Radioactivity in America

Lawrence Badash

RADIOACTIVITY IN AMERICA
Growth and Decay of a Science

THE JOHNS HOPKINS UNIVERSITY PRESS
BALTIMORE AND LONDON

This book has been brought to publication with the
generous assistance of the Andrew W. Mellon Foundation.

Manufactured in the United States of America

The Johns Hopkins University Press, Baltimore, Maryland 21218
The Johns Hopkins Press Ltd., London

Library of Congress Catalog Number 78-20525
ISBN 0-8018-2187-8

Library of Congress Cataloging in Publication data
will be found on the last printed page of this book.

To
Derek J. de Solla Price
and
Alfred Romer

A tribute of my respect and admiration

Contents

Illustrations

Preface

A number of years ago Derek de Solla Price thrust a roll of microfilm into my hands and suggested simply that I might find it of interest. No cloak and dagger tale this; yet while far from containing classified documents the film was nevertheless "hot" in another sense: it was of letters exchanged between Ernest Rutherford and Bertram Boltwood, and in the frankest of terms gave a superb insight into an important corner of early twentieth-century science. Rutherford, as most people with even a slight acquaintance with modern science well know, was widely regarded as the greatest experimental physicist since Faraday. But who was Boltwood, this man on such intimate terms with him? Investigation soon showed that Boltwood was a highly talented radiochemist at Yale University during the first few decades of this century. His association with Rutherford began when the latter was a professor at McGill University, in Montreal, and continued when he moved to Manchester University in 1907. Further study revealed that Boltwood made some excellent contributions to the understanding of radioactivity, perhaps unexpected at a time when—except for such rare geniuses as J. W. Gibbs, Albert A. Michelson, and Henry A. Rowland—American science was held to have been of little consequence.

Now intrigued both by Boltwood's research and the environment in which he worked, I embarked upon the project which has produced this book. Along the way I edited the correspondence just mentioned,[1] and located Boltwood's friend and heir, Dr. Lansing V. Hammond, of the Commonwealth Fund, New York; other friends and colleagues of his, including Mrs. F. R. Hammond, of Hancock Point, Maine; Dr. Loomis Havemeyer, of Yale University; Miss Eleanor Cooper, of North Haven, Connecticut; Prof. Ellen Gleditsch, of the University of Oslo, Norway; the late Prof. Otto Hahn, of Göttingen, Germany; Prof. John A. Timm, of Simmons College; Dr. Hubert Vickery, of the Connecticut Agricultural Experiment Station, New Haven; and the late Prof. Edwin Bidwell Wilson, of Boston; relatives, friends, colleagues, and students of others mentioned in this work, such as Prof. Norman I. Adams, of Yale University; Prof. James S. Allen, of the University of Illinois; Mrs. Wells Barney, of Old Lyme, Connecticut; Prof. Charles D. Coryell, of the Massachusetts Institute of Technology; the late Prof. H. M. Dadourian, of Trinity College, Hartford, Connecticut; Mrs. Harry W. Foote, of New Haven, Connecticut; Miss Helen Goebel, of Hamden, Connecticut; Mr. Arthur W. Goodspeed, Jr., of Springfield, Pennsylvania; Prof. Henry N. Harkins, of the University of Washington; Prof. Lucy Hayner, of Columbia University; the late Prof. Leonard B. Loeb, of the University of California, Berkeley; Mr. William T. Lusk, of Tiffany and Company, New York; Mrs. Charles D. McLendon, of Shreveport, Louisiana; Prof. Benjamin Nangle, of Yale University; Dr. Zalia J. Rowe, of New Haven, Connecticut; Miss Esther Schlundt, of Purdue University; Dr. Bert Stagner, of Los Angeles; Mrs. William E. Stevenson, of Manilla, Philippines; and Mrs. Hans Zinsser, of New York; and still others who shared with me valuable information and insights, including Dr. William H. Byler, of the U. S. Radium Corporation; the late Prof. Kasimir Fajans, of the University of Michigan; the late Dr. Eduard Farber, of Washington, D. C.; Dr. Richard Fink, of Yale University; Dr. E. R. N. Grigg, of Champaign, Illinois; the late Dr. Samuel C. Lind, of the Oak Ridge National Laboratory; Mr. Stephen T. Lockwood, of Buffalo, New York; Prof. A. O. Nier, of the University of Minnesota; Dr. Nathan Reingold, of the Smithsonian Institution; Prof. A. Norman Shaw, of McGill University; and Miss Elizabeth H. Thomson, of Yale University. To all of these individuals I owe much for their unfailing generosity.

No less am I indebted to numerous institutions for the use of their resources and the kindness of their staffs. These include Ms. Joan N. Warnow and Dr. Charles Weiner, of the Center for History of Physics, American Institute of Physics; Miss Jane Hill, former librarian of the Yale Memorabilia Room; Miss Beth Taylor, Yale Chemistry Department librarian; the late Mrs. Dorothy Gordon, Yale Physics Department librarian; the Cavendish Laboratory, Cambridge, England; Mr. A. E. B. Owen, of the Cambridge University Library; the Library of Congress; the physics department of the University of Cincinnati; and the libraries of the City College of New York, McGill University, Oxford University, Princeton University, the University of Chicago, the University of Missouri, the University of North Carolina, the University of Pennsylvania, the University of Pittsburgh, and Yale University.

During the years spent on this project I have at times had support from Yale University and the University of California. To both I am grateful, as I am to the National Science Foundation, which enabled me to spend two years in Cambridge, England, on a related project, which inevitably reinforced this one.

Gail Clark typed a major portion of the drafts and final versions of these chapters, and I am grateful for her skill, intelligence, and unfailing good humor. Henry Y. K. Tom, G. Tilghman Hollyday, Mary Lou Kenney, and Barbara Kraft shepherded the manuscript through the bookmaking process at The Johns Hopkins University Press with professional grace that must be admired as well as appreciated.

My deepest thanks go to Prof. Alfred Romer, of the St. Lawrence University, from whom I have learned much about the history of radioactivity and from whose reading this manuscript has benefitted. The late Prof. Kasimir Fajans criticized chapters thirteen and fourteen, and I regard my conversations and correspondence with him one of the joys of studying the recent past.

Parts of chapter two have been published in an article entitled "Radium, radioactivity, and the popularity of scientific discovery," *Proceedings of the American Philosophical Society*, 122 (June 1978), 145–54, and I appreciate the editor's giving permission for their appearance here. For the merits of this book these good people are in no small measure responsible; for its inadequacies I alone am to blame. Finally, to Derek Price I am forever grateful for that choice roll of microfilm, his probing questioning of an earlier version of several of these chapters, and his warm encouragement of my interest in the history of science.

The Radioactive Decay Series

THE URANIUM SERIES

Radioelement	Corresponding Element	Symbol	Radiation	Half-Life
Uranium I ↓	Uranium	^{238}U	α	4.51×10^9 yr
Uranium X₁ ↓	Thorium	^{234}Th	β	24.1 days
Uranium X₂* ↓	Protactinium	^{234}Pa	β	1.18 min
Uranium II ↓	Uranium	^{234}U	α	2.48×10^5 yr
Ionium ↓	Thorium	^{230}Th	α	8.0×10^4 yr
Radium ↓	Radium	^{226}Ra	α	1.62×10^3 yr
Ra Emanation ↓	Radon	^{222}Rn	α	3.82 days
Radium A 99.98% | 0.02%	Polonium	^{218}Po	α and β	3.05 min
Radium B	Lead	^{214}Pb	β	26.8 min
Astatine-218	Astatine	^{218}At	α	2 sec
Radium C 99.96% | 0.04%	Bismuth	^{214}Bi	β and α	19.7 min
Radium C′	Polonium	^{214}Po	α	1.6×10^{-4} sec
Radium C″	Thallium	^{210}Tl	β	1.32 min
Radium D ↓	Lead	^{210}Pb	β	19.4 yr
Radium E ~100% | 2×10^{-40}%	Bismuth	^{210}Bi	β and α	5.0 days
Radium F	Polonium	^{210}Po	α	138.4 days
Thallium-206	Thallium	^{206}Tl	β	4.20 min
Radium G (End Product)	Lead	^{206}Pb	Stable	—

The Thorium Series

Radioelement	Corresponding Element	Symbol	Radiation	Half-Life
Thorium ↓	Thorium	^{232}Th	α	1.39 × 10^{10} yr
Mesothorium I ↓	Radium	^{228}Ra	β	6.7 yr
Mesothorium II ↓	Actinium	^{228}Ac	β	6.13 hr
Radiothorium ↓	Thorium	^{228}Th	α	1.91 yr
Thorium X ↓	Radium	^{224}Ra	α	3.64 days
Th Emanation ↓	Radon	^{220}Rn	α	52 sec
Thorium A ↓	Polonium	^{216}Po	α	0.16 sec
Thorium B ↓	Lead	^{212}Pb	β	10.6 hr
Thorium C 66.3% | 33.7%	Bismuth	^{212}Bi	β and α	60.5 min
Thorium C′	Polonium	^{212}Po	α	3 × 10^{-7} sec
Thorium C″	Thallium	^{208}Tl	β	3.1 min
Thorium D (End Product)	Lead	^{208}Pb	Stable	—

THE ACTINIUM SERIES

Radioelement	Corresponding Element	Symbol	Radiation	Half-Life
Actinouranium ↓	Uranium	^{235}U	α	7.13 × 10^8 yr
Uranium Y ↓	Thorium	^{231}Th	β	25.6 hr
Protactinium ↓	Protactinium	^{231}Pa	α	3.43 × 10^4 yr
Actinium 98.8% | 1.2% ↓	Actinium	^{227}Ac	β and α	21.8 yr
Radioactinium	Thorium	^{227}Th	α	18.4 days
Actinium K	Francium	^{223}Fr	β	21 min
↓ Actinium X ↓	Radium	^{223}Ra	α	11.7 days
Ac Emanation ↓	Radon	^{219}Rn	α	3.92 sec
Actinium A ~100% | ~5 × 10^{-4}% ↓	Polonium	^{215}Po	α and β	1.83 × 10^{-3} sec
Actinium B	Lead	^{211}Pb	β	36.1 min
Astatine-215	Astatine	^{215}At	α	~10^{-4} sec
↓ Actinium C 99.7% | 0.3% ↓	Bismuth	^{211}Bi	α and β	2.16 min
Actinium C′	Polonium	^{211}Po	α	0.52 sec
Actinium C″	Thallium	^{207}Tl	β	4.8 min
↓ Actinium D (End Product)	Lead	^{207}Pb	Stable	—

SOURCE: Samuel Glasstone, *Sourcebook on Atomic Energy*, 3rd ed. (New York: Litton Educational Publishing, Inc., 1967). Reprinted by permission of Van Nostrand Reinhold Company.

Radioactivity in America

Introduction

As a science, radioactivity is important for its major contributions to atomic physics and for its parentage of nuclear physics. As a subject, it offers a useful vehicle for examining the practices, orientations, and styles of scientists at a particular time, as well as the reciprocal influences between science and society.

This work touches upon a number of aspects of the history of radioactivity. Its scientific origins in Europe are described briefly, then its early exploration in the United States. Because the phenomenon had some startling characteristics, public interest was aroused and the popularization provided an exciting episode—if one of questionable value to the scientists pursuing radioactivity. The investigations progressed, first by the amateurs (even comic characters), then the stumblers, and finally the sure-footed ones. In particular, two chemists, Bertram Boltwood and Herbert McCoy, made such significant contributions that America for a time was in the vanguard of international efforts to explore the radioelements. Their research on the genetics of the decay series, and other chemical work, is described in some detail, with reference given to related achievements abroad.

These accomplishments raised the study of radioactivity in America to a level of maturity, not the dictionary definition, for humans or fruit, of fully grown and developed, but a condition in which the direction the science would take could be seen, impressive results were already in hand, patterns of behavior or relationships were emerging, and a community with related interests was forming. Apart from the science, yet benefitting from this condition of maturity, were two applications of radioactivity that maintained at high intensity the public fascination originally created by the phenomenon alone. These were the medical uses of radium, primarily as a possible cure for cancer, and commercial applications, notably in the manufacture of luminous paint. As part of this story the mining and processing of uranium ore is necessarily discussed.

Further American work in radiochemistry shows the continuing domination of Boltwood and McCoy as they attempted to tie together numerous loose ends, but includes the investigations of several others and an increasing amount of trans-Atlantic communication. Both leaders, whose work contributed to recognition of the next significant problem, then effectively departed the research front and left the concept of isotopy and the group displacement laws to be developed abroad. These foreign stages are nevertheless presented here extensively since confirmation of the ideas fell largely to the American atomic weight expert, Theodore Richards.

Physical research is also described at length, but the lack of an easily identifiable goal makes it a body of science without a clear story line. American study of alpha, beta, and gamma ray phenomena simply did not pay off with first-rate discoveries and pregnant, new lines of inquiry. But, at the time, this work was felt to be valuable, and in retrospect we can see that it was indeed necessary. Without it there would have been gaps in the understanding of certain interactions, blank regions in which unexpected results might have been found. So the physicists performed a useful task, if one that brought them little glory. Yet, description of this competent "normal science," which infrequently survives in historical accounts, provides a better feeling for the human aspect of research; the errors and frustrations and workaday results impart a truer picture than do the triumphs.

Portrayal of the physical work is also important for the overall view of radioactivity as social activity. The chemists were largely ignored by their colleagues; the chemical, as well as the physical aspects of radioactivity, therefore, had closest ties with the community of physicists. These contacts enabled a number of Americans to study abroad, and it appears that international communication was stronger than domestic. On the national scene a few universities created "schools" in which radioactivity flourished, and these genetic relationships between teacher and pupil are as important as the genetic connections between radioelements.

Radioactivity as a science presents some interesting aspects, since the lack of any prehistory to its discovery (no near misses and no closely related phenomena) permits unusually clear definition of its development. Additionally, the radiochemical half of the subject succeeded in answering virtually all the questions it had to ask by the end of World War I, and the physical side evolved into atomic and nuclear physics, leaving little traditional radioactivity to pursue. Thus, in the first two decades of this century we can view the growth and decay of a science, and examine its latter stages with as much care as we customarily give to the origins of such an enterprise.

By restricting the investigation to the United States, it is further possible to describe radioactivity's role in the establishment of the general level of science, especially physics, in this country. America is seen as a developing nation with a sprinkling of genius, but an average competence inferior to that of Western Europe, then the "center of gravity" of worldwide science. The styles of science were similar, and countries on the periphery even led those at the center on occasion, but for the most part the leaders were merely emulated.[1] As a stage in American history the radioactivity efforts are well worth describing. Even more, for a better comprehension of the history of science these efforts offer insights that undercut any view of manifest destiny, of inexorable progress towards national greatness in science. In fact, the picture of the scientific community's growth shows the same frustrations and dead ends, as well as successes, as normal science itself. And neither the subject nor its professors appreciably influenced the generation of the 1920s, which more clearly is taken as the harbinger of our contemporary stature. Nevertheless, this study does describe the process of maturation of both the science of radioactivity and the community of those who toiled at it, necessary though not sufficient foundations for the later accomplishments.

The period covered is approximately 1900 to 1920. Between these dates radioactivity was introduced into the United States, made its mark on the scientific and popular consciousness, and rather abruptly departed the scientific scene. While World War I clearly disrupted scientific life, it is doubtful that the story told herein would have differed appreciably without that conflict. As might be expected, it is impossible to write a history of a twentieth-century science in national isolation. Once there is active communication across borders, whether by journal, correspondence, or travel, the ideas and activities take on an international dimension. Especially, as in the case of radioactivity, if the scientific leadership pursuing a subject is internationally oriented, the other participants will be guided by that standard and endeavor to "play in that ballpark." Hence, in this monograph, no harsh attempt has been made to refrain from describing events abroad. Rather, some work done overseas, by Americans and others, is treated at length in order to establish the context for contributions made in the United States.

The physics and chemistry of radioactivity are the major foci of attention. "Atmospheric radioactivity," meaning radioactivity of the air, soil, water, snow, etc., is treated only in passing, and biological investigations not at all, because they contributed little to scientific knowledge in this period and barely affected the mainstream activity. Even the discussion of chemistry is attenuated, with considerably more attention given to results with interesting radioactivity interpretations than to the wet chemical techniques used to separate the radioelements. This approach is justified because the radiochemical procedures were largely taken from established analytical chemistry, and with rare exceptions were not especially creative. Discussion of reagents, processes, temperatures, concentrations, etc., which normally is absent from the chemical literature of this period (see, for example, the *Annual Reports on the Progress of Chemistry* to the London Chemical Society), was not even then regarded as especially novel, and the value of their description might be likened to that of presenting construction details of the batteries used to charge electrometers.

The opening chapters are in the form of a historical narrative, as are those dealing with the medical and commercial aspects of radioactivity. The other chapters are far more analytical, with extensive descriptions of experimental work, designed to illuminate the progress of this science as well as to impart a feeling for the apparatus, techniques, customs, and concepts of the day. Tables of the three radioactive decay series discussed are provided on pages xvi–xviii to assist the reader in following the narrative. The book's goal, in brief, is to trace the development of a particular science in a particular country, of a new science in a virgin environment. The unfolding of uranium-radium genetics was largely an American story; the physical research was solid but not brilliant, yet not unlike work done elsewhere; public reaction to the science was typical of that abroad; and all these aspects were part of a country's coming of age in science. If it is possible to present new interpretations to a tale never before told, the most interesting conclusions concern the demise of radioactivity, except for applications, by the end of the century's second decade; the circumstance that, in a field not pursued vigorously by many, national characteristics need mean little while personal traits may dominate; and that nonscientific applications of a science may have little bearing on its development, or even be detrimental.

1 European Origins

The Second Scientific Revolution of the Late Nineteenth Century

Speaking in 1913, the great German theoretical physicist Max Planck could look upon his science with a sense of awe. Probably "never," he remarked, "has experimental physical investigation experienced so strenuous an advance as during the last generation, and never probably has the perception of its significance for human culture penetrated into wider circles than it does today." He referred to the great discoveries of wireless telegraphy waves, electrons, x-rays, and radioactivity as exemplars of the new physics, and suggested that their interpretations raised so many problems that the edifice of theoretical physics appeared to be crumbling. But not so, Planck continued, as he proceeded to demolish the straw man he himself had raised. If the great physical principles brought to a fine edge during the nineteenth century conflicted with certain deeply rooted, habitual conceptions, then the former invariably were to be retained. Thus, the conservation of energy reigned supreme over the stability of chemical atoms, the constancy of the velocity of light held primacy over the mutual independence of space and time, and to the principles of thermodynamics the continuity of dynamical effects had to be sacrificed. Rather than bemoan the long-held assumptions now lost, Planck argued, we should be impressed with the strength of the surviving great principles under their new applications.[1]

Planck's approach was wholly appropriate. The discoveries in science during the few decades spanning the turn of the century were so profound that they led to new world views. Yet the accomplishments of the nineteenth century, the classical science which the new work by no means simply displaced, were mightily impressive also. So impressive were they, in fact, that despite the continued outpouring of published research reports and irrespective of the seeming vigor of the discipline, there were those who, paradoxically, feared for the health of science. During the last half of the nineteenth century—before the new discoveries submerged such speculations—a small but vocal group rumored that physical science was near its end.

These were, by and large, not the men who increasingly questioned the very foundations of physical ideas. Had they been of this temper, they would have been aware of numerous unresolved problems that foretold a vigorous future, perhaps even without the turn-of-the-century surprises. For the science of mechanics, so successful in embracing heat and even parts of optics and electromagnetism that all natural phenomena appeared explicable in its embrace, seemed nevertheless to balk in certain cases. Energetics, a science of which energy, not mechanism, was the basis, and the electromagnetic interpretation of matter were examples of this profound questioning. If physical law was felt to be a unifying structure, there was room for more unity.[2]

Superficially, however, there was reason to regard the very real accomplishments of the nineteenth century as heralding the millenium in which the edifice of scientific knowledge would cease to grow. People holding this viewpoint called the phenomenon perfection, for the overall structure of science was seen as virtually complete, with only a few loose ends yet to be tucked in. Physics, chemistry, astronomy, all had been remarkably successful: the major problems were satisfactorily answered and no new fields were seen to conquer.[3]

For example, celestial mechanics, since Kepler and Newton the mainstay of astronomical studies, had been so well pursued that virtually all motion in the solar system could be explained by the law of gravitation, within the accuracy of the observations. Great tables of planetary and lunar positions were published, and the thought that only further polish to the figures lay ahead caused the number of investigators to dwindle.[4] Some who left the field of gravitational astronomy turned to the branches of observational and descriptive astronomy. Yet, here too the star catalogs and computational methods on the one hand, and the cosmological investigations and the tools and techniques of astrophysics on the other, seemed to offer more detailed knowledge but little explanation.[5]

Chemists, having long since purged their science of all vestiges of alchemy and having ordered their impenetrable and indivisible atoms into a periodic table of distinct elements, could look forward only to such activities as the preparation and study of new compounds and the thorough classification of chemical phenomena. While such basic problems as an understanding of the forces which hold molecules together remained, it was felt that the traditional methods of science would suffice to resolve them. The 'age of discovery' in chemistry was over.[6]

In physics the malaise was most pronounced. Such great successes had accrued from close attention to quantitative data that many thought this an end in itself. James Clerk Maxwell, director of Cambridge University's Cavendish Laboratory, saw fit in 1871 to remark that "this characteristic of modern experiments—that they consist principally of measurements,—is so prominent, that the opinion seems to have got abroad, that in a few years all the great physical constants will have been approximately estimated, and that the only occupation which will then be left to men of science will be to carry on these measurements to another place of decimals."[7] While Maxwell went on to claim the shortsightedness of this view, it was his own work which figured large in creating it.

For Maxwell was not only the originator of mysterious equations which described electromagnetic phenomena, he had also contributed mightily to the kinetic theory of gases and other areas of physics, both theoretical and experimental. Put simply, Isaac Newton in the seventeenth century had "explained" mechanics, and Maxwell two hundred years later had "explained" electromagnetics. These subjects, together with the recently developed laws of energy conservation, formed the solid structure of classical physics.[8] Be it a problem in mechanics, heat, light, sound, electricity, magnetism, or the constitution of matter, a formula could be applied to quantitative data and a solution proffered. Even more, science had become "useful," for the accurate definition of the volt, ohm, and ampere was of great value to the fledgling electrical industry, thermodynamics was of increasing importance to mechanical engineers, and mechanics allied with the study of the properties of matter was indispensable to civil engineers.

As late as 1895, the year in which science received its first perceptible shock from the new physics, the *fin de siècle* feeling of contentment (or resignation?) persisted. Queen Victoria was on the British throne and all was well. "Everything essential and possible of knowing was known, and . . . all that remained was mere detail."[9]

With hindsight, of course, we can see that twentieth-century scientists were not destined to be merely the functionaries in an international bureau of standards. Certain problems, earlier considered minor, assumed major importance and provoked the crisis that changed the world picture of science. Interestingly enough, this paradigm change[10] was assisted by the continual effort for refined measurements. This effort, an accepted part of the research in most physical laboratories, was aimed at greater accuracy, not discovery. Yet, the "romance of the next decimal place" *did* lead to discoveries of profound character.[11] Most notable of these are the quantum theory and the existence of the inert gas argon. Max Planck shattered the physics of continuity with his quantum ideas of heat radiation because the data of Lummer and Pringsheim did not fit

Wien's law of temperature radiation closely enough. And Lord Rayleigh found the first element of a whole new group in the periodic table because he was disturbed that his nitrogen sample from air was one part in a thousand too heavy. With a historical look over one's shoulder to Kepler's use of Tycho Brahe's highly accurate data on the position of Mars to prove that planetary orbits are ellipses, we can appreciate the view that if we "look after the next decimal place . . . physical theories will take care of themselves."[12]

True, but not the whole story. Physical theories emerged also from other sources: the crises mentioned above. Since the development of the spectroscope in mid-century, vast amounts of data had accumulated. Each element presented a unique pattern of spectral lines, so the radiant emissions must be due to atomic vibrations. But classical mechanics and traditional models of the atom could not quantitatively incorporate such vibrations, while the ad hoc equations that successfully systematized the spectra also could not explain them.[13]

The luminiferous ether was another problem. It had been burdened with so many properties, some contradictory, that the failure by Michelson and Morley to detect it led ultimately to its expulsion from science by Einstein. Einstein, further, may be credited with extending Planck's quantum notions from heat radiation to electromagnetic radiation, for he thus explained the heretofore puzzling photoelectric effect. In so doing he went beyond Planck's interpretation of a quantized oscillator and quantized the energy being transmitted. The significance of this step is revealed by the circumstance that Einstein was dealing with electromagnetic energy, and after nearly a century of success the wave theory of light was now jeopardized by a corpuscular interpretation.

This wave-particle problem, resolved only by the acceptance of dual properties in the 1920s, had other manifestations. Of foremost significance was the subject of cathode rays. With the mid-nineteenth-century advances in vacuum technology and high voltage sources, a new type of radiation had been seen issuing from the negative electrode of an exhausted tube. This subject was largely pursued in England and Germany and its interpretation was rather along nationalistic lines. The English saw cathode rays as a stream of particles while the Germans viewed them as a wave phenomenon. Since evidence supporting both explanations was found the puzzle was deep indeed before 1897. Even then, when J. J. Thomson announced his discovery of the electron, a particle most likely one thousand times smaller than the smallest atom, and bearing the same charge-to-mass ratio as the radiation produced from ultraviolet light, the Edison effect, Lenard rays, cathode rays, etc., the particle interpretation given to cathode rays lasted only a generation. It too was revised by the later dualistic view of nature, wherein both wave and particle properties for the same phenomenon are acceptable.

The discovery of the electron by Thomson, who followed Maxwell and Rayleigh as head of the Cavendish Laboratory, was profoundly significant as a paradigm changer in another fashion also. As suggested above, this charged particle appeared in a wide variety of effects. Its universal distribution made it likely that it bore the fundamental unit of electrical charge, which in turn implied its extremely small size. The problem for both physicists and chemists then was how to reconcile their concept of the atom, as the irreducible unit of matter, with a universal particle smaller by a thousand times.[14]

For immediate fame, and puzzlement, nothing could surpass the discovery of x-rays towards the end of 1895. Not until the public announcement of the atomic bombs in 1945 would a scientific achievement receive greater attention in the newspapers and magazines. Wilhelm Conrad Röntgen, professor of physics at the University of Würzburg, was working with cathode ray apparatus when he found the more penetrating radiation. The public was fascinated with photographs of human bones and the thought that ladies' clothing was no longer a barrier to prying eyes (lead-lined undergarments were offered for sale), while scientists pondered whether these rays were waves or particles. Even more, since x-rays were so easily produced in cathode ray equipment, and the latter subject had had widespread popularity among physicists for several decades, there was a fear that x-rays had unwittingly affected an unknown amount of scientific work.[15]

But of all these discoveries that changed man's view of the world about him, none was of greater significance than that of radioactivity. This phenomenon challenged the prevailing concepts of the stability of atoms, their uniformity of species, and their construction, and posed profound problems in explaining the source of their energy. In this discovery lay the seeds for greater understanding of the nature of matter, progress which is embodied in the atomic and nuclear physics of today. These subjects, the backbone of modern science, influence not only chemistry, astronomy, geology, biology, medicine, etc., but set the tone of daily life with promises of perhaps Faustian gain through controlled uses of nuclear energy, and forebodings of destruction through military uses of this same power. Just as the Scientific Revolution of the seventeenth century opened man's eyes to the world about him and permitted a new form of rational inquiry, the scientific revolution of the late nineteenth century both initiated another world view and heralded man's increasing control over nature. Those who forecast just a continuation of nineteenth century "normal" science—everyday busy work, however useful, that fell within the range of prevailing concepts—clearly had misjudged the complexity of natural phenomena. The new discoveries and interpretations led to increased scientific activity of an especially creative kind. Yet, these prophets of the completeness of science served a useful function in calling attention to the achievements of classical science, while Planck more specifically emphasized the great classical laws that survived this new scientific revolution centered around 1900. One of these laws, the conservation of energy, would be at the heart of interpreting the phenomenon of radioactivity.

Early Research in Radioactivity

The discovery of radioactivity followed directly from that of x-rays. In the hectic days after Röntgen's announcement, physicists the world over repeated his experiments. One point widely noted was that the x-rays originated in the region of the vacuum tube struck by the stream of cathode rays. Since the cathode ray bombardment caused this part of the glass tube to phosphoresce, it occurred to some that perhaps *all* luminescent bodies emit radiation similar to x-rays. In Paris, Henri Becquerel gathered together a variety of minerals and prepared to examine them in the following fashion. He wrapped a fresh photographic plate in black paper such that it was light tight. Next he spread his mineral crystals on this paper, and then he placed the plate on his window sill where the crystals would be stimulated to phosphorescence by sunlight.

As an authority on luminescence in his own right, and son of an eminent physicist famous for studies in this subject, Becquerel had available a good selection of mineral samples, including potassium uranium sulphate.[16] This mineral, he found upon developing his plates, emitted a radiation that, like x-rays, could pass through matter and expose the photographic emulsion.[17] Becquerel had, therefore, successfully tested his hypothesis, finding a luminescent substance that threw off penetrating radiation, just as was done by the glass of the x-ray tube. But when Becquerel exhaustively examined this radiation, problems arose. First he learned that the crystals were equally active whether or not sunlight had caused them to glow.[18] This news he interpreted as evidence of a new phenomenon of *long lived* invisible phosphorescence. Then he learned that *non*phosphorescent uranium compounds also exposed his photographic plates, a fact he found difficult to explain.[19] Finally, further work traced the activity to uranium element, at which point Becquerel could only explain the source of energy as having something to do with light, presumably a new phenomenon of metallic phosphorescence.[20]

These and other errors of early 1896 were accompanied by much substantial work that has stood the test of time, for Becquerel was actually a highly competent physicist, possessing even the spark of genius. His results showing the absorption of uranium rays by various materials and the ability of the rays to make a gas conduct electricity were accepted, while his mistaken proof that the rays could be reflected, refracted and polarized (and thus different from x-rays either in nature or at least in wavelength) was corrected within a few years. As for the fundamental question of what these rays were, illumination required still more time—and of this, we shall say more.

Radioactivity differed from x-rays in ways besides those supposedly found by Becquerel. The latter enjoyed enormous popularity, while the former aroused but slight attention. Part of the reason for this was their varying utility: x-rays immediately were applied in medical diagnosis with striking success, whereas the photographs produced by uranium rays were weaker, more diffuse, and required longer exposures. Of great importance, too, was the scarcity of uranium minerals and the almost complete nonexistence of uranium metal, in contrast to the widespread availability of cathode ray, or x-ray, equipment. While in a decade radioactive materials were used in medical therapy and, indeed, this demand was the motivation for the "radium industry" that arose, there was little in the early years following Becquerel's discovery to suggest this later fame.

Scientific interest, as distinct from popular enthusiasm, also was at a low level. This was due to the fact that, in the plethora of rays being studied in the late nineteenth century, those from uranium did not stand out.[21] Besides the new and powerful x-rays and the old, but still puzzling cathode rays, physicists of this time were concerned with positive rays, secondary rays, Lyman rays, Schumann rays, ultraviolet rays, visible light, infrared rays, Hertzian waves, radiant energy, canal rays, discharge rays, and a host of other more or less questionable radiations. Given this diversity, Becquerel rays, as they came to be called, would have required far more distinctive properties to command greater attention. A measure of the early interest in radioactivity is given by the seven papers on the subject published by Becquerel in 1896, his two in 1897, and *none* in 1898. The discoverer himself felt he had exhausted the field and went on to other things. Other investigators, including those who saw analogous radiations emitted from glow worms and from sugar, contributed about an additional dozen papers in these years. Contrast this with the total of over one thousand articles and books published on x-rays in 1896 alone!

Radioactivity, then, was not riding a wave of excitement when two new scientists undertook its study in 1898. Rather, it was more like a dead horse; there it was, but no one knew what to do with it. However, these two newcomers approached their subject with a fresh point of view. They asked themselves the simple question: Do any materials besides uranium emit similar rays? And both Gerhard C. Schmidt, in Erlangen, and Marie Sklodowska Curie, in Paris, found that among all the known elements only thorium possessed this property.[22] More important still, Madame Curie detected far more activity in uranium ore than could be explained by its content of uranium metal. Joined now by her physicist husband, Pierre, and by a chemically trained colleague, Gustave Bémont, she extracted from the pitchblende ore first polonium and then radium.[23] Their quantities were minute, yet their activities intense. These great discoveries of 1898, for both substances were suspected and eventually shown to be new elements, served to renew interest in radioactivity. The moribund subject was resurrected and interest now never flagged until radioactivity merged into atomic physics and evolved into nuclear physics.

The Rise in Activity

While the Paris group was concerned with the laborious task of extracting and purifying the minute traces of polonium and radium found in ores, and in detecting yet another element, actinium, in pitchblende, research elsewhere was of a more physical character. In September 1898, the inaugural address by the president of the British Association for the Advancement of Science was delivered by Sir William Crookes, the London consulting chemist and owner and editor of *The Chemical News*.[24] It was a wide-ranging talk, covering economics and agricultural productivity as well as the state of science and recent discoveries. Crookes was happy to note that his idea of a fourth state of matter, the *ultra*gaseous, in addition to the common solid, liquid and gaseous states, was being increasingly accepted as the corpuscular view of cathode rays took hold. Somewhat in connection with this he was pleased to offer a possible explanation of radioactivity. Becquerel's idea of a long-lived phosphorescence, though still retained by its author, never met with much approval, and Crookes envisioned instead a process of energy extraction from the molecules of the air. These molecules possess a wide range of velocities. If, somehow, the radioactive atoms are able to extract the energy excess from the faster moving air molecules, the apparent contradiction with the law of energy conservation would be overcome, for an energy source would be defined. "The store drawn upon naturally by uranium and other heavy atoms," Crookes said, "only awaits the touch of the magic wand of science to enable the twentieth century to cast into the shade the marvels of the nineteenth."[25]

Without waiting for the twentieth century, Julius Elster and Hans Geitel, in Wolfenbüttel, waved their own magic wand and disposed of a few explanations of radioactivity.[26] By photographically comparing the power of a uranium compound in air with its strength in a vacuum, they concluded that air molecules play no part in the phenomenon. Next, they attacked Marie Curie's interpretation that all of space is constantly being traversed by rays analogous to x-rays, which are only absorbed by elements of high atomic weight, which elements in turn emit Becquerel rays as secondaries. This hypothesis, which really was no explanation, but only a shift of the unknown from the radioelements to something else, was tested at the bottom of a deep mine shaft in the Harz Mountains. Since this spatial radiation was not attenuated by the intervening rock, for their uranium sample was as active as on the surface, Elster and Geitel simply denied its existence. Nevertheless, both Crookes and Curie persisted in their understandings of radioactivity, perhaps because they felt no better explanation was available. The suggestion by Elster and Geitel, that energy emission might result from a process wherein an unstable atom moves to a stable condition, fell upon unhearing ears, for the transformation theory of Rutherford and Soddy was not advanced until 1902.[27]

Ernest Rutherford (1871–1937) was a young New Zealander who, in 1895, traveled to Cambridge to study under J. J. Thomson. The program of the Cavendish Laboratory, for some years past, was largely an examination of the discharge of electricity in gases. Röntgen's discovery of x-rays, then, was warmly welcomed, for it provided a convenient means of creating carriers of electricity in a gas. A significant result of extensive experimentation in this area was a theory of ionization by Thomson and Rutherford.[28] Since uranium rays also possessed this ability to ionize, or create charged carriers in a gas, Rutherford was led to test this source in turn. Almost immediately he became the foremost investigator in radioactivity, heading research in this field for the next forty years.

Rutherford, a master of experimental technique, took full advantage of his electroscopes, electrometers, and other apparatus to gather quantitative data. In this fashion, both he and the Curies were able to go far beyond Becquerel's essentially qualitative work. Among his contributions in 1898, shortly before he became professor of physics at McGill University, in Montreal, was proof that uranium rays could *not* be reflected, refracted, or polarized. They were thus not electromagnetic in nature, as Becquerel had thought. Rutherford further showed that there were two components to this radiation, one which could scarcely penetrate matter, but which ionized a gas quite well, and the other which had greater penetrating power, but far less effect on an electroscope. These he named alpha and beta, respectively, "for convenience."[29] The intensely penetrating gamma ray, also emitted from radioactive bodies, was found the following year by Paul Villard, in Paris.[30]

These rays formed the subject of many investigations during the

ERNEST RUTHERFORD (1871–1937)

early days of radioactivity. To be sure, many recognized that the same ray issuing from different sources usually had different energies, but it was not until 1904 that William H. Bragg and Richard Kleeman in Australia showed that each emitter had a characteristic range for its alpha particles.[31] By this time Rutherford had already proven that the alpha was a particle of positive charge and atomic-sized charge-to-mass ratio,[32] and begun his long courtship of the ray. Though he strongly felt it to be a charged helium atom, confirmation required some years more of effort.

Beta rays received far more initial attention than the alphas, simply because they were more amenable to exploration. In 1899 they were deflected by a magnetic field,[33] a feat not performed upon their more massive relations until 1902. Immediately following the evidence that the beta was a negatively charged particle came its successful deflection in an electric field.[34] With this information Becquerel showed that the beta particle carried the same charge-to-mass ratio as J. J. Thomson's electron—they were identical.

In the first four years of the new century there were other facts learned about radioactivity, some of which excited men beyond scientific circles. The emission of radiation was independent of external conditions, for experiments at both high and low temperatures did not affect the activity. These fascinating rays carried much energy and could tint glassware a purple color, cause some crystals to scintillate, elevate the temperature of their surroundings, and inflict a serious burn upon the skin. And radioactivity was not restricted to only a scarce handful of new radioelements and two little-used, known elements. Instead, it appeared that radioactivity was all-pervasive; it was widely found in atmospheric air, soil gases, ground water, rain, and even in the mist at the base of Niagara Falls.

But how could this wide distribution come about? The answer lay in the earlier discovery that both thorium and radium produced gaseous products, each with a characteristic half-life, to which Rutherford gave the deliberately evasive name "emanation," since he was not at first sure of its nature.[35] It was this gas, soon to be shown a member of the inert helium group, that diffused into air, water, and soil from the naturally occurring radioactive minerals the world over. In the laboratory, knowledge of emanation helped explain "induced activity," for objects near the radium and thorium sources strangely acquired an activity of their own. Now it was understood that the gas had deposited something upon the objects' surfaces.

There was another peculiarity, at first devastatingly puzzling, then the key to the explanation of radioactivity. Uranium, thorium, radium, and some other radioelements, left undisturbed for a while after preparation, exhibited the same level of activity whenever they were measured. Yet polonium, emanation, and the induced activity (or "active deposit") sometimes increased in activity, but always eventually decreased. And there seemed something very fundamental about this oddity, as the activity for a given radioelement always decayed to half value in the same period of time. This half-life, indeed, was to become the most accurate means of fingerprinting the radioelements.

The problem now was to organize this mass of information into a theory of radioactivity. Some of the data were chemical in nature; most were physical. It should not be surprising, then, that a successful theory sprang from the collaboration of a chemist and a physicist. Rutherford, still at McGill, found that he had more ideas than time to pursue them. He, therefore, began to develop a thriving research "school" about him. As some of the problems required chemical separations of radioelements, he secured the services of a young demonstrator in the chemistry department, named Frederick Soddy. Together they observed the curious behavior of freshly treated uranium and thorium, which in time regained their activities, even though the supposedly active components, uranium X and thorium X, had been removed.

The best way to explain this growth in activity, they felt, was to assume a change of one element into another.[36] Thorium, for example, transformed into thorium X, which in turn changed into thorium emanation, which in its turn became thorium active deposit. Each substance had its own half-life and its own chemical properties, though at that time thorium's half-life was too long to be ascertained, and knowledge of chemical properties was slight. This explanation was fine, except for one thing. Rutherford and Soddy were, in effect, saying that *transmutation of elements* was occurring, in nature, spontaneously and rather widely. Alchemy had long been exorcised from scientific chemistry, and these two were now reintroducing it. Still, the data of radioactivity fit well into this theory and no one could fault them. In fact, the transformation theory, one of the most iconoclastic ideas ever introduced into science and a giant step forward in understanding the nature of matter, encountered surprisingly little opposition.[37]

With this interpretation of the phenomenon, which was advanced in 1902 and 1903, we come to the point where the development of radioactivity in the United States may be discussed. It now was believed that an alpha, beta, or gamma ray was expelled from the atom simultaneously with its rupture, and this event was itself caused by a large store of energy within the unstable atom. This satisfied the law of conservation of energy, but the concept of an atom that was not indestructible and which, in view of the ejected particles, seemed to have an internal structure, required a great change in the thinking of many physicists and chemists.

It was largely the study of radiation that had led to this theory. Physical investigations continued, of course, in the following years, but transformation, being more a chemical phenomenon, naturally suggested that the bodies emitting the rays were at least as interesting as the rays themselves. If one chemical element were transmuting into another, there was much work promised for the chemists who cared to sort the various known and yet undiscovered radioelements into the correct sequence in the correct decay series. Indeed, radiochemistry moved to center stage in the pursuit of radioactivity, and much of the significant work was performed in America.

2 American Reception of Radioactivity

Initial Reaction

Like their European colleagues, most American scientists were slow to recognize the significance of Becquerel's discovery in 1896. Yet, in the context of nineteenth-century science, this was not at all surprising, for this country usually lagged far behind the European developments. What was surprising, however, was that the professor of electrical engineering at the Alabama Polytechnic Institute (now Auburn University), A. F. McKissick (1869–1938), should be one of the very first in the whole world to follow Becquerel into radioactivity. Perhaps he read Becquerel's papers in the *Comptes Rendus;* more likely he learned of the phenomenon in a review article. One such article appearing in America, really on the subject of x-rays, was penned by J. J. Thomson, who called attention to the similarities and differences between x-rays and uranium rays.[1] As McKissick was intently interested in the Röntgen radiation, it is entirely probable that he was led to radioactivity through his x-ray studies.[2]

McKissick seized upon the fact that the rays initially were found in phosphorescent crystals. Hoping to discover other sources of Becquerel rays, he tested "all such [luminescent] substances that are known as available."[3] Nor did he stop with known luminescent materials; a list of the active emitters of radiation he discovered includes lithium chloride, barium sulphide, calcium sulphate, quinine chloride, sugar, chalk, glucose, and uranium acetate. His procedure was to place the object to be photographed (e.g., a coin or key) directly upon the emulsion, cover it with the plate holder, and sprinkle on top of this the materials to be tested, all of which were previously exposed to the sun for two hours. Development 48 to 72 hours later revealed weak but distinct images of the objects. Granulated sugar was pronounced the best source, since with it he had gotten a fairly clear negative through two and a half inches of wood. A colleague, Professor B. B. Ross, suggested the phenomenon might be due to a high *molecular* weight, but McKissick found no satisfactory verification of this.

Except for the uranium compound, McKissick was, of course, completely wrong. His images were more likely results of chemical action than penetrating radiations. Still, he is worthy of attention, since he was very typical of the casual experimenters who added greatly to the confusing array of emitters and emissions at this time. The suggestion by Ross is also interesting, both because radioactivity was later shown to be found in elements of high *atomic* weight, and because it demonstrates that Becquerel's evidence that the rays come from *elemental* uranium (published in May 1896) was not in November regarded as a relevant factor, or perhaps that this information had not yet filtered to the corners of the scientific world. This short paper by McKissick originally appeared in the *Electrical World* (New York), and was reprinted in its entirety in the *Scientific American Supplement* and *The Electrician* (London). All three journals were popular scientific weeklies, which assured wide dissemination of these false results.

McKissick's paper was the only research effort connected with radioactivity that was reported from this scientifically underdeveloped country prior to the discovery of radium. Even information was scarce, and for those who did not follow the foreign journals, detailed knowledge of the subject had to await Oscar M. Stewart's résumé in the still-young *Physical Review* in April 1898.[4] Stewart (1869–1944), an assistant professor of physics at Cornell, wrote a lengthy review of experimentation to date, chiefly Becquerel's, and included all the Frenchman's errors of observation and interpretation. Thus, because uranium rays were reported reflected, refracted, and polarized, and therefore electromagnetic in nature, they were considered even better understood than x-rays, for which these properties were not seen. Yet, because of the similar penetrating ability of both types of rays, x-rays were strongly suspected also of being "short transverse ether waves," an interesting example of a false premise leading to a correct conclusion.

This juxtaposition of x-rays and uranium rays occurred widely. The name "radioactivity" was not coined by Madame Curie until 1899,[5] and did not become a common journal category for some time. Neither were the terms "uranium rays" or "Becquerel rays" popular with editors. Instead, for articles on this subject, the wise reader would turn to "x-rays" or "Röntgen rays" in his index. Not until radioactivity became a more exciting field did it gain its own headings, and then it often swallowed its parent. A case in point is *Chemical Abstracts,* founded in 1907, which placed its x-ray abstracts in the "radioactivity" section.

If, before the discovery of radium, the American scientific profession heard little about radioactivity, the public knew even less. Neither the *New York Times* nor the *New York Tribune* commented upon the subject. Indeed, the only news concerning uranium dealt with ore purchases in Colorado for the French government, which planned to use this heavy element for hardening gun metal and armor plate.[6]

Popularization for a Technical Audience

The news from Paris changed world opinion from indifference to proper concern. Radium, thought to be a new element, would have been significant enough for that reason alone. But with its outpouring of radiation from infinitesimal traces, it bordered almost on the miraculous. Within two weeks of the French discovery newspapers carried vague reports,[7] while just a month later the Curies' paper appeared translated in *Scientific American.*[8] A number of radioactivity "experts" rather quickly appeared, some serious scientists, others more interested in personal notoriety.

Professor George F. Barker (1835–1910), of the University of Pennsylvania, was one of the earliest popularizers, perhaps because he felt that radium was "a more convenient and economical agent for surgical exploration than that which made Röntgen famous."[9] At a mid-December 1899 meeting in Philadelphia, Barker exhibited radium for the first time in America,[10] though the source of his sample is unknown. In the early days of the new century, Barker again lectured on the subject, this time before the New York Section of the American Chemical Society. After a "brief historical sketch" and the exhibition of photographs made in the style of Becquerel, the high point of the evening occurred when the "room was darkened to enable the audience to observe the feeble luminosity of the substance."[11] This property of self-induced phosphorescence was very popular indeed, and fascinated scientist and layman alike.

Another man who specialized in this scientific-historical popular style lecturing was Henry Carrington Bolton (1843–1903). Like most American students in the nineteenth century, he went to Germany for his graduate work in chemistry. Bunsen at Heidelberg and Wöhler at Göttingen were among his teachers, and he received his doctorate under the latter in 1866. In connection with his dissertation on the fluorine compounds of uranium, Bolton prepared and preserved some uranium compounds. His biographer in the *Dictionary of American Biography* thus gratuitously claimed "that he very nearly anticipated the work of the Curies." While operating a private laboratory, Bolton published an index to the literature on uranium in the *Annals of the New York Lyceum of Natural History* (1870). Then, after a few academic positions, the independently wealthy chemist retired and devoted his time to a variety of interests. An edition of some of Joseph Priestley's scientific correspondence (1892) was one product of this leisure.

His early interest in uranium never flagged, and when radioactivity began to interest the American public, Bolton found himself an "authority" on the subject. Speaking before the Chemical Society of Washington on 21 April 1900, he lashed out at those German physicists (chemists would never be so rude!) who were not "gallant enough to leave the enterprising woman [Mme. Curie] a clear field," and who had, therefore, "announced some minor discoveries."[12] Bolton next proceeded to elaborate upon these "minor discoveries" and the major ones, giving a review of radioactivity notable for its emphasis on the *chemical* properties of the new substances (and the adoption of Henri Becquerel as "the French chemist"). A review bearing such emphasis was rare, at this time and even later, and was highly worthwhile because it pointed out the importance of the chemical work. The vast majority of the many résumés printed in these years regarded chemistry largely as a means to an end, the end being the study of the physical properties of the rays and sources. This widespread attitude caused some anguish among chemists (the vitriolic Frederick Soddy, for example), and it is quite true that their contributions have been unjustifiably minimized. This is apparently the danger in a new area that overlaps two established branches of science and later becomes more and more a child of one of them.

With his background material out of the way, Bolton was then free to describe his own recent experiments. The Smithsonian Institution had acquired ten grams of the two types of radioactive substances sold by de Haën, in Germany, and four grams of "polonium subnitrate" from the Société Centrale de Produits Chimiques, in Paris. The Secretary of the Smithsonian, Professor Samuel Pierpont Langley, allowed Bolton, who resided at the Cosmos Club and was therefore in a position to know influential people about Washington, to test these acquisitions in the "standard manner." Since no electrical apparatus was available, this manner was by impression on sensitized, nonhalation, dry photographic plates. Bolton performed no new experiments and reported no new results. It is probable that he merely was trying to familiarize himself with the phenomenon and also test the sources. In one case, however, he thought he had done Friedrich Giesel, of Braunschweig, Germany, one better by noting that radium compounds slowly regain their power of emitting light, after heating had destroyed this property, *without exposure to sunlight.* Other scientists soon ascribed this change in luminosity to the expulsion of gaseous emanation from the compound during heating, and its subsequent regeneration. For, as was later learned, both heat and the illumination of phosphorescent crystals are due to the capture, within the materials, of alpha particles from the decaying radioelements. But it is revealing that Bolton did not realize that the idea of an insolation effect for uranium had been disproven four years earlier, and no one had seriously considered it for radium.

Radioactivity offered much opportunity for eloquence and few resisted the temptation. Bolton asked rhetorically:

> Are our bicycles to be lighted with disks of radium in tiny lanterns? Are these substances to become the cheapest form of light for certain purposes? Are we about to realize the chimerical dream of the alchemists,—lamps giving light perpetually without consumption of oil?[13]

He was right concerning an alchemist's dream, but it was the vision of transmutation, not eternal light.

Less eloquent, but equally typical for his lack of familiarity with important recent events in the field upon which he was lecturing, was Ernest George Merritt (1865–1948). Speaking before the local section of the American Chemical Society on 18 May 1903, the assistant professor of physics at Cornell University, who was later chairman of the department and dean of the graduate school, presented a highly competent review of radioactivity, but included the barest of historical detail.[14] This omission may account for the fact that Rutherford and Soddy were left unnamed. But it does not obscure the fact that a disintegration theory *was* discussed and the proponents of other theories named. Merritt believed in some sort of transmutation occurring as thorium became thorium X, and this in turn became thorium emanation, etc. His concept, however, seemed to be that of an increase in stability and not necessarily a chemical change. In this he was adhering to the ideas of Elster and Geitel who believed radioactivity to be a manifestation of molecular or atomic change in which an active substance slowly moved from an unstable condition to a more permanent form. But half a year after Rutherford and Soddy had published their transformation theory in the world's premier physics periodical, the *Philosophical Magazine,* and a year after it had appeared in the pages of the *Transactions of the Chemical Society* (London), one might expect Merritt to have been aware of its existence.

Perhaps the most flamboyant of the self-styled experts on radioactivity was William J. Hammer (1858–1934), of New York City. A man of

WILLIAM J. HAMMER (1858-1934)

considerable achievements, he was also something of a showman. After attending the University of Berlin and the Berlin Technische Hochschule, at the age of twenty-one he became an assistant to Thomas Edison, and had charge of tests and records on incandescent lamps in the Menlo Park, New Jersey laboratory. The following year he was appointed chief electrician of the first Edison Lamp Works, also at Menlo Park. His career continued apace, and in 1881 he became chief engineer of the English Edison Company. There he was actively involved in building one of the first stations for incandescent electric lighting in the world, at Holborn Viaduct, London. He also installed twelve Edison dynamos at the Crystal Palace Electric Exposition, in 1882. The year 1883 saw him the chief engineer of the German Edison Company. Other positions in still other Edison companies followed until 1890, at which time he entered into private consulting practice, in New York City. Many electrical plants in this metropolis were installed under his direction.

In view of his familiarity with electricity, Hammer's interest in radioactivity probably stemmed from the ionizing action of the rays. We first see him delivering an address before the 159th meeting of the American Institute of Electrical Engineers, on 3 January 1902.[15] His review of radioactivity covered what were by this time the usual high points, but his remarks about the cost of active samples are of interest.

He exhibited a sealed glass tube containing about one gram of a radium salt, which a friend had obtained for him from Paris at a cost of ten dollars. At this price the degree of concentration must have been rather slight. Even less pure samples were said to be on sale in America at $4.50 per gram, or about two thousand dollars per pound. Though Hammer did not indicate the manufacturer, it is certain these too were imported. Corresponding to further purification of radium by the Curies and by Giesel, the two major suppliers, a preparation was soon expected on the market here for thirty-five dollars per gram (probably of barium chloride or bromide, containing traces of radium).

In an appendix to the published paper, Hammer noted that since delivering the lecture he had visited Paris and called upon Pierre Curie, who estimated that only five or six hundred grams of radium, of all grades, had thus far been prepared in the world. Though this totaled little more than one pound, and scientists eagerly awaited increased production and purification, radium's properties could be easily investigated because its activity was one million times greater than an equal weight of uranium. Curie's own sample of pure radium chloride was quite small, about the size of a buckshot, and had "any value one wished to give to it," but twenty thousand dollars could not buy it. Regarding the hazards of radium, Hammer recorded Curie's comment that "he would not care to trust himself in a room with a kilo. of pure radium, as it would doubtless destroy his eyesight and burn all the skin off his body, and probably kill him"[16]—a statement that was quoted extensively in the popular literature.

While on this trip abroad, Hammer also learned that all the radium manufactured in France was made under Pierre Curie's supervision and was tested and classified by him at the Société Centrale de Produits Chimiques. The best commercial salts from Germany at this time possessed an activity of only 300 (by Curie's test, not by German claims), while the best French product, at one hundred dollars per gram, boasted an activity of 7000. With Curie's permission, Hammer was allowed to purchase some radium and polonium compounds.

A year later (17 April 1903), Hammer again addressed the members of the A.I.E.E., this time at a joint meeting with the American Electrochemical Society, held in the chapel of the City College of New York.[17] His treatment of the subject was, in general, similar, though now, after his travels, he could and did indulge in a bit of name dropping. Intimacy was claimed with such as Pierre Curie, Henri Becquerel, Sir William Crookes, J. J. Thomson, Lord Kelvin, and a gentleman who had sat between Becquerel and Kelvin at a dinner.

Hammer also was able to exhibit the nine different preparations of radium and two of polonium which he brought back from Paris. Although he admitted that neither polonium nor actinium had been obtained in sufficient quantity or purity to be certified as elements, he boasted of having "here perhaps the only sample in this country at present of metallic polonium."[18] Since metallic polonium has probably never been isolated, what he really meant was that he had a piece of metallic bismuth containing minute traces of polonium.

Again, the information of historical value concerns the price of sources. M. Boulay, the director of the Société Centrale de Produits Chimiques, informed Hammer that they intended soon to market a chemically pure (or nearly so!) preparation of radium salt, at a cost of six thousand dollars per gram. Previously, all radium of activity higher than 7000 was retained by the Curies. Almost all the radium in the United States at this date was of German origin, which means that it was supplied by either Giesel or de Haën. The latter was selling his preparations at ten to thirty shillings per gram and, judging from such prices, Pierre Curie was not in error in claiming no German sources possessed activities greater than 300. Curie was unaware, however, that Giesel had just begun to market pure radium bromide. Finally, Hammer indulged his audience's taste for macroscopic or gross orders of magnitude by estimating that it would require five thousand tons of uranium residues to produce a kilogram (2.2 pounds) of radium, at a cost of about two thousand dollars per ton.

With this appealing combination of scientific fact and exciting speculation in his lectures, coupled with numerous lantern slides and demonstrations, Hammer was an extremely popular speaker and much in demand. For several years his talks were scheduled by the J. B. Pond Lyceum Bureau, in New York City, and he traveled from Pittsburgh to Providence, Denver to Duluth, and New Britain to New Brunswick. His enthusiastic audiences packed churches, music halls and high school auditoriums, and the press, in what sometimes appeared to be "canned" articles, hailed "America's recognized authority on radioactivity." Indeed, through 1904, Hammer *was* the outstanding American expert, not only through his numerous lectures and newspaper exposure, but also through his dabbling with the medical applications of radium and his authorship of possibly the first book on this rare substance.[19] He was one of the few able successfully to speak to both scientific and lay audiences, an effort the *scientist*-popularizers seem not to have made.[20]

Radium and Pickled Sheepskins: The Radium Craze

By the middle of this century's first decade scarcely a person in the civilized world was unfamiliar with the word "radium" or the name of its discoverer ("Our lady of radium"). The spectacular properties of this element and its envisioned uses were heralded without restraint in newspapers, magazines and books, and by lecturers, poets, novelists, choreographers, bartenders, society matrons, croupiers, physicians, and the United States Government.

Following the radium burns received by Becquerel and Pierre Curie in 1901, and the early therapeutic application of x-rays to dermatological problems, the power and potential of radium for curative purposes was cause for much speculation. Both because of the element's scarcity and the nonscientific testing by physicians, results during the first few years were equivocal. Newspapers carried reports of success, failure, and uncertainty in killing bacteria, curing blindness, turning the skin of Negroes white, determining the sex of unborn children and, most important, curing deep-seated and skin cancer. The technique suggested for stomach cancer, for example, was the ingestion of a radioactive drink, for the "liquid sunshine" would bathe the affected parts. In a related matter, William J. Hammer advised that "if I were the kaiser's physician I would gargle his throat with radium."[21]

By about 1904 greater supplies of radioactive materials were being imported and some serious physicians were bringing order to radiology. Minor disaster then struck, for the Austrian Government placed an embargo on the export of pitchblende ore and residue, and American Customs declared a twenty-five per cent duty on foreign-manufactured radium. A contemporary observer interpreted the Austrian move as a protection of the national glass industry of Bohemia, which used uranium as a pigment,[22] but more likely it reflected the government's displeasure that much of this material intended for scientific purposes was "ultimately and most unfortunately turned into commercial channels and put upon the market for sale."[23] The Austrian Academy of Sciences subsequently was given control of the mine and works, and under its supervision relatively large amounts of radioactive materials were made available for experimental purposes.

But the importing house of H. Lieber & Company, which previously had brought some forty thousand dollars of radium into the United States, was shocked when informed that a fifty milligram shipment, worth about one thousand dollars, would be taxed twenty-five per cent by Customs. While the official appraisers declared the material to be a chemical compound, the importer claimed it should be free of duty as a crude mineral substance and prepared to go to court (where he lost). Journalists, meanwhile, poked fun at the Dingley tariff schedules whose interpretation transmuted frogs into dressed poultry and snails into wild animals, while raising great obstacles to the importation of pickled sheepskins.[24]

News of this sort simply added to the public's fascination with radium. By 1903, this "natural Roman candle," since it continually emitted radiation, had become the inspiration for a variety of efforts in other fields. Miss Loie Fuller, the American "serpentine lady," created a number of "radium dances" in which monster moths and iridescent halos were formed using fluorescent salts from pitchblende residues,[25] and at the Casino in New York radium dances and the wonders of radium aided the plot of a musical comedy entitled "Piff! Paff!! Pouf!!!"[26] Yet another form of entertainment benefitted from this property of luminescence: "radium roulette" became the rage in wide-open New York. A roulette wheel was washed with a radium solution, such that it glowed brightly in the darkness. "Amid ghastly silence" an unseen hand cast the ball on the turning wheel and sparks marked its course as it bounded from pocket to glimmering pocket.[27]

After the exertion of moving luminous chips about, the weary gambler might desire some liquid refreshment. Radium beverages were available, generally "cocktails of fluorescent liquids rendered radioactive and glowing in the dark when lifted to the lips."[28] Radioactive drinks were, of course, taken more seriously by those seeking a "cure" at numerous fashionable spas during this period. Radium, in addition to all the other beneficial minerals contained in these waters, made such resorts as Hot Springs, Arkansas, and Carlsbad, Germany, more popular than ever.[29]

Fashionable hostesses, in something of the salon tradition, reinforced the radium fad by providing lectures with their afternoon teas, and it quickly became the smart thing to carry in one's pocket Sir William Crookes' newly invented spinthariscope.[30] This was a small brass tube fitted with a lens at one end, through which could be seen flashes of light caused by a speck of radium mounted at a small distance from a phosphorescent screen. Radium became somewhat commonplace even in sporting and literary circles, with one American requesting Madame Curie to allow him to baptize a race horse with her name,[31] and another writing a poem in her honor:

> The All-Master sealed a symbol of His might
>   Within a stone, and to a woman's eye
> Revealed the wonder. Lo, infinity
>   Wrapped in an atom—molecules of light
> Outshining centuries! No mortal sight
>   May fathom in this grain the galaxy
> Of suns, moons, planets, hurled unceasingly
>   Out of their glowing system into the night.
>
> O Man, thou scheme so marvellously planned
>   Of passions, hopes, desires, imaginings,—

Think not they burn within thy blood alone!—
They radiate from the eye, the lip, the hand,
In look, word, deed—that, fading, soon are gone
Or flash unto the eternal verge of things.[32]

The energy of radioactive decay was a source of continual amaze-
ment, and many mental efforts were made to convert this microscopic
phenomenon to macroscopic proportions, such that the layman could
conceive its magnitude. These "gee whiz!" calculations usually included
such thinking as the following: If a lump of coal is burned in oxygen, the
energy released is sufficient to lift itself against the force of gravity
through two thousand miles; hydrogen burned in oxygen releases
enough energy to rise eight thousand miles; radium, in contrast, could
travel three hundred and fifty million miles.[33] A variation, perhaps less
associated with the space age and more with the age of motor cars and
battle fleets, was the eye-opener that the energy stored in one gram of
radium would suffice to raise five hundred tons a mile high, and an
ounce could drive a fifty-horsepower vehicle at thirty miles an hour
around the globe.[34] Visions of the entire British fleet atop Mount Blanc
amused not a few. When, less than fifty years later, information about
the atomic bomb project was declassified, no such editorial magnification
was necessary for one to grasp its enormity.

Suggested uses for the miracle element ranged from the ridiculous
to the deadly serious. One farmer speculated that radium mixed with
chicken feed might have interesting results. The radio-eggs would either
hard boil themselves upon being laid, or would hatch the chicks without
need for an incubator.[35] Radium, additionally, was seen as useful in
detecting fake diamonds, for only the real gem phosphoresces in its
presence. The penetrating ability of the radiations could also be applied,
for making radiographs of iron and steel castings, and for discovering
pearls in oysters—a task for which x-rays then were used. Various time-
keeping, heating and illuminating devices were contemplated, while military enthusiasts rejoiced over luminous gunsights and potential means
of exploding from a distance the powder in an enemy's magazine.[36]

But if radium was so scarce, how would all these benefits mate-
rialize? The answer to this lay in the frequent assertion that new sources
had just been discovered. Indeed, the eastern states vied with the mining
states of the West in claiming extensive deposits of radium. Toward the
end of 1903, for example, a Boston mineralogist named C. P. King
created a sensation by reporting that "under the State of Connecticut,
from Bridgeport north and east to the Massachusetts line, there is a vast
bed of radium of sufficient power and value . . . to make or unmake the
United States."[37] Scientists at Yale greeted this with skepticism, pointing
out that almost all ground water contains some measurable trace of

radioactivity, but that this does not necessarily indicate the presence of commercial grade ores.[38] Yet, such claims of success were repeated so often that one could easily get the impression that many mining millionaires were newly being made. In fact, few made their fortunes in this field, and these few succeeded in the century's second decade, with the extensive carnotite fields of Colorado and Utah.

The American public was further exposed to information about radioactivity through the efforts of several lecturers, whose chief characteristic seems to have been the absence of serious research into the phenomenon. William Hammer, as already mentioned, was extremely popular, but others, such as Samuel A. Tucker of Columbia University, Theodore J. Bradley of Albany Medical College, George R. Southwick of Boston, and W. B. Patty, seem also to have had eager audiences.[39] Additional evidence of this almost insatiable hunger for radium information is the great success of the twice-daily lectures given by the United States Geological Survey at the 1904 Universal Exposition. Indeed, the lectures and the exhibit of radioactive preparations and minerals were considered the outstanding attractions of the fair.[40]

St. Louis was the location, and the centennial of the Louisiana Purchase the motivation, for this Universal Exposition. Its organizers envisioned more than a world's fair, however, and allied events were planned. Thus, in September 1904, foreign and American scholars in all branches of learning gathered on the west bank of the Mississippi River to participate in the International Congress of Arts and Science. Knowledge was divided and subdivided into numerous categories which ranged from language to religion, from belles-lettres to utilitarian science, from gynecology to the wild races of the world, and from painting to physics.[41] Woodrow Wilson, then president of Princeton University, lectured on history; Robert Koch discussed "The theory of serum treatment"; Jane Addams, Jacques Loeb, William Ramsay, and Michael Pupin spoke on their individual specialties, and Sir William Crookes held forth on a subject dear to his heart: "Telepathy." Special pride was taken in the great number of distinguished foreign scholars who accepted the invitation to attend the Congress. These notables included James Dewar, Henri Becquerel, Dmitry Mendeleev, Jakob van't Hoff, Rudolf Fittig, Otto Backlund, Archibald Geikie, Henri Bergson, Wilhelm Ostwald, Ludwig Boltzmann, and Henri Poincaré.[42]

The papers presented in physics concerned the modern aspect of this science.[43] The American Carl Barus reviewed "The progress of physics in the nineteenth century," and then Paul Langevin, from Paris, and Ernest Rutherford, from Montreal, read long reports on the electron and radioactivity, respectively. Rutherford, in particular, deserves much credit for instigating the serious study of radioactivity in the

United States. While he was especially effective among the scientists, as will be discussed in the next chapter, his efforts to reach a wider audience furnished the relatively few lectures and articles that contained sober and accurate information on the subject. In January 1904, *Harper's Monthly Magazine* carried an article by Rutherford on radioactive disintegration.[44] His May 1904 lecture at the Royal Institution of London, in which he proposed that the heat evolved from radium indicated a much greater age of the earth than Lord Kelvin "permitted," was widely reported in the American press.[45] This idea, in fact, seemed to be the source of his booming reputation; his far more profound transformation theory with Soddy, over a year older, was less prominent in the public mind. Interest in the age of the earth was universal and *Harper's Monthly Magazine* commissioned Rutherford to write an article on it, which appeared in February 1905.[46]

What public interest there was in transformation was laid at the feet of Sir William Ramsay, who traveled across America in 1904 as a conqueror. Ramsay was one of the foremost chemists in the world, having been largely responsible for the discovery of an entirely new group of elements in the periodic table, the so-called inert gases. It was, therefore, understandable that his interest in radioactivity should be intense when the emanations appeared also to be inert gases, and when the alpha particle was thought to be helium. In early 1903, Frederick Soddy left McGill University and Rutherford, and began a year of collaboration with Ramsay in London. While Soddy was with him, excellent results emerged from the laboratory. But once he departed, Ramsay's research was incompetent, for he had never really taken the trouble to learn much about radioactivity.[47] And even before Soddy had gone, Ramsay's public statements were misleading, for he was prone to exaggeration and self-glorification.

Yet, such statements are the grist of headlines: "Radium to disappear,"[48] "Ramsay turns glass into lead,"[49] "Elements may be traced to origin,"[50] and "Sir William Ramsay, discoverer of radium."[51] The basis for journalistic consideration of Ramsay as one of the world's greatest authorities on radioactivity was genuine enough, however, for with Soddy he had shown that the gaseous radium emanation transforms itself into helium.[52] Helium, detected decades earlier on the sun, but only recently by Ramsay on earth, excited the public's imagination, and the concept of a related transmutation added to the interest. Thus, at once, Ramsay was *the* expert on transformation, an interpretation he honestly advocated. And because of his far greater fame among nonscientists, the names of Rutherford and Soddy, who authored this theory, were generally omitted from published accounts.

For Americans, in particular, Ramsay's prominence was heightened when the British Society of Chemical Industry's first annual meeting to be held abroad took place in a succession of eastern and mid-western cities. As president and most prominent member of the hundred-man English contingent, Ramsay gave numerous interviews and public lectures, including one at the Congress of Arts and Science in St. Louis, and everywhere he was asked about radium.[53]

Literary magazines abounded around the turn of the century and most made efforts to educate their readers in matters radioactive, much as the *Saturday Review, Reporter, Atlantic, Harper's,* and *Readers' Digest* have in more recent years printed articles on fallout. Thus, we find a large selection of notes and full-length discourses by famous, unknown, and anonymous authors in *Current Literature, Overland Monthly, Harper's Weekly Magazine, Harper's Monthly Magazine, Littel's Living Age, McClure's Magazine, Cosmopolitan, American Monthly Review of Reviews,* etc. Technical journals for the public were also in vogue and *Popular Science Monthly, Scientific American* and *Scientific American Supplement* were, in general, excellent sources of factual, nondistorted information. These latter three often reprinted in full the original research papers by the leaders in radioactivity abroad. Review articles also appeared, and we may cite, for example, one by Robert A. Millikan in the *Popular Science Monthly,*[54] and another, following a chapter of Jack London's *The Sea-Wolf,* by Ernest Merritt in the *Century Magazine.*[55] For the variety of reasons already stated, the subject of radioactivity firmly engaged the public's attention by 1903 and far more articles were printed in that year than in the three preceding years combined. Newspapers shared in this trend, the *New York Daily Tribune,* for example, running thirteen editorials and twenty-three stories on radium during 1903, while printing virtually nothing on it earlier.

American authors (and publishers) were equally anxious to exploit this widespread interest. Charles R. Stevens gave the title *Radio-Activity,* which, curiously, was exactly the title of Rutherford's own book in the same year, to a highly philosophical layman's primer on science.[56] Stevens' volume abounded in statements such as, "knowing and loving man and Nature is the only true way of knowing and loving God,"[57] and it was only in the last of ten chapters that radioactivity was discussed, there to be treated as depending on the tone of ethereal motion.

There were other works of a less theological nature, but they too suffered from defects. William Hammer's printed lecture,[58] though it was voted in 1904 as one of the fifty best books of the past year desirable for a village library in New York State, was neither comprehensive nor systematic. The same, and more, may be said about *Radium and Radio-Active Substances; Their Application Especially to Medicine,* by Charles Baskerville.[59] Baskerville, about whom more in the next chapter, was a chemist with no expertise whatever in the book's area of emphasis. His presentation of medical developments, therefore was haphazard. Nor did the physical side of radioactivity receive much better treatment, as Rutherford's friend Bertram Boltwood noted when he reviewed the book.[60] Even more, Boltwood felt that Baskerville was a plagiarist, and a sloppy one at that, for he found numerous examples of extracts from Rutherford's *Radio-Activity*. The book review, however, contained no mention of this, for it appears that Boltwood had to tone down his criticism in order to have it printed at all. Rutherford was also forced to remain silent, for he had accepted from Baskerville the honor of the dedication.[61]

A somewhat better book also appeared in 1905, entitled *The New Knowledge; a Popular Account of the New Physics and the New Chemistry in Their Relation to the New Theory of Matter*.[62] The author, Robert Kennedy Duncan (1868–1914), was professor of chemistry at Washington and Jefferson College, in Washington, Pennsylvania. He early became convinced that there was a great need for clear but accurate interpretations of the latest scientific advances for the layman. Thus, in about 1900, he became a contributor to a few New York periodicals. This venture being successful, *McClure's Magazine* sent him to Europe to study the field of radioactivity in the summer of 1901. The next year *Harper's Monthly Magazine* published an article by him, one of the earliest, substantial popularizations in this country, which was reprinted in *Current Literature* and *Overland Monthly*.[63] In 1903 the publishing firm of A. S. Barnes returned him to the continent to gather material for *The New Knowledge*. Following this interpretive type work, Duncan sought other means of furthering the cause of science. In later years, as a professor of industrial chemistry, he devoted much time to establishing industrial fellowships at both the University of Kansas and the University of Pittsburgh; out of his efforts at the latter school grew the Mellon Institute.

Duncan was a talented writer and his book was highly readable as well as accurate. But, like Baskerville, he knew a good source when he saw it, and, thus, Rutherford confided to Boltwood:

> I am getting quite accustomed to having my views and remarks and sections copied wholesale without ever a thankyou. If you read Duncan's book, you will find many of my paragraphs reproduced & ideas & everything else including even tables of radioelements.[64]

Yet, aside from its means of fabrication, *The New Knowledge* seems to have been a worthy product and it went through at least five editions. It was a substantial account of radioactivity, well-directed toward the "intelligent layman." For the "final word" on the subject, however, Rutherford's text[65] was recognized as a classic upon its appearance, while others who similarly pursued the science also advanced some works of good quality.[66]

The numerous examples cited throughout this chapter testify to a widespread interest in radioactivity. This interest, understandable on the basis alone of the impressive properties of the new element radium and its industrial potential for heating and illumination, had still a more fundamental foundation. Science had finally become a worthy subject of conversation among nonscientists. Its value now was recognized in the electrical and chemical industries, as well as in medicine, and the prestige of the newly founded Nobel Prizes was immense. Radioactivity research was honored in 1903, when the prize in physics was shared by Becquerel and the Curies. But why was radioactivity considered of such importance? The answer is three-fold: The phenomenon was more mysterious than other physical processes because there were virtually no ties with the older, classical science. Hertz's work with electromagnetic waves, Röntgen's discovery of x-rays, Moissan's achievements with the electric furnace, Rayleigh and Ramsay's isolation of rare gases, and Dewar's liquefaction of hydrogen were exciting and significant, to be sure, but these feats were more comprehensible because the links with the past were more obvious. With radium, what mysteries of the nature of matter would be solved? What new regions would be opened?

To this enigma must be added the recognition that the phenomena of radioactivity touched numerous other sciences as well as other specialties within physics. Be it geology, medicine, biology, astrophysics, cathode rays, x-rays, relativity, etc., there were connections with radioactivity; its influence was ubiquitous.[67]

The final feature of our explanation is more simple and more profound. Radioactivity was a new property of matter, fit to take its place beside gravitation, magnetism, electricity, light, and heat.[68] Its significance rested upon its membership in such an exalted group.

3 First Research Efforts

During the first years of the twentieth century few serious efforts were made in the United States to pursue radioactivity investigations. There were a number of "dabblers," such as A. F. McKissick who found granulated sugar more active than uranium, but though these men may have followed their studies resolutely, to a large extent they were working in the dark. Not only were their experiments unstructured and unsystematic, they were also ignorant of the essential literature on the subject.

The journals, both domestic and foreign, and the populizers who spoke to technical audiences, described in the preceding chapter, therefore served a necessary function of raising the intellectual level to the point where "serious" research could flourish. The distinction between serious work and dabbling becomes subjective at the borderline of the two, and the erection of ironclad boundaries is a fruitless task. Nevertheless, a working definition of the serious efforts to be described would have it that the investigation followed a planned program, that the apparatus and materials were believed sufficient for the purpose, and that the results were taken seriously by the scientific community. Whether the results are considered right or wrong in the light of current knowledge is immaterial; their reception at the time is, however, highly significant as a measure of the developing science.

Radioactivity investigations in America followed the general pattern of these studies abroad, and may be divided into the following categories: the sources, the radiations, effects of the radiations, atmospheric radioactivity, and theory. While the contributions of some individuals were wide ranging, most restricted their work to a single group. Thus, for example, a chemist would logically examine the radioactive sources, a physicist the radiations, and someone with a taxonomic bent the radiation effects.

CHARLES BASKERVILLE (1870–1922)

Sources of Activity

The first research on radioactivity, worthy of the name, to be performed in America, was initiated without any thought of the phenomenon in mind. Charles Baskerville (1870–1922), an alumnus of the University of North Carolina and by 1900 professor of chemistry there, encountered unexpected difficulties in precipitating all of a quantity of thorium in solution.[1] This partial separation puzzled him sufficiently that he attempted the operation with thorium samples from many different locations. He even requested a pure thorium salt from the noted Professor Bohuslav Brauner, of Prague, but never received a reply. Imagine his surprise, said Baskerville, when he then saw an article by Brauner in a 1901 issue of the *Proceedings of the Chemical Society* (London), in which evidence of the complexity of thorium was presented. Brauner termed the constituents "thorium alpha" and "thorium beta." Baskerville thereupon took the first opportunity—a meeting of the North Carolina Section of the American Chemical Society, on 23 April 1901—to make public mention of the work he had begun in 1896 (before thorium was known to be radioactive). He claimed that "as soon as the unexpected properties were noticed almost five years ago, I indicated the differences by terming one Th and the other Th(X)."[2]

Baskerville's claim of priority made particular sense at this time for it was widely felt that neither pure thorium nor uranium was radioactive. The existence of these elements had been known for seven and eleven decades, respectively, and it was hard to believe that this property could have been overlooked so long. Was it not more likely, the argument ran, that the radioactivity is embodied in another material that is invariably found with these two elements? Not long before, Sir William Crookes had extracted the "active component" uranium X from uranium, and now was looking for an analogous thorium X.[3] Baskerville's work, which began solely as a chemical problem, had therefore been given more meaning by the radioactive property of thorium. When he placed some of his thorium "components" upon a photographic plate for 72 hours, he was pleased to note that the plate was exposed. While not ready to assert that his new substance differed from Debierne's actinium, or indeed that it was a new element, Baskerville nevertheless kept his foot in the door by writing:

> on account of the extensive occurrence, in this state (North Carolina), of the monazite sands from which the original material was obtained, if the investigation give a successful issue, I should like to have the element known as Carolinium, with the symbol *Cn*.[4]

Rutherford and Soddy announced discovery of their thorium X the next year, but it is almost certain that it was not Baskerville's carolinium.[5] By this time both Becquerel and Crookes were aware that their uranium X preparations had inexplicably lost their activities, while the uranium from which the X material was separated had its activity restored. Upon learning this the two McGill scientists immediately tested their thorium and thorium X and found the same results: the X material was virtually inactive, while the thorium had regained its normal activity. More than any other evidence, the results of this experiment led Rutherford and Soddy down the path to the transformation theory of radioactivity. Each radioelement, they reasoned, was genetically related to others and would be formed and would decay at fixed rates indicated by the half-lives.[6] In this particular case, thorium was the parent of thorium X. When freshly separated, the thorium X continued to decay at its usual rate, but since no more was being formed here its quantity decreased and, correspondingly, so did its measured activity. The activity of parent thorium was seen as a mirror image, for it was very low when initially deprived of its daughter products, but it rose as additional thorium X and subsequent products were formed.

Characteristically, for chemists, Baskerville used the qualitative photographic technique in testing the radioactivity of his Th(X) or carolinium. Physicists generally employed electrometers and electroscopes which gave quantitative data, from which half-lives could be calculated. His chemical procedures offer no clue to the identity of the material he separated, while some odd specific gravity measurements reported and the lack of decay period information further confuse a reconstruction. In Baskerville's defense it may be said that his investigation was primarily chemical, not radioactive, and that there was good precedent for suspecting that thorium was a complex substance. It was found widely, in a confusing variety of compounds, but usually associated with elements of the rare earth group. One of the rare earths, yttrium, had already been separated into four elements, and Baskerville felt that thorium's difficult chemistry could well be a result of similar complexity. With the clear vision of hindsight, we can see that the problems were caused by thorium's spontaneous disintegration into a series of other elements which, even if separated chemically, begin immediately to accumulate with the thorium once more. Baskerville, no more than a competent chemist, had unwittingly drifted into two areas—rare earths and radioactivity—where skills of a higher order were required. Further, he had logically followed the standard chemical tradition associated with the Curies, that of separating measurable quantities of apparently permanent radioactive elements. He could hardly be aware that the most fruitful line of investigation now was the Rutherfordian pursuit of transient activities and the genetic connections between such sources.

In his work thus far, Baskerville had simply indicated that he believed common thorium to be more than a single element; he had not attempted to apportion the radioactivity between a new elemental thorium and his carolinium or Th(X). In February 1902, Karl A. Hofmann and Fritz Zerban, in Germany, introduced a new twist into the question by asserting that thorium possesses a secondary activity, induced from uranium, but is not intrinsically active.[7] George Barker, the University of Pennsylvania popularizer of radioactivity, mentioned in the last chapter, objected vigorously to this conclusion and embarked upon a series of experiments in which he exposed photographic plates to numerous thorium samples.[8] In order to disprove the Germans' contention, however, he required thorium free from traces of uranium. Hearing that Baskerville had prepared an exceedingly pure thoria (thorium oxide), Barker requested and obtained loan of a sample to confirm his work. As the thoria darkened a photographic plate after 96 hours exposure, he felt justified in concluding that "thorium does not owe its activity to an induction from uranium existing in the minerals from which it was obtained."[9] Curiously, no one suggested that the induction might have been long-lasting, dating from the time thorium and uranium coexisted in the mineral. But perhaps such explanations were out of date, for even Becquerel had dropped his long-lived-phosphorescence interpretation of radioactivity. More likely, the question of thorium's

GEORGE B. PEGRAM (1876–1958)

primary or secondary radioactivity was considered of minor importance. Far more significant was the question whether thorium had any activity at all, for this involved the relationship with thorium X and a host of other genetic problems.

Probably the first American to undertake research on radioactivity whose name will be familiar to contemporary scientists was George B. Pegram (1876–1958). Though an accomplished physicist, his fame rests more on his administrative skills in such posts as dean of the School of Mines, Engineering, and Chemistry, dean of the Graduate Faculties, and vice-president of Columbia University (during Dwight Eisenhower's presidency of the university). To a larger audience he is known as the man who first informed the United States Government of the possibility of an atomic bomb.[10] This contact was made in March 1939, after Enrico Fermi and his colleagues, working then at Columbia, obtained encouraging results in fission experiments.

While a graduate student at Columbia, Pegram in 1901 exhibited a broad interest in radioactivity by writing a popular, review-type article for *Science*.[11] Very likely, he was thrown into this field by the physics department chairman, Ogden Rood, who had devised a new electrometer and encouraged its use. Pegram probably had no objections to this strategem, however, for he not only found several new radioactive minerals,[12] he chose his Ph.D. dissertation topic in radioactivity. Reflecting a widespread conviction (even at that early date), Pegram wrote that "the study of the radio-active substances will surely lead to a better knowledge of that which is the subject of much of the physical research of today, the intimate structure of matter."[13]

His approach to this fundamental problem was through the electrolysis of radioactive solutions.[14] This was a relatively virgin area, pioneered by Ernst Dorn and Willy Marckwald in Germany, which offered another means of separating radioactive constituents. Rutherford had shown that thorium emanation, a gas, deposits a temporary activity on nearby bodies, and others had found that in air this activity will concentrate preferentially upon a negatively charged wire. In contrast, Pegram's electrolysis of a thorium nitrate solution placed the greater activity upon the anode. Chemical analysis of the anode deposit showed it to be a lead oxide, while electrical measurements revealed a half-life of eleven hours. This corresponded to the period of Rutherford's active deposit from thorium (later named thorium B, and now known as the isotope lead-212), and the circumstance of its attraction to the anode was due to the simple chemical fact that the lead-oxygen ion in question is negatively charged—though Pegram was not then clear about this. Similarly, he was silent about the distinction between an ion formed by the detachment of an electron from an *atom* in the transformation process, and one formed in an electrolyte by the splitting of the bonds holding a *compound* together.

The material from which his salt was prepared was Brazilian monazite, imported from Germany. The commercial purification either had not removed the lead, or had been done long enough before to permit decay of the thorium into measurable amounts of new radioactive lead. When Pegram then tested exceedingly pure thorium salts, obtained from Baskerville, among others, the anode deposit's mass and activity were both much smaller than before, and the activity's decay period was only one hour. Had he been a chemist, or had the supply of pure salts been larger, Pegram might have attempted the isolation and study of greater amounts of this substance, which became known as thorium C (bismuth-212). As it was, this ended his electrolytical experiments and his later work in radioactivity was decidedly physical.

While Pegram's data fit into no theoretical framework of the time, and, like Baskerville, the stage of his study inclined him toward trying to identify substances rather than relate them genetically, his investigation, nevertheless, illustrates a few characteristics of the science around 1902–1903. The radioelements with eleven hour and one hour half-lives were but two among a large and increasing number of such materials. For the most part, their chemical identities were as yet unknown and they presented a confusing array of apparently unrelated substances. The Rutherford-Soddy transformation theory of radioactivity brought order out of randomness by designating decay series and sequence within series. But it was to be the work of the next decade and more to place the radioelements properly.[15] Pegram was strongly inclined to accept the transformation theory, though it is questionable whether he and others fully understood the transmutational intent of the authors. His statement that "radio-activity is a property of some kind of matter derived from the thorium which in some way becomes closely attached to the lead peroxide"[16] reflects the prevailing uncertainty. It took ten years more for the phrase "closely attached" to lead to the concept of isotopy, for there was an understandable reticence to proclaim boldly that the "derived matter" was merely another form of the common chemical element to which it adhered. This reticence derived naturally from the paucity of evidence, but also from the feeling that, while transformation consisted of a subatomic chemical change, this change somehow occurred *outside* the periodic table of elements. Only in the case of the production of helium from radium, as proven by Ramsay and Soddy, was a real transmutation widely and vocally credited.

What little overt opposition to the transformation theory there was came from scientists not working on radioactivity. Presumably, there were many more who could not accept the idea of transmutation, but whose views were not put into print. Baskerville seems to have fallen into this group, for he was unable to link the rise and fall of activities with the growth and decay of radioelements. In a letter to Pegram, who earned extra money by testing radioactive minerals, Baskerville wrote:

> I confess I am not altogether able to understand this matter at all. If only the Alpha rays are given off and they are driven off at red heat, then where in thunder do the new Alpha rays come from? . . . I still am inclined to the opinion that the thorium itself is not strictly speaking radio-active, but contains a constituent which may be distributed through the several fractions which we have secured.[17]

By the spring of 1904, Baskerville's chemical investigations had progressed to the point where he felt that "the preponderance of evidence is favorable to the assumption of the existence of two new members of the family of chemical elements."[18] Though data were entirely lacking on the two criteria considered by chemists essential for such proof—spectral lines and atomic weights—Baskerville named his new elements "carolinium" and "berzelium." He also took the occasion to side with Hofmann and Zerban in the running controversy with Rutherford, Soddy, Curie, Schmidt, and Barker over the intrinsic activity of thorium.

Baskerville was, in fact, confused by the many thorium decay products, and his carolinium and berzelium now share a showcase with Fermi's transuranium elements in that mythical Museum of False Discoveries. At the time, however, his work was accepted eagerly by his chemical colleagues and he became famous overnight. The *New York Times* headlined "Dr. Baskerville the Only American Who Ever Found a New Element," and the thirty-three year old former football star at North Carolina was hailed as an all-American aristocrat.[19] Baskerville received job offers from many "prestigious northern universities," the newspapers reported, and accepted the department chairmanship at the City College of New York. From 1904, however, his activities fell more and more in the areas of industrial research, consulting, administration of his new laboratory and expanding department, and in being a public figure, such that carolinium and berzelium were quietly forgotten.

Yet, for a short time in his new post, Baskerville continued his study of thorium. Indeed, he was joined by Hofmann's pupil, Fritz Zerban, who had received his degree from Munich in 1903, and was enabled to come to the United States on a Carnegie assistantship. Zerban earlier had found traces of uranium in supposedly pure thorium ore and this was the basis for his belief that thorium's activity was actually due to other substances.[20] With Baskerville, he continued this line of work, and the two reported a new source of inactive thorium in "a rock from South America."[21] Typically, the quantity under observation was too small for several desired tests and plans were announced for further investigation. These erroneous results underscore the prevailing situation: standard chemical analyses were highly unreliable. They may have been satisfactory when the presence of traces of an element was of no chemical importance, but now far greater precision was required, for radioactive elements generally appeared in minute quantities. Baskerville and Zerban no doubt saw no radioactivity in the rock because the weak radiation from the small amount of thorium was unable to darken their photographic plates appreciably. Buoyed by this "success," and noting that someone had found radium in the thorium ore, monazite, they then overlooked Zerban's own finding of uranium in this ore and postulated that "the radio-activity of thorium from uranium-free minerals may be explained by the presence of radium in them."[22] In so doing, they also overlooked the transformation theory. Radium's descent from uranium had long been suspected, and Boltwood had just published strongly persuasive evidence for it: radium did not occur without uranium.[23] And even if the mineral was indeed uranium free, the transformation theory suggested the possibility that radium might be found as a decay product of thorium, although, in fairness, the chemical identity of thorium series members remained uncertain for a few years more.

Radiation Properties

From the time of Becquerel's discovery of radioactivity to perhaps a year after the transformation theory's publication, the radiations formed the core of radioactivity research. But even after chemical investigations of the radiators began to supplant physical studies, the rays continued to receive significant attention. By the time Americans entered this field the basics were well established: Becquerel had shown the equality of the beta ray and electron in early 1900,[24] Paul Villard had discovered the highly penetrating gamma ray shortly thereafter,[25] and in 1903 Rutherford had proven that the alpha particle bears a positive charge and is of atomic dimension.[26] The American work did not contribute materially to the determination of the alpha and gamma radiations as charged helium atoms and electromagnetic radiation, respectively, but other problems of the rays were attacked with creditable results.

After his work on the electrolysis of thorium solutions, George Pegram selected another interesting topic for investigation. With Harold Webb, a graduate student at Columbia, he attempted to measure the energy liberated by thorium. The previous year Pierre Curie and Albert Laborde had shown that radium maintains a temperature higher than its surroundings.[27] This had given rise to much speculation about unlimited sources of energy, and it was to be expected that someone would try eventually to show this energy emission in other radioactive bodies. Since, weight for weight, radium was the most active substance known, the determination for thorium required greater delicacy, but it was a case of finding something known to be there.

Detection of the minute effect was announced in April 1904, at a meeting of the New York Academy of Science.[28] The following month Pegram told the same group of recent experiments concerning the emission of electricity by radium.[29] In effect, he confirmed Soddy's conclusion that when positive alpha particles are expelled from radium there is no corresponding negative charge left behind. Since, however, the negative charge might have been removed in the emission of beta particles by some of radium's decay products, Pegram suggested a similar test using radium bromide which had recently been in solution. The freshly prepared radium would have none of its daughter radioelements mixed with it and the radiation would therefore consist only of positive alphas.

Before he reported any such results, William Duane (1872-1935) obtained definite indications of residual charges and presented a clear explanation that showed how much the science had progressed in the intervening two years.[30] Like Alexander Dallas Bache (1806-1867), who was superintendent of the United States Coast Survey and first president of the National Academy of Sciences, Duane was a descendent of Ben-

WILLIAM DUANE (1872-1935)

jamin Franklin who became an eminent scientist. After study at the University of Pennsylvania and Harvard, Duane spent two years in Göttingen and Berlin, earning his doctorate in the German capital under Nernst. Until 1907 he was professor of physics at the University of Colorado, but at that time he went to Paris, becoming the first American to work in Madame Curie's laboratory. Six years and seventeen good papers on radioactivity later, Duane returned to Harvard, where he held a joint position in the physics department and on the Cancer Commission. Now his interests slowly shifted to the biological applications of radioactivity and to x-rays. In fact, he achieved his greatest fame for his x-ray work: the Duane-Hunt law, which shows that the short wavelength limit of the continuous spectrum varies inversely as the voltage across the tube. This law had particular merit, because from it a highly accurate value of Planck's constant could be determined. Fame of a different sort also came to Duane for his heated denial of the Compton effect. To his credit, though, he examined his discrepancies jointly with Compton and then freely admitted his error.

In the 1906 paper on radioactivity, Duane wished to show not only that he could detect a residual charge, but that here was a new method of studying the rate of decay of the excited activity deposited by radium emanation. He was able to observe an accumulation of negative charge for the first five or six minutes, which after ten minutes changed to a net positive charge. The explanation was simple: radium A emits only alpha particles,[31] while radium C emits both alphas and betas.[32] After ten minutes the radium A had virtually disappeared, and the effect then was due to radium C, in which beta decay predominated. When the experiment was repeated in vacuo, however, the electrometer indicated a continual accumulation of positive charge from the start. Apparently, the total beta emission was now masking the alpha emission from radium A. After the first ten minutes the rate of positive charge accumulation was seen to increase in a manner very similar to the theoretical curve given by Rutherford for the sum of radium B and radium C activities. Formerly it had been believed that radium B was a rayless product, meaning that it decayed into its daughter without the emission of any radiation. Duane now concluded that radium B emits about as much negative electricity as does radium C; its beta rays are relatively weak and their absorption in air had prevented them from reaching the grounded part of the circuit. Duane's work thus led to a reexamination of the so-called rayless transformations.

His interpretations, however, came to be modified by the recognition that ionization phenomena were far more complicated than believed. In particular, emission of secondary electrons, from the atoms of the radioactive sample as they were struck by radiation from their disintegrating neighbors, caused a variety of effects depending on air pressure and applied electric field. It was not a simple case of subtracting from the source a charge of $+2$ for each alpha particle expulsion and -1 for each beta particle. Secondary rays were widely examined, the American contribution being a determination by Henry A. Bumstead that, *if* the secondaries continue to leave the metals once the provoking beta and gamma radiation is removed, the emission lasts no longer than nine thousandths of a second.[33] Bumstead performed this work in 1904, in the Cavendish Laboratory, Cambridge, England, while on sabbatical leave from Yale University. It was, in fact, an extension of experiments by the Cavendish professor, J. J. Thomson, who maintained a lively interest in radioactivity while pursuing his "corpuscles" (electrons).

Canadian-born Samuel J. Allen (1877–1966) embraced a whole range of beta radiations, from primary to quartenary, as might be expected of a former student of Rutherford. He had been trained at McGill University, where his efforts were directed toward atmospheric radioactivity, and then at the Johns Hopkins University, where he turned to a study of beta rays. In 1906 he began a lifelong career at the University of Cincinnati, his research there marking him as perhaps the most productive physicist in America in the field of radioactivity.

Difficulty in obtaining the expected magnetic and electrostatic deflection of beta rays led Allen to examine them more critically.[34] Further refined experimental geometry, replacement of his electrometer with an electroscope which could be better shielded from electrostatic influences, and placement of his apparatus in a vacuum permitted him to achieve the bending of not only the primary betas, but secondaries and tertiaries as well. It was these latter radiations that had caused the initial problem. Walter Kaufmann, a few years before in Germany, had shown that beta particles, traveling at nearly the velocity of light, seemed to exhibit an increase in mass.[35] Allen, in 1906, confirmed this and extended the observations to secondary betas in his measurements of velocity and charge-to-mass (e/m) ratio. The subject was of considerable interest at the time, for, because of what later became known as the electron's relativistic mass, a number of leading physicists inclined toward the view that the particle's *entire* mass was electrical in nature. Electromagnetic theory suggested not only this, but the further extreme position that *all* mass in the universe was of electrical origin. Allen declined to speculate, being content to present his experimental findings. In this he was typical of most new world physicists, who were not especially strong in theoretical matters, particularly those of a mathematical type. A genius such as Yale's J. Willard Gibbs (1839–1903), for example, was unique, for his isolation from other mathematical physicists as well as for his brilliance.

The American tendency to repeat and confirm work done abroad (with, of course, the hope of discovering something new) extended also to alpha radiation. Not only Rutherford in Montreal, but William H. Bragg in Adelaide, was finding the alpha a fruitful subject in this century's first decade. With Richard Kleeman, Bragg showed that this particle loses its energy by ionizing the material through which it passes and that it has a definite range in each material.[36] Further, the Australians found that each alpha emitter imparts a different energy to its radiation; thus the particle from radium C, for example, will always travel a greater distance than that from radium. Whereas Bragg and Kleeman had performed their experiments with an electrometer to measure the ionization, Edwin P. Adams (1878–1956) sought to repeat portions of their work with a scintillating zinc sulphide screen.[37] Adams, who spent his entire career at Princeton University, corroborated their empirical relation that the loss of range as an alpha particle passes through a gas is proportional to the sum of the square roots of the weights of the constituent atoms. But neither he nor the Australians were able to explain the square root in the equation which, it was learned only decades later, was but a first approximation, though an accurate one, to a more precise understanding of the interaction of alpha particles with matter.

Another Princetonian who wrote on alpha particles was Owen W. Richardson (1879–1959), the winner of the 1928 Nobel Prize (as an Englishman), who is more famous for his work on electronic phenomena. Richardson had been brought to Princeton in 1906 from the Cavendish Laboratory, as part of an effort by the school's president,

Woodrow Wilson, to transform that university from a provincial institution to a cosmopolitan center of excellence. Like many others, Richardson was concerned with the true nature of the alpha particle.[38] Was it really an atom of helium (as Rutherford and Royds had yet to prove), or could it be perhaps an atom of hydrogen? Would it be possible, he asked, to have an alpha that behaved cyclically, losing its charge upon collision with a gas molecule and regaining it at the next collision? With such a system, in which the particle was charged only half the time, the *measured* charge-to-mass ratio would be in reality not e/m but e/2m. "On this view," he wrote, "Rutherford's measurements would indicate that the α particles are hydrogen atoms with the normal charge instead of helium atoms with twice that charge."[39] But, Richardson pointed out, this would leave helium, which had been observed, with no place in the picture.[40] If then the alpha really is a helium atom, he suggested, the idea of an alternating charge might somehow explain why ionization ceases at a certain range. Thus, besides examining the arguments for the alpha's elementary identity, Richardson's speculations were an effort at understanding the ionization phenomenon.

Radiation Effects

From the time of the earliest radiation burns experienced in Germany and France, the diagnostic and therapeutic effects of radioactive materials had been studied. It was quickly demonstrated that x-rays produce a sharper photographic image in a shorter exposure, so the application of alpha, beta, and gamma rays was limited largely to therapeutic efforts, in such cases as dermatology, tumors, and cancer. Radium also was hailed as the universal curative, and in the period before World War I many businessmen, ranging from honest to ignorant to unscrupulous, offered the public a variety of products. Radioactive mineral waters, bath waters, mud packs, and inhalation devices for the gaseous emanation competed for the faddists' money, few of whom were injured, fortunately, since the products contained vanishing quantities of active substances. The detailed story of medical and biological research and applications is beyond the scope of this volume, which deals with physical and chemical research, though a brief survey is given in a later chapter. The two cases of medically oriented investigators, discussed below, are cited to illustrate the thinking in some quarters on *scientific* topics.

In the humanitarian spirit that guided many professional men at this time, William Rollins (1852–1929) was anxious to find the most efficient radiation "to aid in the relief of human suffering."[41] A Harvard-trained Boston dentist, he was a pioneer in the development of x-ray protection methods.[42] He had experimented with the Röntgen rays since within a half year of their discovery, but in 1900 wrote:

> A distinguished American physicist [G. F. Barker] has expressed the opinion that the radio-active substances would be so intensified as to act as substitutes for the complicated and troublesome apparatus now required for producing X-light for medical purposes. Having a high regard for his opinion I abandoned experiments with X-light to work with these substances.[43]

When Rollins made the test, however, he found that the radiation diffused so much in the tissues that not even the bones of the hand were visible. This had long been known to occur with uranium sources, but apparently better results were anticipated with the more intense radium. Still hoping to use radioactivity for diagnostic purposes, Rollins then proposed that a vacuum tube be constructed with a radioactive cathode, which would eliminate the need for an electrical generator. Since Becquerel rays consist partly of the same particles found in cathode rays, and their velocities are approximately the same, all that would be needed for the production of x-rays (by the cathode rays which stimulate their emission) would be the proper degree of evacuation. In truth, it is an ingenious idea; in practice, it neglects beam intensity, range, and collimation considerations.

Myron Metzenbaum, an assistant in the anatomy department of the Cleveland College of Physicians and Surgeons, became interested in radioactivity at a time when numerous sources besides radium were known to exist. In 1904, he reasoned that the disadvantage in reduced activity of one of these sources might be balanced by its lower cost and more convenient form of application. To obtain such material, Metzenbaum suspended tubes of radium in or next to various solutions, solids, and powders, but, strange to tell, they acquired no activity.[44] Like many other physicians, he knew little about the science and had assumed that activity could be induced in other bodies by mere proximity. After all, was not the active deposit left by radium and thorium emanations called "induced activity"? Indeed it was, but the terminology was misleading, for the induced activity was mechanically deposited by the emanation, not caused mystically. It would be a few years yet before the medical profession obtained the services of full-time physicists, an event that corresponded with the use of "radium bombs" in hospitals.

Still, in this first decade of the century, many people wanted to know, simply, what these rays could do. The obvious way to gain satisfaction was to expose numerous materials to the radiation and look for effects. If the effects were spectacular, this would suffice to satisfy most onlookers. It happened that diamonds phosphoresce under the radiation's stimulation, and the combination of radium and jewels was a journalist's dream.

This combination was also of fascination to the Tiffany and Company gem expert, George Frederick Kunz (1856–1932). During his youth in New York City, Kunz showed such a remarkable ability for geological research and a knowledge of precious stones that he was employed by the famous jewelry store as a specialist when he was but twenty-three years old. Here he earned a reputation which brought him appointment as representative of the United States Government at several international expositions, and such was his authority, that he became a pundit on fashions and novelties in jewelry and gems.

In 1903, even before Baskerville became famous for his alleged discovery of two new elements, Kunz invited him to New York for the purpose of examining a large number of minerals under the influence of x-rays, ultraviolet light, and radium radiation. The minerals, some fourteen thousand specimens, comprised the great Morgan-Bement collection in the American Museum of Natural History, and had been acquired some years before by J. Pierpont Morgan through the efforts of Kunz, who was an honorary curator of the museum. Another museum benefactor provided a radium source of 300,000 activity,[45] enabling Kunz and Baskerville to proceed with the most eye-catching aspect of their work. Since they also examined the priceless Morgan-Tiffany jewel collection, strict security precautions were taken, leading to headlines such as: "Testing priceless gems in a New York secret strong room; Scientists, armed and guarded, work amid famous collections; World awaits result; They are seeking to determine the radio activity of every known material; Task is first attempted."[46]

They were, in fact, not testing the minerals for radioactivity. Their project was to expose the many samples to the three types of radiation mentioned above and to classify them according to whether they phosphoresced to all three radiations, only two, only one, or none. They succeeded in this laborious task, which had been initiated because of Crookes' discovery of alpha ray-caused scintillations on phosphorescent screens.[47] But their classification scheme, a taxonomic *tour de force*, was of no particular value, and their decision to study the physical effects of the radiations, rather than the radiations themselves or the sources, led them into a not very promising area at the time.[48] Their demonstrations of the phosphorescence produced in diamonds were, however, enthusiastically received. In fact, the paper read by Kunz and Baskerville before the New York Academy of Science, in October 1903, drew an audience estimated as the largest to attend a meeting in many years.[49]

For the most part, the two men were disinclined to speculate upon the mechanism of phosphorescence. Yet, the problem faced them of explaining why certain materials exhibited the effect and others did not. "Is it not possible," they asked, therefore, "that we have here . . . a new substance perhaps elementary, which acts as a radium foil. . . ?"[50] This research led to the discovery of no new element, but it is characteristic of Baskerville's repeated publication of incomplete investigations with the apparent hope that he could thereby claim priority if a new element were eventually discovered. Within a few months other work led him to add carolinium and berzelium to the list of elements, but, as already mentioned, his eagerly sought fame was transitory.

In other applications, several authors examined the well-known phenomenon of the violet coloration of glass by rays from radioactive sources, without expressing any detailed understanding of the effect.[51] The action of the rays was tested also on perhaps more interesting objects. Bergen Davis and C. W. Edwards, of Columbia University, found small amounts of water produced from hydrogen and oxygen gases,[52] while F. C. Brown and Joel Stebbins, of the University of Illinois, observed that the electrical resistance of a selenium cell decreases with exposure to the rays.[53] These investigations need not be considered trivial, for studies of this nature were necessary to explore the effects and applications of radioactivity. Yet, in the development of the *science* of radioactivity, such applications played a minor role. Indeed, this is equally true for the area of medical and biological research, where many applications met with great success.

Atmospheric Radioactivity

The discovery in 1901, by the German physicists Elster and Geitel, that radioactive substances are distributed in the atmosphere explained the century-old puzzle of leakage of an electrically charged conductor—at least until cosmic rays offered an additional explanation. With the recent work on ionization by such men as J. J. Thomson, Ernest Rutherford, and C. T. R. Wilson, the radiation was seen to be creating charged particles or ions through collisions with air molecules, and these ions carried away the charge on the conductor in question. It was precisely this process that caused the divergent gold leaves of an electroscope to fall together, slowly when measuring the instrumental "background leakage," and faster when measuring the activity of a radioactive source.

Many scientists the world over quickly moved to investigate the extent of distribution of radioactivity in the air, water, soil, rain, snow, etc., which we shall cover with the rubric "atmospheric radioactivity." This excitement was mirrored in America in the numerous newspaper stories of local successes and in the far fewer contributions to scientific periodi-

HENRY ANDREWS BUMSTEAD (1870–1920)

cals. These latter were to a large degree stimulated by the extensive Canadian work under Rutherford at McGill and J. C. McLennan at Toronto, and in one particular case by J. J. Thomson in person. In the spring of 1903, Thomson voyaged to New Haven to inaugurate the now traditional Silliman Lectures at Yale University. As part of a wide discussion on "recent developments of our ideas of electricity,"[54] he called attention "to the work done in the Cavendish Laboratory on a radioactive gas found in waters coming from deep levels."[55]

At his request, Henry Andrews Bumstead (1870–1920) and Lynde Phelps Wheeler (1874–1959), both of Yale's Sheffield Scientific School physics department, began a search of the neighboring Connecticut area. "J. J." could not have had a more willing follower than Bumstead, who had studied under Henry A. Rowland at the Johns Hopkins University and received additional training during two years employment as an assistant in the great physicist's laboratory.[56] Then he went to Yale, where he earned his doctorate in 1897 for a dissertation on the theories of electrodynamics. Until 1906 he was an assistant professor in the Sheffield Scientific School, being then appointed to a full professorship in Yale College and the directorship of the Sloane Physics Laboratory (after the position had been turned down by Rutherford).[57]

His ties with England, and with the Cavendish Laboratory in particular, were always strong. When x-rays were discovered, Bumstead was one of the first in America to study them. Equally intense was his interest in Thomson's work on the electron and gaseous discharges. It was, in fact, largely due to his efforts that Thomson came to Yale in May 1903 to inaugurate the lecture series. Bumstead was an active traveler and, in addition to summer trips abroad, he spent the winter of 1904–1905 at the Cavendish Laboratory. He returned to England for another extended period toward the close of World War I, serving in London as scientific attaché. Bumstead enjoyed the esteem of his colleagues and was elected president of the American Physical Society, vice-president of the American Association for the Advancement of Science, a member of the National Academy of Sciences, and a member of several other learned societies. It was during a term as chairman of the National Research Council that he died, while on a pullman returning from a meeting in Chicago.

His colleague, Wheeler, received his Ph.D. from Yale in 1902 and remained on the faculty in New Haven for about thirty years. He later was involved in research on radio wave propagation and served as chief of the technical information section of the Federal Communications Commission (1936–1946). He is perhaps best known as the biographer of Yale's outstanding theoretical physicist, J. Willard Gibbs.

While he had not investigated radioactivity himself, before Thomson's suggestion, Bumstead's interest had been stimulated by "shop talk" with Rutherford, following the Christmas 1902 Physical Society meeting in New York.[58] In his work with Wheeler, Bumstead obtained water from a fifteen hundred foot deep spring near New Milford, collected the trapped gases by boiling, and tested them in an electroscope.[59] The rate of leak (activity) was approximately three times that for normal air. They further found that the water did not have to come from great depths to be radioactive. The gases contained in the water from one of the New Haven reservoirs, which was fed entirely by surface drainage, caused the electroscope to discharge in one-twelfth the time normally required. Whether this water came through the city pipes or was collected at the lake, Bumstead and Wheeler got similar results. A notable point was raised when they learned that the water from which the gases were extracted had not recovered its activity in over two weeks. This could not, therefore, be a case of an emanation being produced by a radioactive substance *dissolved* in the water. Additional proof of this was found when the water's mineral residue was shown to be inactive. The active gas was radium emanation, as they proved by measuring its half-life of about four days. Besides finding it in surface and subsurface waters, Bumstead and Wheeler were able to draw it out of the ground.

The whole science of radioactivity was tending toward greater accuracy and away from qualitative results. As participants in this trend, the two Yale physicists, aided by Bumstead's student H. M. Dadourian, plotted decay curves for the gases found in water and soil.[60] They even substituted a more precise, specially constructed quadrant electrometer for their gold-leaf electroscope, but still found only radium emanation in their samples. This puzzled them, for Rutherford and S. J. Allen in 1902 had obtained an active deposit, on a negatively charged wire suspended in *air*, which exhibited a half-life longer than that of radium emanation's deposit,[61] and Allen, shortly before he left McGill a year later to be a graduate student at Hopkins, reported that his deposit's decay curve was not entirely regular.[62] Both findings clearly suggested that the radium active deposit was mixed with some other active material and the measured decay curve was a composite of both. Yet, Bumstead and Wheeler could not understand why the mixture should be found in air but not in water or soil.

First, Bumstead decided to confirm that the New Haven air contained the additional material. In an extended series of experiments, involving hundreds of meters of copper wire suspended above the ground, he succeeded in obtaining sufficient active deposit on the wire to measure its decay over a period of up to a day.[63] This ability to continue activity observations after the more powerful deposit from radium emanation had virtually decayed away, was of the greatest importance. For, upon comparing this curve with one produced from the active deposit of a sample of pure radium emanation, Bumstead saw that

> it is evident that they do not entirely agree within the limits of experimental error; indeed a mere inspection of the curves ... shows that, toward the end, the air wires were falling off at a slightly slower rate than the one exposed to radium emanation. It seemed probable that this might be due to the presence of a small proportion of some form of activity decaying more slowly than that due to radium, and which would, therefore, show itself in a more and more marked manner as time went on.[64]

More extensive tests allowed him to record a half-life for this longer-lived substance of about ten or eleven hours. These times corresponded to the accepted half-life for thorium excited activity,[65] leaving little doubt of its presence. Bumstead, therefore, not only confirmed in 1904 the mixed nature of radioactive materials in New Haven air, he proved that it contained thorium emanation as well as radium emanation.

Availing himself of his professor's experience, Dadourian then proved that the underground air in New Haven also contained thorium emanation.[66] It had been missed in the earlier studies by Bumstead and Wheeler because the level of activity, at the time when its presence would have become evident, had fallen to background level. Dadourian further proved that the radium and thorium emanations were not only necessary, but sufficient to account for the decay curves obtained. Comparisons of the activities due to air, and to the two emanations (from chemical compounds) showed that other gases, for example, actinium emanation, were present in undetectable amounts, if at all. In work a few years later he calculated that the amount of radium emanation in air was thirty thousand to fifty thousand times as great as the amount of thorium emanation.[67]

Since radioactive materials seemed to be so widespread, albeit in minute quantities, there was a certain amount of interest in the extent to which humans were in contact with radioactivity. This contact was probably greatest in the use of water for bathing and consumption, so the proprietors and visitors at health spas and the bottlers and purchasers of mineral waters were curious whether the benefits derived could be attributed to radioactivity. Of course, in the middle of this century's first decade, with radium touted as a specific for most illnesses, nothing but benefits could be seen from radiations.

Although routine, commercial-type chemical analyses were "not as exciting or interesting" as research work, Bumstead's good friend Bertram Boltwood felt himself obliged to "do a little 'pot boiling' occasionally."[68] At the request of the United States Department of the Interior, Boltwood, about whom more in subsequent chapters, examined the waters on the government reservation at Hot Springs, Arkansas, and found all forty-four of his samples radioactive.[69] This activity was due entirely to radium emanation and not to any dissolved salts, but its distribution was a perplexing problem. "All of the hot springs are situated on a narrow strip of land about 500 yards in length," he noted, yet the strongest sample was over five hundred times more active than the weakest. No connection could be "discovered between the location of the springs and their radio-active properties,"[70] nor could any relationship be established between activity and temperature, flow, or chemical composition.

In all this work, Boltwood advocated and used as a standard "the quantity of radium emanation set free when a known weight of uranium in the form of a natural mineral is dissolved in a suitable reagent."[71] Like a few others of vision, he was eager to establish reproducible standards which would allow workers in different laboratories to compare different samples. Two University of Missouri professors who adopted his suggestion were Herman Schlundt and Richard Moore, both of whom also reappear later in this narrative, when they expressed the activities of several deep wells in the neighborhood of their school.[72] Boltwood, in fact, tested their mineral sample of uraninite in his standardized electroscope, confirming its chemically determined uranium content within about a quarter of one per cent. Such work testified to the increasingly accurate efforts being brought to bear upon the problems of radioactivity. It must be recognized, however, that these examples were the quantitative prominences among a mass of qualitative reports of radioactive springwater that appeared in newspapers across the nation.

Theoretical Thoughts

Few Americans ventured explanations of radioactivity; those who did were united in their innocence of experimental experience. This is not to say that a theoretician must also work in the laboratory, but merely that he must be well familiar with the phenomenon. The insights offered appear not very novel and were little different from explanations originating across the Atlantic. The most curious aspect, perhaps, is that they were seriously discussed a year and more after the first publication of the Rutherford-Soddy transformation theory. But, perhaps again, this was not so unusual, for the famous Curies themselves were slow in accepting this theory over their own.

Thomas A. Edison, outstanding inventor and, to the public, the perfect image of a great scientist, when called upon to explain the everlasting power of radium, said that it is constantly being replenished by the air, sunlight, and other things.[73] Indeed, replenishment or transformation[74] of some sort was required if energy within the atom was denied. A somewhat more precise theory was contributed by Hudson Maxim, brother of the inventor of an efficient machine gun and a well known inventor and explosives expert in his own right. But his belief that radium's heat is derived from ethereal waves seems to be a cross between the Curie and the Crookes hypotheses and not particularly original at a time when the luminiferous ether was being invoked to explain almost all inexplicable phenomena. When, in September 1903, the newspapers carried articles reporting the suggestion by Lord Kelvin that radioactive bodies are capable of converting etheric vibrations into vibrations of a lower pitch, corresponding to heat and light, Maxim wrote a strong letter to the editor of the *Scientific American Supplement* claiming priority.[75] This same journal carried another letter within a year from J. C. Featherstone, who also argued that radioactive materials change the frequencies of the invisible waves, but whose unique contribution was the placement of the source of these waves in the interior of the earth.[76]

External Pressures

These first research efforts, while neither qualitatively nor quantitatively outstanding, have been described in some detail because they are accurately representative of the type of radioactivity investigations pursued by competent scientists throughout the world. This is the background, during approximately this century's first half-decade, out of which came the contributions that made European eyes turn westward, and not always then toward Montreal. The achievements which advanced the science significantly were, of course, relatively few in number, and were made by a small group of people. Indeed, radioactivity always seemed to be a small corner of physics and chemistry, with a corresponding magnitude of practitioners. This being so, it is not always possible to detect trends and themes running through its history. To a large degree, the story of radioactivity is one of individuals rather than of movements.

Nevertheless, examination of the scientific leaders and those of the second and third ranks leaves the strong impression that lines of communication and influences were more international than national. This will be shown in individual cases in following chapters. The effect is not surprising, since any American with good results in his research would want to share them with colleagues abroad, for that was where the "action" was. American science was still far from the center of gravity of the world scientific community.

The *overall* rise of interest in radioactivity was also due to foreign influences, and not only because of the obvious circumstance that the discovery was made abroad, but in particular the activities of Ernest Rutherford. Rutherford served as something of a catalyst, imparting enthusiasm wherever he spoke and encouraging others to pursue this scientific specialty. After all, he had already made some outstanding contributions by studying radioactivity, proving that the phenomenon was a fertile one. He and his students attended meetings of the American Physical Society each Christmastime, presenting many papers on radioactivity. John McLennan of Toronto and his students also contributed to these sessions, making the total of Canadian work in this subject impressive indeed. Michael Pupin, the noted professor of electromechanics at Columbia University, in discussing the newly formed (1899) American Physical Society in his Pulitzer Prize winning autobiography, wrote that "Rutherford's wonderful discoveries in radioactivity were reported regularly by himself at the meetings of the society, and I often thought that these reports alone, even without the many other good things which came along, amply justified the existence of the society."[77]

Rutherford further was in demand as a speaker at universities and other places where scientists or interested laymen might meet. His Yale University Extension Lecture in the spring of 1904 led to an invitation to return to New Haven the following year to deliver the series of Silliman Lectures.[78] Another city to which he returned was St. Louis, where he was an invited speaker before the American Association for the Advancement of Science in 1903,[79] and the International Congress of Arts and Science, held as part of the 1904 Universal Exposition.[80] At Dartmouth College, where he lectured in March 1904, Rutherford carried with him, as he often did, a small lead box containing about twenty milligrams of radium bromide, for use in a demonstration experiment. Gordon Ferrie Hull, the physics professor there, many years later recalled that "before the lecture he desired to get the radium salt into a small glass tube in order to pump off the 'emanation'. . . . He was given a piece of good paper about two inches square, slightly creased. He poured the radium on to the paper and coaxed it into the glass tube, using a pen knife to scrape in the final particles." Hull retained the paper, which became the "radium supply of our laboratory for these nearly 40 years."[81]

Rutherford became quite accustomed to rail travel in the United States, speaking at Ohio State University,[82] the University of Illinois,[83] Clark University in Worcester, Massachusetts,[84] the Franklin Institute in Philadelphia,[85] and teaching during the 1906 summer session at the University of California, Berkeley.[86] So much did he seem to be away from his laboratory that Oliver Lodge, one of the elders of British science, felt constrained to write: "I trust you will not waste your time in lecturing but will go on with your experiments and leave lecturing to others."[87]

American institutions not only wished to hear Rutherford, they wished also to honor and to hire him. The universities of Pennsylvania and Wisconsin conferred honorary degrees upon him, and job offers came from Yale, Columbia, Stanford, and the Smithsonian Institution.[88] Columbia, in fact, set its sights extremely high, with simultaneous invitations to Rutherford and J. J. Thomson to join its faculty in 1902.[89] Another measure of Rutherford's influence on American science is the extensive correspondence he conducted from Montreal. These letters contained both information and encouragement, vital commodities to his friends south of the border.

It was not a one-man show, however. There were other influences on the progress of radioactivity in America, from foreign and domestic sources. Baskerville, whatever the quality of his work in this field, tried sincerely to advance it. In addition to providing space in his laboratory for Fritz Zerban, as mentioned earlier in this chapter, Baskerville did the same for Leo Guttmann,[90] who came from Heidelberg via Ramsay's laboratory in London, and he attempted unsuccessfully to tempt Otto Hahn to the United States.[91] Hahn, who was working under Ramsay at the time, shortly became famous for discovering radiothorium. Before returning to Germany, he decided to spend the 1905–1906 academic year at McGill, learning more about radioactivity from Rutherford.

The instigation of Bumstead and Wheeler's research on atmospheric radioactivity by J. J. Thomson has already been described. No doubt there were other cases of such influence, for many American scientists had done their graduate work abroad and kept in touch with European colleagues by mail and by attendance at international meetings. In addition to this personal contact, the exciting advances in radioactivity were reported in professional journals to which most university libraries subscribed. Thus, the periodicals, such as the *Philosophical Magazine, Nature, Chemical News, Proceedings of the Royal Society, Comptes Rendus, Physikalische Zeitschrift, Berichte der Deutschen Chemischen Gesellschaft, Le Radium,* and *Jahrbuch der Radioaktivität und Elektronik,* which published the greater part of research in radioactivity, were reasonably accessible to American scholars. Two new library tools helped to simplify the task of feeping abreast of current developments: *Science Abstracts,* which began in England just before the turn of the century and specialized in physics, and *Chemical Abstracts,* an American product which began publication in 1907.

Among domestic journals, the *American Journal of Science, Physical Review, Journal of the American Chemical Society,* and *Science* carried original research papers. It is entirely likely that the common practice of simultaneous publication in a foreign periodical encouraged American scientists to furnish copy to the local editors. The *Physical Review,* in addition to printing abstracts of papers presented at meetings of the American Physical Society, carried in 1898 and 1900 two early and impressive reviews of worldwide research in radioactivity, written by Oscar M. Stewart of Cornell University.[92] Other journals, as well as some of those mentioned above, frequently either reprinted articles that had appeared elsewhere or abstracted them. Though second-hand, the information that was disseminated in this fashion, in such prestigious books as the *Annual Reports of the Smithsonian Institution,* and in such weekly newspapers for the layman as the *Scientific American* and its *Supplement,* figuratively increased the level of radioactivity in the United States.

While this foreign "fallout" predominated, there were lines of communication developing within America. As just indicated, some domestic journals were supported by domestic contributors. There were also direct personal contacts, for the American Physical Society meetings were small, "clubby" affairs, reflecting the size of the profession at that time. Even the American Chemical Society, though larger because chemists then were able to find employment in industry as well as academia, was still relatively intimate.

Communication, thus, was the key to the growth of radioactivity research in America. Conventionally, the journals perform this function, but this requires scientists able both to recognize significant papers and to initiate their own investigations. For some Americans, especially those trained in centers abroad, entering this research tradition was no problem. Others never developed the frame of mind, mastered the experimental skills, or acquired the ability to distinguish between the important and the commonplace. Since American universities in this period had come to encourage scientific research, the fairly common examples of graduate training in Europe and hiring of foreign scholars may be seen as part of an effort to develop a domestic research tradition.

Communication to those who had not been taught as apprentices how to conduct research was better performed by the popular and scientific lecturers, who could direct newcomers to the important current work. Most valuable, however, was personal contact with a scientific "impresario," be it through visits, conversation at scientific meetings, or, for newcomers to the field, direct supervision for a time in his laboratory.

This, then, is the background against which the major American contributions to the mainstream of radioactivity research may be seen. For the period before American world leadership in science, it seems typical of the manner in which a given scientific subject unfolded. As such, it offers a valuable case study of the development of science, a study whose merit is enhanced by the circumstance that radioactivity had virtually no ties to earlier science. There were no precursors or near-misses before Becquerel's discovery in 1896, and recognition of the phenomenon cleared up no long-standing, puzzling mysteries. Because of this relative isolation we may more clearly see the influences on scientific progress: personal, professional, and institutional.

The radioactivity research described in this chapter illustrates an evolution of research ability and orientation. This early period comprised a number of men who embraced the subject but briefly, and then departed as the research front eluded their grasp. Scientific opportunists also were attracted, with Baskerville persistently generating a mixture of misinformation and trivia, highlighting his lack of critical judgment by sensational claims. Then sober, serious investigators appeared, of the sort of Pegram, who went as far as they could with a problem but failed to resolve it. And next came the well-trained physicists, Allen and Bumstead among them, who were quite capable of initiating, conducting, and completing research projects. These men, in fact, were typical of physicists for the next decade and a half. Their major handicap was a lack of intuition; they performed quite competent research on topics that history has found to be relatively uninteresting because they led to no profound understandings of nature. These men, in effect, ran into blind alleys. There were, however, a few scientists touched with something—insight, luck, or genius—and the research they performed both resolved and raised major issues. The next few chapters are about their work.

4 Boltwood and McCoy

Perhaps the first foreign research student to work in Rutherford's laboratory at McGill was Howard L. Bronson, a physicist who was born in America but spent almost his whole career in Canada. His entry into radioactivity studies was fortuitous, as he noted in some published reminiscences:

> Early in the spring of 1904, I had the good fortune to attend a Yale University Extension Lecture by Rutherford and it proved to be a turning point in my life. I was hoping to receive my Ph.D. in June and was working on a somewhat trivial problem that I found neither interesting nor creative, and the future did not look very promising.[1]

> This was only shortly before E. R. was to go to London for his Bakerian and Royal Institution Lectures and he was full of his subject. His enthusiasm was contagious and his subject appealed strongly to me. This was just what I needed. I now knew what I wanted to do. I therefore wrote to him at once asking if I could come to Montreal and work under him.[2] I had been a Teaching Fellow at Yale and I asked for and was given a part-time demonstratorship at McGill. So began three most rewarding years of research under Rutherford's guidance before he left for Manchester in 1907.[3]

Rutherford returned to New Haven in 1905, as already noted, to deliver the series of Silliman Lectures, and was accompanied on this trip south by Bronson. At that time Yale was considering the thorny problem of combining the physics departments of Yale College and the Sheffield Scientific School into one university department and

> Rutherford was urged to head up the whole undertaking, practically on his own terms. He discussed with [Bronson] the pros and cons on the way back to Montreal, but his reaction was—"Why should I go there? They act as though the University was made for the students."[4]

59

Rutherford declined the offer, as related earlier, and the chairman-ship of the Yale College physics department went to Bumstead. Con-cerning this appointment, Yale's President Arthur Twining Hadley wrote to Secretary Anson Phelps Stokes that

> After leaving the matter open until the last minute, Rutherford finally decided that he could not come here for the Sloane Laboratory. The posi-tion is to go to Bumstead. Bumstead has during the last year been doing really distinguished work. We are going to associate Boltwood with him as his assistant in the Laboratory. I think that this will make a good combina-tion. Boltwood is a man of great mechanical ingenuity, and an untiring experimenter. His vivacity will counteract the effects of Bumstead's se-riousness, and Bumstead's concentration will prevent Boltwood from scat-tering his brains loose around the table. It is delightful to see the en-thusiasm and ambition with which the two men are going into the work. I predict great things for them.[5]

President Hadley was a shrewd judge of men. Bumstead's notable career was sketched in the last chapter, while Bertram Borden Boltwood (1870–1927), who has been so unceremoniously introduced here, soon became the foremost authority on radioactivity in the United States, and contributed significantly to this subject.

Boltwood's Education

Boltwood's ancestors on both sides were among the early European settlers of America.[6] The paternal line came to New England in the mid-seventeenth century and lived there as farmers, millers, and blacksmiths for several generations. Bertram's grandfather, Lucius, de-parted from this tradition to become a lawyer in Amherst, Mas-sachusetts, where he was active in the founding of Amherst College, serving as its secretary from 1828 to 1864, and where he was also active in politics, being a candidate for the governorship in 1841. Bertram's father, Thomas Kast Boltwood, graduated from Yale College in 1864, and received a degree from the Albany Law School two years later. He practiced his profession for but a short time, however, until his untimely death in 1872. His son, an only child, was two years old at this time and thereafter was raised entirely by his mother, Margaret Van Hoesen Boltwood, in her native village of Castleton, New York. The Van Hoesens, of Dutch stock, were among the pioneering families in Rens-selaer County during the seventeenth century.

Despite his father's death, Boltwood's family was well provided for, and he received his secondary education at the excellent Albany Academy. Although he rarely, if ever, saw his cousin Ralph Waldo

BERTRAM BORDEN BOLTWOOD (1870–1927)

Emerson, the latter's fame made intellectual achievement of importance to the young "Bolty." Of even greater influence, perhaps, were the mineralogical and chemical interests of uncle Charles Upham Shepard, a professor at Amherst College. Yet, Boltwood was by no means a one-sided, serious youth, and his love of practical jokes and of mechanical devices (both traits which lasted his entire life) caused his family some concern about his scholarly future. Nevertheless, he was admitted to the Sheffield Scientific School of Yale University in 1889.

Although registered in the chemistry department, Boltwood received a freshman prize for excellence in physics. As a very pleasant reward for this success, he was taken upon a trip to Europe during the summer of 1890. This was the first of a number of visits to the British Isles and the Continent and established his warm feelings for these places. Upon completion of the three year course, Boltwood took highest honors in chemistry, and then departed for two years of further study at the Ludwig-Maximilian University in Munich. The famous Adolf von Baeyer was among his instructors there, but Boltwood seems to have profited most from his contact with Alexander Krüss, who taught him special analytical techniques and exposed him to the chemistry of the rare earths, both of which were to prove valuable in later years.

In 1894, Boltwood returned to Yale as a graduate student, supporting himself with a position as assistant in analytical chemistry. He returned to Germany for half a year in 1896, and spent part of this period working in Wilhelm Ostwald's laboratory in Leipzig. At this time the controversy over energetics was raging around Ostwald, and Boltwood seems to have absorbed some of this philosophy. After his return to America, he presented a paper to an informal student group on the "Indivisibility of Matter or the Atomic Theory Exposed,"[7] which, from the title, appears to follow Ostwald's antiatomic views, though we have no indication whether Boltwood fully agreed with the concept that energy, not matter, is the fundamental constituent of the universe. Whatever his commitment at this time, however, in following years Boltwood often spoke out against the energetics philosophy. This was, in fact, one of the relatively few instances when he concerned himself with theoretical or philosophical ideas in science, but his later work in radioactivity could hardly have permitted him to be an antiatomist. Boltwood's student days at Yale ended in 1897, when his dissertation entitled "Studies on chlorides" was accepted and he was awarded his doctoral degree.

Professional Activity

A year before his graduation, Boltwood was appointed instructor in analytical and physical chemistry, a post he held until 1900. In addition to his teaching responsibilities, he translated two chemistry texts from the German[8] and made himself a storehouse of information on laboratory techniques. He also devised new apparatus, for example, an automatic Sprengel pump,[9] a lead fume pipe for the Kjeldahl nitrogen determination apparatus,[10] and, later, "Boltwax," a low melting point wax with fair mechanical properties, useful in vacuum work.

Sometime during 1898 or 1899, Boltwood began his research on radioactivity, having had his curiosity aroused by the Curies' papers describing polonium and radium that appeared translated in the *Chemical News*.[11] No publication resulted from this and the details of the work are unknown; we are aware only that he separated some of the radioactive constituents from pitchblende, presumably following the Curies' procedures.[12] Boltwood also was adviser to a student named Clifford Langley, who in 1899 wrote a very short bachelor's thesis describing the chemical separations of the Curies.[13] Langley repeated these separations and came to the conclusion that there existed in the pitchblende another radioactive element differing from uranium, radium, and polonium.[14] But the work was dropped upon his graduation in June, and in October 1899, Debierne announced from Paris the discovery of what was later called actinium. Boltwood thereupon tested Langley's active substance and became convinced that it was identical with Debierne's product.

Not until early 1904 did Boltwood return to the study of radioactivity.[15] But in the five years since he last courted the subject a significant change had occurred in his career. Apparently at odds with Russell Chittenden, director of the Sheffield Scientific School, whose policy, he felt, was "to grind everyone down to the lowest possible limit," and who had "succeeded in antagonizing all of the younger members of his faculty,"[16] Boltwood left his Yale position in 1900 and conducted a private laboratory in New Haven for the next six years. For part of this time he was associated with a schoolmate under the title of "Pratt and Boltwood, Consulting Mining Engineers and Chemists."

Joseph Hyde Pratt was a geologist who did much field work in the Carolinas, and who later became professor of economic geology at the University of North Carolina. Thus, by one of those fortuitous circumstances, Boltwood was called upon to analyze numerous ore samples, such as monazite, many of which contained thorium or uranium. His original interest in the rare earths was now supplemented by extensive practical experience, since the rare earth elements were usually found in radioactive ores. With this background it was perhaps inevitable that he should resume the investigation of radioactive minerals, of which he must have had a sizeable collection.

It is also likely that Boltwood's close friendship with Henry Bumstead, who was then working on radioactivity, had some influence upon this choice of research area. Yet, if one overriding influence is desired, it was probably that provided by Ernest Rutherford, when he delivered the 1904 lecture at Yale, earlier mentioned in the Bronson quotation. Boltwood actually began his work shortly before this visit, but he received great encouragement when Rutherford visited his laboratory on Orange Street to discuss the relative proportions of uranium and radium in natural minerals.[17] Boltwood soon sent Rutherford word of his experimental results, beginning a correspondence that extended for the next two decades, and which developed into a warm friendship.

This new direction in Boltwood's life may have made him desire a return to academic surroundings. Another factor may have been the appointment of Pratt as North Carolina state geologist in 1906, and the probable dissolution of their partnership. Bumstead's selection as chairman of the Yale College physics department, and a recommendation by Rutherford, elicited an offer in 1906 of an assistant professorship in physics,[18] which Boltwood accepted even though he considered himself a chemist, because it "offered increased opportunities for the continuation of [his] scientific work."[19] This is an interesting indication of the "problem" created by radioactivity because it did not fit neatly into one of the standard categories of science—it was not entirely physics nor entirely chemistry. Another element of probable importance was that his appointment was in Yale College, leaving him relatively free of the influence of the director of the Sheffield Scientific School. Boltwood retained this position until 1910, when he was made professor of radiochemistry, in recognition of his notable accomplishments.

Rutherford, who possessed no small measure of laboratory genius, found himself impressed with Boltwood's "power of anticipating experimental difficulties which were likely to arise," and his skill in "arranging his apparatus and methods to overcome them." Commenting on a joint research effort, Rutherford noted that "every detail of the complicated apparatus and arrangements was so carefully thought out beforehand that not a single change was required for the successful conclusion of the measurements."[20] The research upon which Boltwood's abilities were brought to bear lay in the chemistry of the radioelements and in tracing their genetic connections. Within the exciting field of radioactivity, this particular area was the most fruitful during the period of his greatest productivity: 1904–1911.

Except for a year's leave of absence, in 1909–1910, Boltwood remained at Yale for the rest of his life. The one year away was spent as a John Harling Fellow at the University of Manchester, to which he had been invited by Rutherford. It was an exciting and productive sabbatical leave, marred, however, by the death of his mother, who had accompanied him. This unhappy association with Manchester may have been the reason he declined Rutherford's offer of a permanent position.[21] Then too, Yale University did not wish to lose one of its best known scientists, and President Hadley was most persuasive. "I have heard with some anxiety," he wrote, "of the efforts which the Old World is likely to make to detain you upon its shores for an indefinite period, and thereby take you away from us."[22] He professed to be very sad at this prospect, especially since a generous alumnus was about to give the money for a new Sloane Physics Laboratory. Hadley, therefore, continued, "I write to offer you promotion, on your return here, to a full professorship of physics in Yale University," at a salary of $2,500 a year.

Apparently, Boltwood was reluctant to take the position in physics, for Hadley next wrote to him that the professorship could be in radioactivity, or radiochemistry, or any other descriptive name.[23] It should not be in chemistry, he felt, because Chittenden and some other (possibly antagonistic) members of the faculty might have to be consulted, and, further, it was desirable for this subject to maintain a connection with the physical laboratory. Hadley had not brought the nomination before the faculty, taking advantage of a clause in the statutes which allowed him to appoint University Professors. His course of action was justified by the speed that he felt was required and by the prudent realization that the creation of a new professorship in the chemistry department of either Yale College or the Sheffield Scientific School would upset several vested

interests. On the other hand, it seemed likely that the physics departments of "Y. C." and "Sheff" would soon be combined into a single
department upon construction of the new laboratory, and Bumstead,
Boltwood's friend and the present head of the college department,
would be chairman of the new university department.

Boltwood accepted the professorship of radiochemistry at the
suggested salary and returned to New Haven in the summer of 1910.
President Hadley was highly pleased and assured Boltwood that Mr.
Sloane was likely to give an ample endowment for research, in addition
to the laboratory costs.[24]

Scientific Inactivity

It appears, however, that research funds were available in homeopathic
doses, if at all. During the first three or four decades of this century Yale,
and other institutions, made either a minute profit or erased deficits
using unrestricted endowment funds.[25] Under these circumstances, research expenses continued to be met in the "time-honored" manner—
out of the professor's pocket, or from whomever he could beg, borrow,
or steal.

In Boltwood's case it is hard to say whether lack of funds influenced
his withdrawal from research. Certainly, he had pursued his work
strenuously, both when self-employed and when an assistant professor.
Very likely, he then met the expenses himself, for, though he apparently
was not wealthy, he was a bachelor and able to live quite comfortably.
Upon his promotion, he received a much larger salary and should have
been able financially to continue his research. Yet, once he returned to
New Haven his scientific work virtually ceased. The few research papers
he published as late as 1911 discussed the investigations pursued in
England. Though he later guided a small amount of graduate research
on radioactivity, and supervised the efforts of a Norwegian scholar who
came to America to work under him,[26] Boltwood accomplished no more
publishable research.

This, however does not mean that he was idle. Indeed, his other
activities were the main reason for this change in his career. Upon his
return from Manchester, Boltwood was the acknowledged leader in
radioactivity studies in the United States. He was elected to the National
Academy of Sciences, the American Philosophical Society, and several
other prestigious organizations. His correspondence for this period
bulges with inquiries from miners, mining engineers, chemical refiners,
industrialists, physicians, the Bureau of Standards, etc.—all asking for
information pertaining to their own interests. Boltwood replied to them

BERTRAM BORDEN BOLTWOOD (1870–1927)

all, frequently undertaking to devise a chemical process or test a mineral sample for a fee. In essence, he resumed his former role as a commercial consulting chemist, though retaining his post on the Yale faculty. Not that this was unethical or uncommon, for academic scientists then had frequent contacts with industry; it is the pervasiveness of this practice in current times that offers a quantitative contrast. Though Boltwood's efforts helped greatly to stimulate the uranium mining and radium processing industries in America, such that this country's output led the world by the middle of the century's second decade,[27] this activity left his basic research potential untapped. Physicists, perhaps, were more "fortunate" than chemists at this time, for neither government nor industry had yet learned of their usefulness. Boltwood, further, was called upon as an expert witness in fraud trials, particularly relating to the sale of radioactive mineral waters, and the required testing of evidence for activity consumed much laboratory time. And, finally, Boltwood was in demand as a speaker, giving, for example, a series of five lectures to the members of the Brooklyn Institute of Arts and Sciences during the winter of 1910–1911.

But his primary activity was, of course, in the Yale physics department. Here he continued teaching one of the few courses in radioactivity available to undergraduates and graduate students in this country, and, as the only full professor besides Bumstead, shared in the daily task of running the old laboratory and in the designing and supervision of construction of the new one. This burden was increased first by

Bumstead's ill health,[28] and then by his complete absence from the department during the 1913-1914 academic year,[29] when Boltwood became acting director of the laboratory.

With the outbreak of World War I in Europe, basic research ground to a halt.[30] Many young scientists abroad, with no call for their special skills, felt the patriotic urge to enlist in the army. A few were assigned to artillery sound-ranging and submarine detection projects, but a great many wound up in the trenches. Rutherford considered the death at Gallipoli of his brilliant pupil, Henry Moseley, a national tragedy, and urged the British Government not to expose its talented citizens on the firing line.[31] The pall over scientific research extended even to neutral America, for with the flow of periodicals, books, and correspondence often curtailed, and even more, with the world center of science no longer generating new ideas and discoveries, there was little inclination to continue peacetime investigations.

All these circumstances probably played a part, along with the most important and intangible one—personal inclination—in Boltwood's scientific inactivity after 1911. His academic career, however, continued apace. In 1918, he transferred his allegiance to the Yale College chemistry department, being appointed acting director of its laboratory. In this position he was instrumental in the consolidation of the chemistry teaching and facilities of Yale College and the Sheffield Scientific School. Then, as chairman of the new university department of chemistry and later as its faculty representative, Boltwood devoted more and more time to the planning, construction, and equipping of the Sterling Chemistry Laboratory. This involved a great amount of work with the architects and scientific-supply manufacturers, but resulted in a modern and well-furnished chemical laboratory by 1922.[32] Tragically, the strain of this effort caused a breakdown in his health, from which he never fully recovered. Severe periods of depression alternated with the more customary cheerful spirits and resulted finally in his suicide during the summer of 1927.

McCoy's Education

Like Boltwood, Herbert Newby McCoy (1870-1945) made major contributions to the study of the radioelements. Again, like Boltwood, he was born in 1870, descended from pioneer stock, lost his father while a boy, worked for a while as an industrial chemist before returning to an academic career, and had close ties with the radiochemical industry. But, unlike his contemporary at Yale, McCoy also pursued chemical fields other than radioactivity, with outstanding success, and though he often sent his papers for publication in foreign journals, he seems not to have had significant contact with colleagues abroad.

His father, James Washington McCoy, who was born in Ohio, had his health shattered in the Civil War, and worked as a tailor in the industrial town of Richmond, Indiana, where Herbert was born.[33] Sarah Newby McCoy, born in North Carolina of English Quaker extraction, necessarily taught her son to be serious, frugal, and hard working. Thus, for example, at the age of thirteen, Herbert was earning three dollars a week by riveting buckles on straps at a roller skate factory. Neither this nor numerous other jobs, however, interfered with his education and he graduated from Richmond High School in 1889, first among its thirteen senior class members.

A lifelong interest in birds inclined him first to study zoology, and he planned to work under David Starr Jordan at the University of Indiana. But Jordan left to become president of Stanford, and McCoy's thoughts then turned to another local institution, Purdue, where John Ulric Nef was making a name for himself in organic chemistry. This was a subject that also appealed to McCoy, and that was indeed quite fashionable in the late nineteenth century. Yet, once again, McCoy was the victim of academic mobility, for when he arrived at Purdue he learned that Nef had just left for Clark University, then a strong center of science in America. McCoy, with little funds, was far less mobile, so he remained at Purdue, where he fortunately encountered the enthusiasm for research of Göttingen-trained Winthrope Stone. Even with a year of absence from the campus to restore his depleted bank account, he finished the course in three years, and was hired for the following year as an assistant, while he worked for his master's degree.

Having earned this in 1893, he took employment as an analyst for Swift and Company meatpackers in Chicago. A year later, though offered the post of superintendent of a fertilizer works in Kansas City, McCoy chose instead to be a chemistry instructor at Fargo College in North Dakota. By 1896, however, the desire for further graduate study prevailed, especially since Nef was now at the University of Chicago. So once more McCoy enrolled in an institution . . . and found his professor gone. But while this time it was only for the summer (McCoy started in the summer session), it was enough to place him under the wing of another professor, Julius Stieglitz, who suggested to him his dissertation research topic. This he completed in 1898, then remained a year in Chicago as an assistant, and next served for two years as an assistant professor at the University of Utah. He was called back to Chicago as an instructor in 1901, and during the next decade climbed the academic ladder to a full professorship.

Academic Accomplishments

Though McCoy had trained and had published in organic chemistry, his teaching position was in physical chemistry and soon his research moved to this area also. Physical chemistry at the turn of the century was a new and exciting subject and was just establishing itself as an authentic discipline alongside inorganic, analytical, and organic chemistry. One of the leaders of this new science, Wilhelm Ostwald, in 1901 proposed a theory of indicators that interested McCoy. The German suggested that indicators were either very weak acids or bases, and therefore capable of ionization. If the ionization constant could be determined quantitatively, a rationale would be brought into the selection of indicators in acidimetry, for the constant is a measure of the hydrogen ion concentration at the point where the change of color occurs. McCoy, who had been led to this study through extensive work on the equilibria between carbonates and bicarbonates, in which he used phenolphthalein, published in 1904 the ionization constant of this indicator.[34] This served as the foundation for the practice of determining the pH of solutions, but the subsequent work was largely left to others.

For even before McCoy's paper on phenolphthalein appeared, he had begun in 1903 his investigations of radioactivity. The Rutherford-Soddy transformation theory, presented and refined in 1902–1903, had attracted the attention of numerous chemists, for the transmutation of one element into another was something they could understand and hope to examine. McCoy, no less than Boltwood, was intrigued with the idea, though he later hinted humorously that some hostility underlay his curiosity: "What right had physicists to tell chemists their atoms could disintegrate!"[35] The evidence initially presented by Rutherford and Soddy in support of their theory was highly impressive and sufficient to convince all but a few skeptics. The most significant next step, however, would be to show a definite connection between radium and uranium, since the new element was always extracted from uranium ores. This was not a strict requirement of the theory, for the two elements were then looked upon as beginning distinct decay series, but if a relationship could be found it would serve as additional support for the iconoclastic doctrine. As will be described in the next chapter, McCoy was one of those who succeeded in establishing that radium existed with uranium in a definite percentage, and hence their proximity could not be simply a chance occurrence.

He continued the study of the radioelements, as did Boltwood, helping to organize them in proper sequence in the correct decay series. But, unlike Boltwood, he also trained graduate students, at least half a dozen of whom made their own contributions to radioactivity. Still, McCoy could not be confined to one area of interest. Before Rutherford's nuclear atom concept was published in 1911, J. J. Thomson's atomic picture of concentric shells of electrons revolving in a sphere of positive electrification held considerable prominence. It was felt that this interpretation would lead to an understanding of valence, periodic properties, and other important properties of matter. To McCoy, it gave a particularly satisfying picture of the metallic state, with its mobile or free valence electrons. In analogy to the electrolysis of salt solutions, in which a metal is deposited at the cathode, as metal ions pick up electrons, he wondered if an *organic* metal could be formed by the electrolysis of an organic salt solution. With his student, William Moore, McCoy succeeded in 1911 in preparing tetra methyl ammonium amalgam and in demonstrating that it is a heavy, crystalline solid, with a metallic luster and good electrical properties.[36]

Industrial Career

In 1917 McCoy resigned his professorship at the University of Chicago and devoted his efforts to an industrial career. Though his relationship with his former teacher and now colleague, Julius Stieglitz, was always thoroughly cordial, McCoy, whose major pursuit from 1903 to 1919 was radioactivity, may have felt something of a scientific outcast in the rather conventional chemistry department. Though there was no research on radioactivity in the physics department at that time, there was enormous interest in the subject among Millikan's students who were working on modern problems, and there were even some radioactivity experiments in the elementary course on electricity, magnetism, and light.[37] But there is no evidence that McCoy enjoyed significant contact with this department.

He may, moreover, have begun to feel uncomfortable in the somewhat pious academic atmosphere, following his divorce, especially when he then began to court the daughter of an eminent history professor. Ethel M. Terry, herself a Ph.D. chemist and an assistant professor at Chicago, coauthored a popular textbook on general chemistry with McCoy before marrying him in 1922. Yet, it is most likely that McCoy's resignation came about because of the challenge offered by industry, coupled with the opportunity to exceed his professorial income, which would enable him to fulfill his goal of early retirement.

Just as Boltwood established close contact with the Welsbach Company in New Jersey, from whom he obtained many valuable samples of thorium ores and concentrates, McCoy found the Lindsay Light Com-

HERBERT NEWBY McCOY (1870–1945)

pany of Chicago a source of great help. These relationships, of course, were not just one-sided, for the chemists reciprocated by pointing out to the companies any commercially valuable radioelements in their stockpiles that could be separated economically. Both of these companies manufactured mantles which, when placed over a gas burner, furnished more satisfactory illumination than an ordinary gas flame. This attachment continued long in service while the carbon filament electric lamp slowly established its superiority and electrical supply cables were installed in urban areas. The mantles consisted of gauze impregnated with thorium containing a small amount of cerium. Hence, these companies often found themselves in a position to be of service to chemists interested in radioactivity or the rare earth elements.

In 1909, McCoy received from the Lindsay Company a gift of two pounds of the ash from scraps of gauze left over from their mantle production. After extracting the mesothorium present for his own research needs, he returned the remainder to the company in the form of thorium nitrate. This courtesy was appreciated by the manufacturer, for the compound was of prime importance in its procedures. Three years later McCoy suggested to the Lindsay Company that it consider obtaining its thorium nitrate from the monazite sands readily available in the Carolinas, Brazil, and India. He was hired as a consultant to plan the process, this work being completed in July 1914. When, just a few weeks later, the British blockade cut off American gas mantle companies from their source of thorium nitrate in Germany, a plant was built for McCoy's process. This was placed in operation in November of that year, and supplied about half of the Allied demand for this compound during World War I. McCoy, characteristically, also developed commercial methods for extracting mesothorium from the wastes of the thorium nitrate process, for this radioelement was useful for luminous watch dials.

In this same period that he was devising commercial procedures for the thorium series, McCoy developed comparable interests in the uranium-radium series. He took out a patent in 1914 for a new method of preparing radium from its ores, and the following year was engaged as a consultant by the Carnotite Reduction Company, newly formed to produce radium bromide, uranium nitrate, and ammonium meta-vanadate from the Colorado ores. Upon his departure from the University of Chicago in 1917, he became president of this company, a post he held for three years, at which time the company was sold to another ore processor. It is unlikely that McCoy had advance knowledge of the circumstance, but within a few years the "bottom dropped out" of the industry, as radium from the Belgian Congo captured the American market.[38]

About the time he severed his connection with the radium business, McCoy became vice-president of the Lindsay Company, a more diversified chemical manufacturer, and retained this affiliation until his death in 1945. Typically generous to students with his own funds, he carried the inclination over to employees, with company funds. His weekly dinners for Lindsay chemists succeeded in their goal of promoting good fellowship, and were much appreciated.[39] Like Boltwood, therefore, McCoy was an instrumental figure in the rise of an American radiochemical industry, perhaps even more important because his graduate student, Charles Viol, became chief chemist for the Standard Chemical Company of Pittsburgh, the country's largest manufacturer in this field. In 1927, McCoy moved to Los Angeles, both to enjoy semiretirement and to be closer to the tropical bird life he loved. Annual trips to Chicago kept him in touch with Lindsay Company activities, but his influence in the firm necessarily lessened.

His interest in chemistry, however, continued unabated. McCoy's early efforts in radioactivity had given him a strong acquaintanceship with the rare earth elements, and the fact that the Lindsay Company dealt with monazite, the only rare earth ore worked commercially, meant that he had easy access to these intractable substances. For some years he examined the more abundant rare earths, extracted from the Lindsay residue heap, and by the mid-1930s had fixed his attention upon europium. While this element had been discovered at the close of the nineteenth century, its scarcity and lack of industrial application explained the meagerness of information about its behavior. McCoy added greatly to the knowledge of its chemical and spectral properties, preparing essentially the world supply of europium in the process. For the fundamental, pioneering contributions made in his several areas of endeavor, McCoy was awarded in 1937 the Willard Gibbs Medal, the American Chemical Society's highest honor. He continued active to the end of his life, becoming a consultant to Arthur H. Compton's branch of the Manhattan Project, the metallurgical group at the University of Chicago, during World War II.[40] In this fashion, he returned to the subject of atomic and nuclear science after many years' absence.

McCoy and Boltwood, coming, respectively, from lower-middle and upper-middle class backgrounds, were typical of the two trends in graduate education. One worked his way through school and completed his training in a domestic university; the other polished his graces at an elite institution and perfected his talents abroad. Both, however, were fortunate to train in schools where a research tradition existed, Boltwood in Munich and McCoy, under European-educated Stieglitz, in Chicago. Both came out of fields where numerical data mattered, Boltwood from analytical chemistry and McCoy from physical chemistry, and they knew how to get numbers. And both were keenly independent, acquiring their interest in radioactivity after their formal education was over. Boltwood clearly benefitted from his contact with Rutherford, but he was always a colleague and never an apprentice; McCoy mastered the subject alone. How well they both succeeded in formulating their research problems and employing their laboratory skills to see them through is first apparent in their work on uranium-radium genetics.

5 Uranium-Radium Genetics

The Beginnings: Boltwood, McCoy, Millikan

Despite much remaining uncertainty and confusion, by early 1904 there was established a considerable body of information about the radioactive elements. A number of them had been identified chemically, including uranium, thorium, radium, actinium, polonium, and radiolead. Even more, family relationships were proposed, incorporating these and still other radioelements, as yet unidentified. Rutherford and Soddy were primarily responsible for establishing the genetic chain based on thorium: Th – ThX – ThEm – ThA – ThB; Rutherford alone was the key figure in defining the series based on radium: Ra – RaEm – RaA – RaB – RaC; and Becquerel and Crookes were credited with uncovering the transformation of uranium into uranium X. Most important, these relationships were grounded in the Rutherford-Soddy transformation theory of 1902 and 1903. According to this concept there is a continuous production of fresh radioactive matter. For example, radium spontaneously decays by the expulsion of an alpha particle to form the new substance emanation, which in turn ejects another alpha particle to become radium A. But the activity of none of these radioelements is permanent: it decreases in time following a precise mathematical law.

This decay would be noticeable if a particular substance were isolated, while the time for it to drop to half its activity would be recognized as characteristic for that particular radioelement. When several related bodies are together, as in a mineral, and are left undisturbed long enough for a condition of equilibrium to be established in which as many atoms of a given element are formed as decay in a specified time, then the quantity of each member of the chain stays constant and the total activity measured also registers constant. This will be the case if the first member of the series has a half-life too long to be measured directly. The rate of disintegration of each substance is represented by the number of its atoms present (N) multiplied by its decay constant (λ). For products in equilibrium the number, for example, of radium atoms decaying per second (λN) (which always equals the number of radium emanation atoms formed per second) equals the number of radium emanation atoms decaying per second. Thus, for all members of a series in equilibrium, the relationship $\lambda_1 N_1 = \lambda_2 N_2 = \lambda_3 N_3$ holds true. And since the decay constant is inversely proportional to the half-life (T), the equalities $N_1/T_1 = N_2/T_2 = N_3/T_3$ also are valid. Expressed in this fashion it is clear that radioelements with long half-lives of many years, such as thorium, must have many more atoms present than products such as thorium X, whose half-life is only a few days. It is also apparent that radium's great activity results from a half-life short enough to require rapid decay of its relatively few atoms, yet long enough to give separated amounts more than a transitory existence.

Boltwood believed in this transformation theory. He fully subscribed to the interpretation that radioactive atoms are unstable systems which break apart with explosive violence, emitting small particles (the radiations) at great velocity, and leaving larger residues consisting of atoms entirely different chemically and physically from those that had decayed. He further accepted the view that the transformation period of radium was far too short for the quantities currently found to have been created when the earth was formed. Any primordial radium would have long since transformed away. Present-day radium, therefore, must continually be formed, and its most likely ancestor was the uranium in whose minerals it was first found. A rather elegant way to confirm the transformation theory, Boltwood felt, would be to show that the ratio of the amounts of radium and uranium in various minerals was a constant. Such regularity could not be attributed to chance; it would have to be explained as a genetic relationship in secular or long-time equilibrium. If they were related, the relationship $N_1/T_1 = N_2/T_2$ would apply. Since this may be expressed equally well as $N_1/N_2 = T_1/T_2$, and since the half-lives are invariable, the ratio of numbers of atoms must be constant. In other words, when a mineral is old enough to have reached the equilibrium in which (for all members except the first and last of the series) as many particles of a radioelement decay as are formed in a unit time, the amount of one radioelement compared to another is in a constant ratio. Uranium, though the first member of this decay series, and therefore actually decreasing in quantity, was relatively so abundant that for practical purposes its quantity could be regarded as constant.

Five different uranium ores, four from North Carolina and one from Branchville, Connecticut, were dissolved in acid and the gaseous radium emanation collected.[1] After introduction into an air-tight electroscope, the rate of fall of the gold leaf, due to the dissipation of its charge by the Becquerel ray-produced ionization, was measured through a microscope which had a scale mounted in the eyepiece. This rate of leak was a reasonable measure of the amount of radium contained in the minerals, since emanation is strictly proportional to radium, and a three-hour delay before taking readings allowed the excited activity deposited by the emanation to reach a steady value. The ratio of the measured rate of leak to the analytically determined weight of uranium gave, in fact, a ratio of the amount of radium to the amount of uranium in these minerals. In his arbitrary units, Boltwood got results of 211, 212, 181, 207.2, and 218. The agreement was striking, since the low of 181, for uranophane, could be explained as the result of a gelatinous silica formed in the chemical reaction, which occluded a portion of the emanation.

In early May 1904, Boltwood rushed this information to *Nature*, where it appeared within three weeks as a letter to the editor, and to the *Engineering and Mining Journal*. The choice of *Nature* was logical, since it was (and is) one of the few periodicals having such a short publication delay, and it was a European journal (London) sure to be read by a good many of those pursuing radioactivity abroad. It was also an appropriate place to refute some shoddy conclusions by R. J. Strutt, son of the eminent physicist, Lord Rayleigh, and a physicist of growing reputation himself.[2] Strutt had heated several minerals and drawn off the emanation, whose activity he then measured. This method detracted nothing from his proof that the minerals contained radium and thorium, for decay period and not quantity mattered here. But when he discussed the *amount* of emanation in a sample, he went astray. Boltwood attributed this to poor experimental technique, which he, as a chemist, was more able to understand and resolve:

> These results show a direct variation from those obtained by Mr. Strutt ... which may perhaps be explained by the fact that he secured the emanation by heating his minerals. Experiments which I have made show that on heating samarskite to a low redness only 10 per cent. of the total emanation is given off, and that heating to bright redness releases only 20 per cent. of the total emanation obtained when the very finely powdered mineral is completely decomposed by heating with concentrated sulphuric acid.[3]

This technique of complete dissolution of the mineral, which Boltwood adopted from the start, was the only means of releasing *all* the emanation occluded in the solid, and, therefore, the desired means of obtaining valid quantitative information. Boltwood understood that when dealing with such small quantities of radium the only reliable method of determining its amount was to measure the emanation, which, being a decay product, was directly proportional to the amount of radium. Radium, with various chemical properties of its own, was difficult to separate and test quantitatively, while emanation is an inert gas requiring only mechanical separation.

Boltwood designed much of his own apparatus, including odd-shaped glass tubing which allowed his chemical operations to proceed in sealed vessels, and an electroscope for measuring emanation activity. For many years electroscopes were manufactured with glass windows to allow the leaves to be seen and yet protected from drafts. With emanation studies there arose a need for air-tight electroscopes, and this need was filled mainly in the laboratory workshops. Boltwood's first device, a detailed account of which was given in the *American Journal of Science* of August 1904, was based upon a design of C. T. R. Wilson, who was noted for his improvement of this instrument as well as for his cloud chamber. Brass stopcocks served as the gas ports, hot sealing wax made it air tight, a sulphur rod insulated the leaf from the charging cap, and charging was effected by drawing an iron wire hanging from the cap into contact with the leaf support by a magnet. His preference for an electroscope to the more complex and exasperating electrometer probably reflected his training as a chemist. Yet, accuracy did not suffer, for, as Rutherford noted many years later, "this method in Boltwood's hands became a weapon of precision."[4]

Boltwood's published conclusion stated that "the quantities of radium present in the uranium minerals which have been examined are apparently directly proportional to the quantities of uranium contained in the minerals," and that this affords strong evidence that "radium is formed by the breaking down of the uranium atom."[5] The completion of this work gave Boltwood an opportunity to initiate a correspondence with Rutherford. In his letter he looked to the future:

> From the extreme sensitiveness of this method I am inclined to believe that it may be possible to actually measure the rate of formation of radium from uranium. The normal air leak of the electroscope was only about 0.01 of one division per minute and as will be noted the actual quantity of uranium taken was only about 0.1 gram. With about 500 grams of uranium in solution it ought to be possible to get some idea of the rate of production. I am preparing to carry out some experiments in that direction.[6]

Rutherford replied encouragingly that "I would not lay any especial stress on negative results of attempts to grow radium as in Soddy's letter to Nature. It is quite likely that products (possibly non-radioactive) of very slow rate of change may intervene between the uranium and radium."[7] This casual remark began the work for which Boltwood became most famous.

Herbert McCoy, who was then an assistant professor at Chicago, also reported his results on the uranium-radium ratio in the spring of 1904.[8] Unlike many other American chemists, he had never made the grand tour of European universities, but he recognized the value of publishing abroad and sent this communication to the German Chemical Society. Radium must have a parent, he maintained, since estimates of its loss in mass per year suggested that all radium on earth would disappear within a short geological period if the supply were not replenished. And since radium was first found in a uranium ore, might not this be the logical connection to try to establish? McCoy, therefore, prepared powders of a dozen different pitchblende, gummite, and carnotite ores and tested their activities with an electroscope. This method was not as elegant as Boltwood's, since he was measuring activity due to uranium, actinium, polonium, and the radium decay products, as well as the radium itself, but the ratio of activity to the analytically determined amount of uranium was still quite strikingly constant. Of course, this meant that "we are . . . led to the conclusion that radium is one of the successive decomposition products of uranium."[9] Further, the mean value of his activity coefficient was 22.1 for ores, while only 3.86 for half a dozen pure uranium salts. "Hence the ores are 5.7 times as radio-active in proportion to the uranium they contain as the pure salts."[10] This was a result whose usefulness will be seen below.

Yet another American investigated this same relation between radioactivity and the uranium content of minerals during the spring of 1904. Robert A. Millikan (1868–1953), who was to receive the 1923 Nobel Prize in physics for his work on the elementary charge of electricity and on the photoelectric effect, was then an assistant professor of physics at the University of Chicago. Oberlin College had granted him his first two degrees (A.B., 1891; A.M., 1893) while the Ph.D. degree was awarded to him by Columbia in 1895. Following a year in Berlin and Göttingen, he moved to the shores of Lake Michigan to begin a twenty-five year association with the University of Chicago's physics department. This was terminated in 1921, when Millikan was appointed director of the physics laboratory and chairman of the executive council of the California Institute of Technology.

In March 1904, Millikan, assisted by H. A. Nichols, the assistant curator of geology at Chicago's Field Columbian Museum, began examination of half a dozen uranium minerals in order to determine whether

ROBERT A. MILLIKAN (1868–1953)

the radioactive substances found in these ores were indeed decomposition products of uranium.[11] "If such be the case the ratio between the uranium content and the [total] radioactivity . . . ought obviously to be constant, in case the assumption may be made that the active products of the decomposition are not washed out of the mineral by percolating water or other agencies."[12] Awareness of this latter possibility was important, since some minerals, such as gummite and uranophane, are formed from uraninite by the action of percolating water.

Millikan used an electrometer to measure the activities of his samples, and though there was as much as a thirteen per cent departure from the mean ratio, he felt that he had provided additional evidence in support of the view that uranium is the parent of radium. By the time he announced his results, Boltwood and McCoy had already published their papers and the agreement of these three authors must have been impressive.

Like McCoy, Millikan measured the *total* activity of powdered samples of the ores, and used the fairly constant ratio of this activity to uranium content as proof of a genetic relationship between uranium and the radioelements found with it. Boltwood, on the other hand, chose to determine the radium-uranium ratio only. His was perhaps a less modest plan, but it was capable of greater definition, and his techniques capable of greater accuracy. Precision suffered somewhat in both McCoy's and Millikan's experiments because the powdered minerals they tested not only occluded varying amounts of emanation, but they absorbed some of the alpha radiation that went to create the ionization current. One may indeed wonder whether these two men, both at the University of Chicago, were aware of each other's work prior to McCoy's publication. Even after that event there is no evidence of contact, let alone collabora-

tion, except for some chemical analyses McCoy performed for Millikan.[13] Yet, this need not be regarded as unusual, for in this period it was
customary that a scientist work alone, or perhaps with his research students. A team effort by men of equal rank was an approach to science
found rarely, although within a generation or two it would become widespread. The circumstance of several men working independently on the
same topic is encountered not infrequently in the history of science.
Such simultaneous discovery strongly marks an idea whose time has
arrived.

Testing Ores of Low Uranium Content

Throughout the period 1904 to 1907, the Boltwood-Rutherford correspondence maintained a high level of intensity. Suggestions for investigations, prepublication reports of results, and requests for information
and radioactive samples passed back and forth rapidly. Soon the saluta-
tions "Dear Professor Rutherford" and "Dear Mr. Boltwood" gave way
to the less formal "My dear Rutherford" and "My dear Boltwood." A
warm professional amity developed, and was given opportunity to grow
into personal friendship when they saw each other at meetings of the
American Physical Society, and when Rutherford came to New Haven
for the 1905 Silliman Lectures.

Rutherford early suggested that "results of most importance would
be obtained from analysis of radioactive minerals with a small quantity of
uranium."[14] He had found that orangite and thorite, both thorium ores,
contained radium in moderate quantities, although they were supposed
to contain only traces of uranium. If it could be shown that the Ra-U
ratio was constant even in essentially nonuranium minerals, then the
family connection would be as strongly proven as it could be, except by
the direct growth of radium from uranium.

Boltwood had, in fact, already examined some ores of low uranium
content, but now he greatly extended this investigation.[15] With new minerals from Joseph Pratt, H. S. Miner, chemist of the Welsbach Light
Company, and S. L. Penfield, Yale professor of mineralogy, he encountered new problems. "One decided difficulty," he wrote, "is going to be
the analytical separation of the uranium when present in very small
amounts. One has to take a lot of material to begin with and it is difficult
to separate the uranium precipitates completely from other things which
persist in coming down with it."[16] Monazite, in particular, gave him
much trouble. Boltwood easily detected the presence of *some* uranium in
various samples of the mineral, though "the routine methods . . . either
fail altogether in giving even a qualitative test or else result in the separa-

tion of only a few hundredths of a per cent." He knew that there should be much more uranium, since emanation measurements gave him the radium equivalent of from 0.3 to 0.4 per cent uranium. "These results were very disconcerting," he continued, "until I found that when I added 0.4% of uranium to the original mixture the analytical results came out just as before, that is, only a few hundredths of a per cent of the uranium present could be separated."[17]

Obviously, a new method of analysis was required, whereupon Boltwood developed one which depended "on the separation of the greater portion of the rare earths by simple recrystallization from a solution containing a large excess of sulphuric acid and the chemical separation of the uranium from the residual solution." With this technique he succeeded in getting amounts of uranium of the approximate magnitude expected. The difficulty in separating uranium from monazite was attributed "either to the presence of some rare element in the monazite which combines with the uranium and changes its ordinary chemical behavior," or, more likely, "to the possibility that the large quantity of phosphoric acid in the mineral is the disturbing factor." Phosphoric acid has a very strong affinity to uranium, and monazite is practically a pure phosphate. Boltwood's chief difficulty in the first method had been, in fact, the *complete* separation of uranium from "the great excess of phosphoric acid which remains in the solution after the removal of the other constituents."[18]

With an evident sense of satisfaction and confidence, Boltwood concluded:

> I have found this problem of the separation of the uranium from monazite a most difficult one and have already devoted weeks to it. It is, however, most fascinating to me since I believe that it is the first case where quite positive indications (You will note my conviction!) of a quantitative physical character have demonstrated the pitiful crudity of established chemical methods. When I have established the presence of the indicated quantities of uranium in monazite, which I mean to do even if it takes years to accomplish it, I am going to drop this line of research right there and rest on the assumption that the radium-uranium ratio is constant in ALL MINERALS, leaving it for others, who may so desire, to attempt to PROVE the opposite.[19]

A week later he sent Rutherford the following new year's greeting:

> I feel sure that you will be interested to learn that I have succeeded in separating uranium equivalent to 0.38 per cent from N.C. [North Carolina] monazite. I do not believe that there can be any doubt as to the accuracy of the analytical method employed or any question as to the identity of the substance separated.
> Uranium is most certainly the parent of radium.[20]

Refining the Techniques

Now began a long series of papers by Boltwood on uranium-radium genetics. McCoy also contributed significantly, though in lesser quantity, while it appears that Millikan decided to leave the field to the chemists. This is noteworthy in itself, since it was symptomatic of radioactivity studies at that time: the exciting work was largely chemical, in the tracing of the uranium and thorium decay series. To be sure, there were significant physical advances in the middle of the century's first decade, such as the measurement of alpha ray ranges by Bragg and Kleeman; the counting of particles, based on a scintillation method discovered by Crookes and perfected by Rutherford and Geiger, and on an electrical method developed by Rutherford and Geiger; and work by Rutherford on alpha ray charges and scattering, but it would seem that many physicists were disinclined to enter into radioactivity investigations.

On 10 February 1905, Boltwood read a paper before the New York Section of the American Chemical Society, entitled "The origin of radium."[21] He had now tested twenty-two samples of twelve distinct mineral species and further confirmed the proportionality between uranium and radium. Of these twenty-two, six contained less than one per cent of uranium, yet the Ra-U ratio was in line with the other values. McCoy's work was criticized by Boltwood because of his assumption, on the basis of activity measurements roughly proportional to uranium content,

> that the radium contained in each mineral is directly proportional to the uranium present, a conclusion which, under the circumstances, is only justified on the assumption of the further hypothesis that all of the other radioactive constituents (polonium, actinium, radio-lead) are also directly proportional to the uranium. This latter relationship still lacks experimental confirmation. McCoy's method is moreover quite unsuitable for minerals containing any notable quantities of thorium, as he himself acknowledges.[22]

Boltwood also found fault with some of his own methods and promptly corrected the error. Heretofore he had neglected the activity of the emanation that diffused from the minerals at ordinary temperatures and was lost, but now he was able to include this factor. Weighed portions of each sample were finely pulverized, introduced into glass tubes between plugs of cotton-wool, and subjected to a slow current of air which removed any emanation clinging to the solid. The tubes were then sealed in the flame of a blowpipe and allowed to stand for a number of days. Careful testing showed that it was unnecessary to wait until equilibrium was established; the seal could be broken and the newly released emanation introduced into an electroscope[23] after a much shorter period of time, and the maximum activity calculated by the standard growth and decay equations. Now Boltwood combined the activity value of the spontaneously evolved emanation with that for the emanation released from the solid by dissolution in acids, and obtained the total or equilibrium activity value of the emanation, corresponding to the radium contained in one gram of the mineral. Besides showing the constancy of the ratio of this figure to the uranium content in one gram of mineral, Boltwood calculated the "percentage of the total emanation lost by diffusion from the cold mineral (the emanating power of the mineral at ordinary temperatures)."[24] The range, from 0.2 per cent for aeschynite to 26.0 per cent for xenotime, both from Norway, and the variation in samples of the same mineral, pitchblende, from different localities, suggested clearly that the amount which diffused before the mineral was dissolved could not be neglected.

Fruitful Suggestions for Further Investigations

Two other important observations were announced in this paper before the American Chemical Society. Boltwood was impressed by the persistent appearance of lead in uranium-radium ores and suggested that it was one of the final, inactive products of the disintegration of uranium. Appreciable quantities were found in all his samples but one, and this exception was the youngest in geological age, a fact which tended to confirm rather than disprove his hypothesis.

The second announcement involved his seven month effort to grow radium directly from uranium. This was the work he projected when he first wrote to Rutherford, and which the McGill professor encouraged. But, like Soddy's trial, Boltwood's attempt was unsuccessful, the reason, he felt, being:

> It is therefore highly probable, as suggested by Rutherford [in the 1904 Bakerian Lecture before the Royal Society], that one or more intermediate changes exist between the uranium atom and the radium atom, although the identity of these intermediate products has not yet been established. The suggestion by Rutherford that actinium may be such a substance is of interest since the position of actinium in the family of radioactive elements has not yet been determined. Thorium appears to be quite out of the question because of the total lack of proportionality between it and the other substances.[25]

In the same month that Boltwood published this paper, an article by McCoy appeared in the *Journal of the American Chemical Society* in which he not only presented an improved method of determining activity, but used this method to *prove* that radioactivity is an atomic property.[26] His earlier work had suggested to him that activity measurements of powders might yield incorrect values due to absorption of the alpha rays in the upper layers of the solid. By depositing thin films of uranium oxide on plates of uniform size, McCoy was enabled to plot a graph of the activity vs thickness of film, and extrapolate to zero thickness. With an absorption coefficient thus obtained, the true activity of a sample (sufficiently thick to allow the maximum alpha ray activity) was a matter of simple calculation. Now that he had proved that "the *total* α-ray activity of any uranium compound is strictly proportional to its percentage of uranium," he felt this to be "a direct confirmation of the theory that radioactivity is an atomic property."[27] Since early belief in this theory was largely based on very similar type evidence, McCoy's work was a quantitative step forward.

The Equilibrium Amount of Radium

The correspondence between Rutherford and Boltwood by this time had evolved into active collaboration. During the spring of 1905, the data mailed between Montreal and New Haven formed the basis for the first of several joint papers.[28] Thus, before he left for his first visit to New Zealand in several years, Rutherford sent Boltwood a rough draft which the latter smoothed into a short but significant contribution.[29]

Since 1898, scientists had known that pitchblende, if not every uranium ore, contained radium. They often discussed the factor by which radium was more active than uranium (a figure in the millions). They even calculated roughly the amount of radium in a uranium mineral, but usually from separations data of the Curies or other large-scale operators. Seven years after the discovery of radium, Rutherford and Boltwood actually determined this ratio by sound methods.

A standard was prepared by dissolving a known weight of pure radium bromide in water and sealing up a portion of this. Some sixty days later, more than time enough for the maximum amount of emanation to accumulate, the emanation was completely removed by boiling and its activity measured in an electroscope. The data obtained for this standard were 24.24 scale divisions per minute for 0.926×10^{-7} grams of radium. For the uranium material, Boltwood chose a North Carolina uraninite, the radium emanation from which gave 193.6 divisions per minute for each gram of uranium present. A simple ratio showed that "*the quantity of radium associated with one gram of uranium in a radio-active mineral is equal to approximately* 7.4×10^{-7} *gram.* One part of radium is therefore in radioactive equilibrium with approximately 1,350,000 parts of uranium."[30]

Several months later Boltwood revised this figure, since he had discovered that the widely accepted textbook value for the percentage of uranium in the phosphate he used was in error.[31] For some reason he had redetermined this figure and found it to be over five per cent off. The corrected amount of radium in equilibrium with one gram of uranium was 8.1×10^{-7} gram, and, as he pointed out to Rutherford, "this illustrates, if further illustration is necessary, the rotten uncertainty of much chemical analytical data."[32]

But alas, the figure was still wrong! Exactly one year after their first paper was printed the two authors were forced to publish a correction, under the identical title.[33] The standard solution had been prepared in Montreal by Rutherford's colleague A. S. Eve and its radium content confirmed by gamma ray tests. About two months after making the standard, a portion was transferred to another bottle and sent to Boltwood, who assumed the radium content to be as stated. But, as Eve found many months later when experiments indicated the standard was defective, a considerable amount of the radium had deposited on the surface of the glass, leaving the solution much weaker than it was originally.[34] Therefore, with some hydrochloric acid present to prevent the standard solution from attacking or adhering to the glass, Boltwood had to repeat his determination in 1906. He then announced with Rutherford that *"the quantity of radium associated with one gram of uranium in a radio-active material is equal to approximately 3.8×10^{-7} gram."*[35]

Growth of Radium

In September 1905, Boltwood delivered a withering criticism of Soddy's most recent and allegedly successful work on the growth of radium from uranium.[36] Boltwood had never met Soddy, but it is not impossible that a little personal animus helped sharpen his pen, since Soddy was already gaining a reputation for his scathing tongue.[37] On the other hand, though Boltwood was capable of more tact than Soddy, and was a highly popular individual himself, he was not in the habit of withholding criticism when he thought it due. Especially since the subject of radium's parentage was one in which he was greatly interested, he could not stand by idly while Soddy's papers were taken as gospel by a large audience. For it must be remembered that Soddy himself was now a rather well-known personage. He had worked with Rutherford and Ramsay, both famous figures, published a book on radioactivity, delivered the prestigious Wilde Lecture before the Manchester Literary and Philosophical Society, and in general had managed to keep himself in the news So when Boltwood wrote that "the experiments described in this paper are considered to indicate that the results obtained by Mr. Soddy are without significance,"[38] he was attacking the competence of a person whose work had heretofore been considered sound.

Boltwood clearly presented the problem:

> If radium is a *direct* product of uranium through the intermediate stage of uranium-X and if the average life of radium is approximately 1,000 years, then it can readily be deduced that, with the delicate methods of measurement at command, the quantity of radium formed in a few hundred grams of uranium salt will be readily detectable and measurable after the lapse of a period no longer than a month. If, however, one or more transition products of a relatively slow rate of change intervene between the substance uranium-X and radium, the production of radium will be so protracted that no quantity of it sufficiently great to permit its detection will be formed within a greatly extended period.

> The difficulties involved in the experimental demonstration of the growth of radium do not appear to be great. Uranium forms no radioactive, gaseous disintegration product, while the radium emanation affords a most convenient means of quantitatively estimating any radium which may be present. A solution of a carefully purified uranium salt can therefore be prepared and can be tested at intervals for radium emanation. If radium is formed from the uranium its existence will be indicated by the presence of radium emanation in the uranium solution.[39]

Then Boltwood discussed Soddy's three papers.[40] He pointed out that "no conclusive evidence is brought forward to show definitely how much or how little radium was present in the uranium solution at the commencement of the experiment."[41] In fact, Boltwood had

> convinced himself at the beginning of his own experiments that the method of procedure followed by Mr. Soddy in testing his solutions for radium emanation is entirely unsuited for the determination in question. A concentrated solution of incompletely purified uranium nitrate containing traces of radium gave up only a fraction of the total radium emanation generated within it when the solution was allowed to stand for days in contact with a small air space and air was bubbled through it. It was speedily found that only by boiling the solution vigorously for about fifteen minutes could the total emanation present be positively separated.[42]

Further,

> it appears extremely possible that the increase in the content of radium which Mr. Soddy believes he has observed in his uranium solution may in fact have been due to the accidental and unconscious introduction of

radium salts during the tests conducted at the end of the twelve months period. According to his own statements these tests were carried out in a laboratory notably contaminated with various radio-active products, and the accidental introduction of the sub-microscopic quantity of material (1.6 × 10^{-9} gram) which was afterwards detected would account for the later positive results. The liability of contamination from an extraneous source is strongly suggested by the behavior of Mr. Soddy's electroscope, in which the normal air leak has risen from 0.048 division per minute to 1.56 division per minute, an increase of over thirty times, during the period covered by his experiments.[43]

And even if the strong possibility of contamination could be ignored, the results were still questionable, for

> although it is stated in this paper [in the *Phil. Mag.*] that observations had been taken occasionally over a period of eighteen months and that these observations indicated a *gradual* growth of the emanating power of the uranium solution, the only definite and directly comparable numbers are restricted to a total period of about three weeks (Dec. 17, 1904 to Jan. 9, 1905) and include only four measurements conducted at the close of the period of observation.[44]

In short, "it would appear that none of these essential conditions," specified by Boltwood as necessary for a successful determination, "has been fulfilled in the experiment described by Mr. Soddy."[45] Rutherford apparently agreed, for he wrote, "you have rather run a tilt at Soddy and I do not think it undeserved considering the uncertainty attaching to his methods and accuracy of experiment."[46]

Boltwood's paper did not contain only an attack on Soddy's work; he also presented his own experimental results. Over a year before he had purchased a quantity of "purest uranium nitrite" from the firm of Eimer and Amend, of New York City. Since he had found radium emanation present, he repeatedly recrystallized a solution of the salt until virtually all the radium was removed. This uranium solution was then sealed for a period of thirty days, at which time all dissolved gases were boiled out and tested. There was *no increase* in the leak of his electrometer, though "an increase of 0.005 division per minute could have been detected with certainty."[47] Since 0.005 division per minute corresponded to a radium equivalent of 1.7 × 10^{-11} gram, less than this amount had been grown in the thirty days.

The solution was again sealed and allowed to stand for six months. Entirely negative results were again obtained. And the same was true after a total of 390 days. Thus,

> the quantity of radium which can have been produced in the given time is therefore less than one two-millionth of the equilibrium quantity and less

than one sixteen-hundredth of the quantity which would be expected from the disintegration theory [using Rutherford's value of the decay constant]. The quantity is furthermore only about one-tenth of the quantity assumed by Mr. Soddy to have been formed from an equal quantity of uranium in his solution during an interval of eighteen months.[48]

Radium was undoubtedly a daughter product of uranium, as Boltwood had proven earlier. Now, these results showed that it could not be the direct daughter of the short-lived uranium X, but that a long-lived member of the decay series lay between them.

6 Useful Decay Series Relationships

The Ultimate Disintegration Products

> In the naturally occurring minerals containing the radio-elements these changes must have been proceeding steadily over very long periods, and, unless they succeed in escaping, the ultimate products should have accumulated in sufficient quantity to be detected, and therefore should appear in nature as the invariable companions of the radio-elements.[1]

In the few years following this statement by Rutherford and Soddy there was a minimum of real work on the subject, accompanied by a fair amount of speculation. Certainly, the discovery of helium in radium, by Ramsay and Soddy in 1903, led many to suggest that this inert gas was the final decay product of not a few elements.

More serious attention was given to the ultimate disintegration products of the radioelements by Boltwood, in 1905. In the early part of the year he ventured the suggestion that lead is one of the final, inactive members of the uranium series.[2] Then, in the next several months, he gathered what other information he could into an admittedly speculative paper for the *American Journal of Science*.[3] Meager as his data were, he wished to publish his hypothesis, hoping to awaken others to the problem of finding the end products. In this respect, his efforts were largely to no avail; yet his own results were quite interesting.

His investigation had failed to find a single primary uranium mineral that did not contain lead. The same was true of the secondary, or altered, minerals, though in at least one case Boltwood was hard pressed to detect the lead. In general, he determined that "greater proportions of lead and helium with respect to uranium are found in those primary minerals which occur in the oldest geological formations."[4] Very ancient Ceylonese thorianite, supplied to him by George Kunz, helped to illustrate this point. These empirical observations were strengthened by theoretical considerations voiced by Rutherford in the Silliman Lectures at Yale just a few months earlier: if the alpha particle consists of helium of atomic weight 4, then the five alpha emitters between radium (atomic weight believed to be 225) and the inactive radium G would require the latter to have an atomic weight of about 205. "This is not far from the accepted atomic weight of lead, namely 206.9."[5]

Other elements that Boltwood suggested as possible end products for the different decay series were bismuth, barium, hydrogen, and argon. The first two were found to appear persistently in uranium minerals, though in small quantities. Since thorium appears so often with uranium, and since the "atomic weight of bismuth differs from the atomic weight of thorium by exactly 24 units, an even multiple of 4,"[6] a connection was thought possible through the emission of six alpha particles. Such reasoning, crude as it seems, was, in fact, the way the holes in the decay series were often filled. Barium was suggested as the final product of the actinium series, since both appear in small ratio to the uranium in natural ores. Boltwood had recently been interested in actinium, for while it seemed to be proportional to the amount of uranium present there was too little of it. He and Rutherford had earlier felt that it "is not a direct product of uranium in the same sense as is radium,"[7] and this belief later led to a more explicit statement of branching.

Hydrogen was proposed not only because the identity of the alpha particle was yet uncertain, but because water had been found in a number of impervious radioactive minerals. The inert gas, argon, was also included because Ramsay and Travers in England had found that most minerals which evolve helium also evolve argon in small amounts.

Boltwood concluded this speculative paper:

> If it can be ultimately demonstrated that lead, bismuth, barium, hydrogen and argon, or any one of them, actually result from the disintegration of uranium, an interesting question which naturally arises will be: Have the quantities of these chemical elements already existing been produced wholly in the same manner? Any discussion of this problem at the present time would certainly be premature, but the time may not be very far remote when this question will deserve serious consideration.[8]

Another paper on this topic was published by Boltwood almost a year and a half later, in February 1907.[9] This was based on far more experimental evidence and was, as would be expected, more sound. Since recent work of his (to be discussed below) indicated that actinium was probably an intermediate product between uranium and radium, and therefore would not require a unique element at the end of its series, "the number of possible ultimate products has been correspondingly reduced." In addition, a careful examination of typical primary uraninites from Branchville, Connecticut, and Flat Rock, North Carolina, and of thorianite from Ceylon, had convinced him that "neither bismuth nor barium can be considered as disintegration products in the main line of descent from either uranium or thorium, at least on the basis of the present disintegration theory."[10]

Boltwood again declared lead to be one of the final products of uranium, and presented impressive data to support this claim. Quantitative analyses of forty-three mineral samples clearly showed a striking constancy in the ratios of lead and uranium percentages for minerals of the same locality, and therefore the same geological age, and an equally striking increase in these ratios among the older minerals, in which more of the final product would be expected. In the case of secondary or altered minerals, e.g., uranophane, all analyses gave lower ratios than for primary minerals from the same locality. Of course, the most common alteration product of uraninite, known as gummite, was excluded from consideration, since lead is one of its natural, chemical constituents.

Thorium's final products were not thoroughly investigated at this time, but Boltwood (erroneously) felt certain that lead was *not* one of them. No proportionality could be found between lead and thorium in several minerals, while the lead-uranium ratio in these same minerals was "proper." The explanation of this was recognized only many years later: lead from uranium generally forms a relatively insoluble compound, while that from thorium forms a more soluble one.[11]

Helium was also suggested as a final decay product of uranium and some data were presented to show the possibility but not the certainty, since an unknown amount of the gas could have escaped from the mineral. Boltwood assumed that all the uranium (atomic weight 238.5) decayed ultimately to lead (206.9) and helium (238.5 − 206.9 = 31.6). In other words, for every 207 parts of lead there should be formed 32 parts of helium. Then he calculated the percentages of helium expected in twenty different minerals by multiplying their lead contents by 32/207, and compared these values with those obtained by analysis. At first glance, the results were not particularly good. But upon consideration, one could see that the observed amounts of helium "found in radioactive minerals are of about the order, and are not in excess of the quantities, to be expected from the assumption that helium is produced by the disintegration of uranium and its products only."[12] In general, also, the minerals of greater density were found to retain a greater proportion of their helium.

Radioactive Dating

Untangling some of the elements of the uranium decay series was a significant accomplishment in its own right. Boltwood, however, went further and developed the first geological dating system. In effect, he reversed the procedure of confirming the lead-uranium ratios by *accepted* geological age, and used these ratios to determine the age. This was a striking application of the science of radioactivity and one that, like the medical applications, could be comprehended by the layman.

The original idea was not his and he readily acknowledged its birth in Rutherford's amazingly fertile mind.[13] But it was in Boltwood's hands that the method was tested, developed, and used. And, of course, it was a logical consequence of his proof that the uranium series has a final, inactive product in lead.

The greatest uncertainty was introduced by the value for the half-life of radium, and, until this number was better determined, all calculated ages would be somewhat uncertain. As described in chapter five, at secular equilibrium the product of the number of particles and the decay constant is equal for all radioelements of a series. Since the decay constant λ is inversely proportional to half-life ($\lambda = 0.693/T$), the ratio of particles to half-life is therefore also constant in each series. The decay constant, or rate of disintegration, for uranium had not been determined by direct experiment because of the slowness of the process, but since this rate had been roughly calculated for radium, it was thereby possible to fix it for uranium.

Not long before, Rutherford, on the basis of the reckoned number of alpha particles emitted from radium each second, had estimated radium's half-life to be about 2600 years. This is equivalent to saying that the fraction of radium undergoing transformation in one year (λ) is 2.7 $\times$ 10^{-4}. The amount of radium in a gram of uranium had also been determined, by Rutherford and Boltwood, as 3.8 $\times$ 10^{-7} gram. This meant that (2.7×10^{-4}) (3.8×10^{-7}) = about 10^{-10} gram of radium is changing each year per gram of uranium. At equilibrium, however, the number of atoms of each element in the decay series changing per unit time is equal, and if the relatively small differences in atomic weights are neglected, the fraction of uranium transformed each year is also 10^{-10}.

The application of this information was extremely simple. Assuming lead to be the end product of uranium decay, Boltwood realized that about 10^{-10} gram of lead would be formed per year from each gram of uranium, and the decrease in quantity of the uranium would be negligible. All he need do was divide the previously obtained lead-uranium ratios by 10^{-10}:

$$\text{Age} = \text{Pb/U} \times 10^{10} \text{ years.}$$

Thus, for example, the 0.041 ratio of Connecticut uraninite showed the mineral to be 410 million years old, and the 0.22 ratio of Ceylonese thorianite proved its great antiquity of 2200 million years.[14]

These results were of particular interest because they indicated that the earth was far older than commonly allowed.[15] In the mid-nineteenth century William Thomson (later Lord Kelvin) had calculated the age of the globe on the basis of its rate of cooling from a molten mass. His enormous prestige and the straightforward physical methods employed caused acceptance of a value of less than 100 million years (it was frequently revised), much to the distress of some geologists and biologists. Formerly, such scientists had been unconcerned about geochronology, only maintaining that the time required for mountain building and changes in organic life were "very long." Indeed, following the pronouncement by James Hutton at the end of the eighteenth century that "we find no vestige of a beginning, no prospect of an end," many geologists simply assumed *unlimited* time for the grand succession of events they studied. Their initial reaction to Thomson's work was disdain at the meddling of a mere physicist in *their* science. But when his calculations were subsequently seen to be irrefutable, more and more geologists advanced data on sedimentation, physical properties of rock, and the salt content of the oceans which served to corroborate an earth age under 100 million years. Thus, by the beginning of this century, earth and life scientists overwhelmingly favored a relatively short time scale.

It came as a shock when in 1903 Pierre Curie and Albert Laborde announced in Paris that radium continually evolves enough heat each hour to melt more than its own weight of ice. This prodigious production of heat is caused by the loss of kinetic energy as alpha particles, ejected from the radium, collide with the atoms of the chemical compound and lose their momenta. The significance of this discovery was not lost upon the scientific community: Kelvin had based his investigations on the assumption that the earth is a steadily cooling globe, with no source of heat other than that remaining from its once molten condition. If, however, radioactive materials were present in the earth in sufficient quantity, as seemed probable, their disintegration might produce enough heat to invalidate all age calculations based on a rate of cooling, and extend this age considerably.

That radioactive bodies were widespread could not be doubted. Witness the measurements made around the world on rain, snow, air, soil, and ground water—a subject which I have called "atmospheric radioactivity." Further studies by Strutt showed, moreover, that the earth's radium could not be distributed throughout its mass, but must be concentrated in a crust about 45 miles thick. Otherwise more heat would have to flow from the earth than was actually measured. Yet, while some popular writers now lauded radioactivity for delivering geology from Kelvin's "tyranny," there seemed to be little professional interest in a revision of geochronology. Even when Boltwood published his results in 1907, few geologists chose to pursue the technique of radioactive dating. Some, no doubt, had a vested interest in the *geological*, not physical, data which supported a short earth-age. Others recalled that their predecessors had once before been deceived by physical evidence that appeared correct, and were unwilling so soon to accept the evidence of radioactivity. In this they were not unwise, for the age calculations were based upon some constants, such as the half-life of radium, and some assumptions, such as that the thorium series does not end in lead, which were later revised. Yet, though accuracy would improve and refinements would be introduced in the future, Boltwood had shown that the lead method of radioactive dating was a viable technique. Widespread acceptance of these procedures came finally in the late 1920s, largely as the result of efforts by the British geologist Arthur Holmes. Though other elements have also been used successfully to date rocks, such as the helium method developed by Strutt, and more recently the rubidium-strontium, potassium-argon, and other techniques, the lead method has not been superseded.

Relative Activities

Another subject of great importance, investigated by both Boltwood and McCoy, was the relative activities of the various constituents of the uranium decay chain. Among the benefits of such information would be a certainty that all members of the chain could be accounted for, the total activity logically equaling the sum of its parts.

For his 1904 paper in the *Berichte der Deutschen Chemischen Gesellschaft,* McCoy had tested twelve uranium ores from different localities and determined that the total radioactivity of each was approximately proportional to the percentage of uranium contained in the sample.[16] He also tested commercial uranium compounds and found them far less active than the minerals, a fact he attributed to removal of some or all of the radium in processing. In general, the activity constant (determined by rate of electroscope discharge and percentage of uranium) was 5.7 times greater for minerals than for pure compounds. This meant that the decay products of uranium emitted alpha particles with 4.7 times the ionizing power of the alpha particle from uranium itself.

Shortly thereafter McCoy proved that absorption of the alpha rays by the source must not be neglected, and developed techniques employing thin films of uranium compounds and ores to determine the absorption coefficients.[17] The greater accuracy resulting not only enabled him to revise the 5.7 activity ratio of minerals to pure compounds, but gave him the opportunity to establish a reproducible standard of radioactivity, something for which the need was beginning to be strongly felt.[18] The absorption coefficient allowed him to correct the observed activity of a thick film to its true activity; the film's known weight permitted calculation of the total activity of one gram of pure uranium compound. If this figure was divided by the percentage of uranium in the compound, the result was the total activity K, of one gram of elemental uranium. When pure U_3O_8 was taken as the standard source, and its activity normalized to unity (i.e., the activity of one square centimeter), McCoy found K to be 791. This meant that the total activity of one gram of uranium was 791 times the observed or surface activity of one square centimeter of a U_3O_8 film sufficiently thick to be of maximum activity. It also meant that McCoy had found a definite area of a thick film to be a more reproducible standard than a definite amount of the oxide. This is understandable, since the film could be of *any* thickness over a certain minimum. Up to this point the alpha activity above the surface would increase with thickness, but beyond it no additional alphas would be able to penetrate the blanket of material.

At the same time he determined the total activity of five different uranium ores, whose average value was 3280. The ratio of ore to compound (3280/791 = 4.15) meant that, for equal uranium content, ores were 4.15 times more active than pure compounds, not 5.7 as found earlier. This figure also gave a means of calculating the relative activity of equal amounts of radium and its products compared to uranium. Rutherford and Boltwood had shown that one part of radium is in secular equilibrium with 1.35×10^6 parts of uranium. Since radium and its products were 4.15 − 1 = 3.15 times as active as uranium in equilibrium amounts, they would be $(3.15)(1.35 \times 10^6) = 4.25 \times 10^6$ times as active in equal amounts. While this value did not agree too closely with a Rutherford and McClung determination in 1900, it was of the same order of magnitude and therefore something of a confirmation.[19]

McCoy's paper had been read before a meeting of the American Physical Society, in Chicago, on 21 April 1905. Boltwood's next paper was presented to the same group, at their 24 February 1906 meeting, held at Columbia University.[20] In all previous statements regarding the activity of radium salts, he noted, nothing indicated that care was taken to preclude the escape of a portion of the emanation, nor were data presented on the proportion of the emanation yet retained in the solid.

For example, in her 1903 thesis, Mme. Curie merely noted that the maximum activity attained by solid radium salts, several months after preparation, was five to six times the initial activity. Since more accurate information was desirable, not only intrinsically but "for the interpretation of other more complicated relations," Boltwood undertook to "determine the relative α-ray activity of radium salts from which all emanation had been removed and the activity of the same salts when the total, equilibrium quantity of radium emanation was retained within them."[21] By preparing films of radium-barium chloride and monitoring the increase in activities during a few days, then determining the percentage of emanation retained, he calculated that a salt containing an equilibrium amount of radium, its emanation, and the short-lived decay products is 5.64 times as active as a pure radium salt.

But what does this value of 5.64 mean? To explain it Boltwood quoted some alpha particle ranges published by Bragg and Kleeman in the *Philosophical Magazine*. By this time it was clear that each radioactive substance which emits alphas ejects them with a precise range (in air or other material), and that this range is a means of identifying the emitter. For radium, emanation, radium A, and radium C[22] (all the alpha emitters except the long-lived radium F, which need not be considered here since little of it would have had time to form) the ranges in air at atmospheric pressure were 3.5, 4.23, 4.83, and 7.06 centimeters, respectively. Their sum, 19.62, is 5.60 times the range of the alpha particle from radium, leading him to conclude:

> The value found for the relative ionization or activities of the different products, namely 5.64, agrees so closely with the ratio of the relative ranges of the same products that it appears highly probable that the α-ray activities of the different products are proportional to the ranges of their α particles. Moreover, since according to the disintegration theory when the parent substance and the products are in radio-active equilibrium the same number of atoms of each are undergoing disintegration per second and the same number of α particles are projected from each exploding atom, it would appear probable that the ionization produced by each α particle is proportional to its range.[23]

Boltwood did not discuss in this article the relative activities of uranium in equilibrium with its decay products and uranium alone. He had, however, been working on the subject and was not entirely happy with McCoy's published value of 4.15. He felt the difficulty was connected with McCoy's disregard of the amount of emanation spontaneously evolved and lost, for he wrote Rutherford:

> I am in correspondence with McCoy and am helping him to get his own results on the activity of uranium minerals straightened out somewhat. He

had sent me a couple of his minerals, one of them, his uraninite No. 1, loses no less than 13% of its emanation; and his carnotite loses *33.6%* of its emanation. As you will perhaps remember, he found that minerals (per gram of uranium) were 4.1 times as active as uranium, while I find the ratio to be about 5.3 when all emanation is retained. I think that I can convince him of his errors and can get him to publish an explanatory paper with my own, so that people will not have to puzzle over the disagreement of our results and say: "McCoy found 4.1, but Boltwood says 5.3. Which is right?"[24]

This projected joint paper was never published and since no pertinent correspondence between the two men has been found the details of one of the apparently rare occasions when they engaged in scientific intercourse are obscure.

Nevertheless, McCoy *did* subsequently correct his results for the lost emanation. In a joint paper with William H. Ross (1875–1947), he recalculated the "specific activity" of uranium, taking into account not only the emanation but several other errors newly uncovered.[25] McCoy's collaborator, Ross, was a student who received his doctorate in 1907. Though born in Canada, he remained in the United States and pursued a long career as an agricultural chemist for the federal government. In the sense that he was later employed in a field different from that of his dissertation, Ross was typical of the early radiochemical students.

The original experiments had been performed in an electroscope having a distance of from 3.5 to 4.5 centimeters between the active film and the charged gold leaf system. However, Bragg and Kleeman showed that both radium A and radium C emit alpha particles with ranges in excess of 4.5 centimeters, which meant that the activity measured had been less than the total amount. A larger electroscope was therefore constructed in which an alpha particle could complete its ionizing path before striking any part of the instrument.

Another source of error was also connected to the range of the alpha rays. McCoy formerly deposited his films of active material on tin-plated jelly glass covers. These were quite suitable since they were cheap enough to discard after use, thereby avoiding contamination problems. But the covers had a rim 0.8 centimeters high which acted selectively on those alphas of greatest range. This, too, was now changed and the films were deposited on flat plates.

Admittedly, both these corrections produced only small changes in the results, but the emanation correction had a larger effect. McCoy and Ross obtained values of 3616 for the specific activity of uranium in minerals and 796 for that in pure uranium compounds. Their ratio showed minerals 4.54 times as active as compounds of equal uranium content, almost a ten per cent change from the previous figure of 4.15.

Combining Boltwood's determination of 5.64 − 1 = 4.64 for the ratio of activities of emanation and its products to radium alone, with the derived value of 0.53 for radium free from its products to the equilibrium amount of uranium in a mineral, the activity of emanation plus its products was calculated: (4.64 × 0.53)/4.54 = 0.54 of the total activity. Direct experiment had given 0.52, a quite good confirmation.

If, as Boltwood suggested, the activity of a member of the radioactive series was proportional to the range of its alpha ray, then radium F (polonium) should be 1.10 times more active than the equilibrium amount of radium alone. Inverting the radium-uranium activity ratio gave 1/0.53 = 1.87 (roughly) for the uranium's activity relative to radium. This continual setting up of ratios may seem highly confusing at first, but it is beautifully illustrative of the manner in which every bit of information concerning the activity of an individual member of a decay series was squeezed out of the gross empirical results.

Happily, however, McCoy and Ross pulled everything together:

> Since D is inactive and E gives only β-rays, uranium together with radium and its products are 1.87 + 5.64 + 1.10 = 8.61 times as active as the radium alone; or 8.61/1.87 = 4.60 times as active as the equilibrium amount of uranium. Our direct experiments gave 4.54 as the ratio of the activity associated with equal amounts of uranium in minerals and in pure compounds. The close agreement of the two values would seem to indicate that the activity of a uranium mineral (free from thorium) is due solely to uranium and radium and their recognized products.[26]

The uranium series, however, was not complete; other products were to be found. Nevertheless, the method of determining relative activities was valid, if not yet sufficiently precise. If at this point it yielded no new radioelements, it did much to further precision measurement in radioactivity, a trend of general value. A particular piece of benevolent fallout was the development of reproducible standards.

Boltwood and McCoy, both quantifiers, helped push the study of radioactivity into ever greater exactitude. Proof of lead's relationship to uranium, the method of radioactive dating of minerals, and the information available from relative activity calculations showed the benefits of their rigorous approach. Great discoveries soon would come from this orientation, particularly the isolation of an important new radioelement and recognition that certain radioelements are chemically inseparable.

7 The Parent of Radium

Actinium's Uncertain Position

According to the Rutherford-Soddy transformation theory, the substance which produces radium must be found in radium-bearing minerals. This was naturally believed to be uranium since radium was first found in uranium ores. Elaboration of these ideas followed two trails: a) Proof of the constant ratio of the amount of radium to uranium in such minerals, such work having been described in chapter five, and b) Search for the immediate parent of radium, uranium being accepted as the ancestral or ultimate parent. The logical procedure in this second pursuit was to attempt to grow radium in a uranium solution from which the radium initially present had been removed. When failure was encountered it was recognized that probably a long-lived product in the decay series intervened to slow the accumulation of new radium. Intimately associated with this search were efforts to determine the relative activities of all the decay products and to measure the half-life of radium. It is, therefore, no surprise that the same people were concerned with all these matters.

During the summer of 1906, Boltwood moved his apparatus into the Sloane Physics Laboratory of Yale University and prepared to undertake his new academic duties. These responsibilities proved more extensive than anticipated, since, owing to illness in Bumstead's family, and as a result of the small size of the faculty, Boltwood was left in charge of the extensive renovations in the old laboratory.[1] Thus, it was over a year after he delivered his strong criticism of Soddy's work on the growth of radium that he returned to the subject. In his first communication as a member of the physics department staff, he presented "strong evidence . . . in support of the assumption that actinium is the intermediate disintegration product between uranium and radium."[2] By this time no one doubted the genetic relationship between uranium and radium, nor the fact that there must be a slowly decaying product between uranium X and radium; the task was to locate this element.

"A considerable mass of experimental data . . . ," Boltwood wrote, "points to the conclusion that the quantity of actinium in a mineral is directly proportional to the amount of uranium present and that, accordingly, actinium is a product of uranium."[3] Suspecting that this might be the elusive parent of radium, he separated the actinium from a kilogram of carnotite ore and sealed the solution in a glass bulb. Two months later the gases and emanation collected in the bulb were boiled out and tested in an electroscope, the amount of emanation corresponding to the presence of 5.7×10^{-9} gram of radium. Again the bulb was sealed and allowed to stand undisturbed, this time for 193 days. When tested now, the emanation corresponded to 14.2×10^{-9} gram of radium, an increase of 8.5×10^{-9} gram, or a production of about 1.6×10^{-8} gram of radium in one year. The amount of radium in equilibrium with 200 grams of uranium (the approximate quantity in the mineral from which the actinium was extracted) was, according to Rutherford and Boltwood's work, 7.6×10^{-5} gram. Thus, the decay constant λ for radium would become $(1.6 \times 10^{-8})/(7.6 \times 10^{-5}) = 2.1 \times 10^{-4}$ per year. The half-life would be therefore the natural logarithm of 2 divided by $\lambda = 0.693/(2.1 \times 10^{-4}) =$ 3300 years, a value of the same order of magnitude as Rutherford had obtained by other means.

This work appeared to show strongly that actinium was, indeed, the parent of radium. Boltwood, however, would only admit publicly to the probability, since "the original content of uranium in the material used, and the completeness of the separation of the actinium, are both uncertain."[4] In private, though, he had not "the least suspicion that actinium is not the true father. You may doubt the sharpness of my chemical nose," he wrote Rutherford,

> but I can assure you that I felt perfectly confident that I should find radium growing from actinium because *there is no final disintegration product* of actinium in the uranium minerals. So I am unwilling to believe that the intermediate product between uranium and radium is merely entrained with the actinium. Besides the latter is altogether too plebeian an idea and wholly discreditable to the parent.[5]

Rutherford was, however, more cautious and also somewhat less aesthetically inclined in his science. Though certain experiments of his own also indicated the growth of radium in actinium solutions, he felt that "the growth of radium observed in [Boltwood's] actinium solution possibly might arise, not from the actinium itself, but from another substance, normally separated from the radio-active mineral with the actinium."[6] He also felt the matter of the relative activities of the actinium products must be settled before conclusive statements about genetics could be made.

Difficulties Presented by Relative Activities

The problem of the relative activities was, indeed, a thorny one. In a letter to the editor of *Nature,* written before he joined Rutherford in Montreal for the 1906 Christmas vacation, Boltwood presented the following puzzle.[7] For a nonemanating mineral, containing no thorium, which had reached equilibrium, the activity of thin films (measured in an electroscope with a large ionization chamber) was about 5.3 times the activity of the uranium in the mineral. This was composed of approximately 0.52 for radium, 2.4 for the radium products of rapid change, 0.55 for radium F (polonium), and 1.0 for uranium itself. The total of approximately 4.5 times the uranium activity, left only $5.3 - 4.5 = 0.8$ times uranium for the activity of the four alpha ray products of actinium.

Other reasoning showed, however, that the activity of the actinium products must be higher. Since at secular equilibrium the same number of atoms transform in unit time for each radioelement, the activity contribution of each element should be directly proportional to the range of its alpha particle, since such range was an indication of the alpha particle's energy. Otto Hahn had measured the alpha ray ranges of the four actinium products, their average being 5.6 cm. According to Bragg and Kleeman, the range of radium's alpha ray was only 3.5 cm. Assuming all alpha particles are similar, then the average actinium product would produce $5.6/3.5 = 1.6$ times the ionization of radium. And since radium's activity was 0.52 times uranium, there would be expected from the actinium series $0.52 \times 1.6 \times 4 = 3.32$ times uranium, while only 0.8 times uranium could be allowed. Knowledge of this approximate number was, in fact, the reason for Rutherford and Boltwood's remark that "actinium is not a direct product of uranium in the same sense as is radium," in their first joint paper in 1905.[8]

On the credit side of the ledger, Boltwood called attention to uranium's activity being about twice that of the radium present. This was in good agreement with a recent claim by Richard Moore and Herman Schlundt of the University of Missouri that there were two alpha ray changes not just one, in uranium. Assuming that the average range of these alphas was about 3.5 centimeters, the ledger balanced well—at this point anyway.

In the debit column, though, the bookkeeping was far more complicated. H. S. Allen, of King's College, London, had written a letter to the editor of *Nature* the previous month suggesting that the alpha particle was only half a helium atom.[9] This possibility had, in fact, been mentioned earlier by Rutherford. But while Rutherford and Royds had not yet *proven* the identity of helium and the alpha particle, most workers accepted their equality on the basis of strong secondary evidence. Boltwood objected to Allen's proposal because, while it would work for a total of six alphas between uranium (238.5) and radium (226.5, recently revised by Mme. Curie from 225), i.e., uranium, uranium X, and the four actinium products, it would fail between radium and lead, where far more changes would be required than were known to exist.

Although he questioned the value of his own speculations, Boltwood could not resist hypothesizing.[10] Assume, he said, the mass of each alpha to be one, not four, and further assume that four alphas are emitted in *each* of the two uranium changes and five radium changes, while only one alpha is emitted in each of the four actinium changes. Thus, the entire series would be as follows: U(238.5), less $2 \times 4 = $ Ac (230.5), less $4 \times 1 = $ Ra (226.5), less $5 \times 4 = $ Ra F (206.5, lead, now called Ra G). Apparently no one, however, including Boltwood, thought enough of this scheme to consider it seriously.

Moore and Schlundt

The two men mentioned above as the discoverers of a second alpha ray change in uranium were, after Boltwood and McCoy, the most prominent American radiochemists; by their activities in university and governmental research laboratories they significantly aided the fledgling radium industry in this country. Richard Bishop Moore (1871–1931) was born in Cincinnati, Ohio, but educated in England, where his parents moved when he was seven years of age.[11] At University College, London, he chose chemistry as a career, and studied rare gases under Sir William Ramsay. In 1895, he returned to the United States and received a B.S. degree after one year at the University of Chicago. He remained another year in Chicago as an assistant in chemistry and a classmate of Herbert McCoy, before accepting a job as an instructor at the University of Missouri.

He held this post, in the oldest state university west of the Mississippi River, between 1897 and 1905, at which time he became professor of chemistry at Butler College, Indianapolis, Indiana. Except for a leave of absence in 1907, to work again in Ramsay's laboratory, Moore taught at Butler until 1911, when the U.S. Government acquired his talents. By

RICHARD B. MOORE (1871–1931)

this time he had established a reputation for his work on the radioactivity of waters and soils, such that he was employed by the Bureau of Soils and then the Bureau of Mines, until 1923.

During this period of federal employment, he surveyed the radium deposits in Colorado, devised methods for concentrating ores, and supervised the preparation of some of the first radium salts produced in the United States. In another field, he pioneered in suggesting the use of helium for balloons and airships, encouraging first the Navy Department and then Congress to conserve this natural resource. Between 1923 and 1926, he worked in private industry, and then he returned to academic surroundings for the last five years of his life. This final post was as professor of chemistry and dean of the school of science at Purdue University.

Moore's colleague was Herman Schlundt (1869–1937), a native of Wisconsin who received his higher education from that state university.[12] While a graduate student he was employed as an assistant in chemistry (1894–1896) at the university, as an instructor of physics and chemistry in the Milwaukee high schools (1896–1899), and was able to spend the last year of the century at the University of Leipzig. He remained at the University of Wisconsin for one year after receiving his

HERMAN SCHLUNDT (1869–1937)

doctorate in 1901, and then began a lifelong association with the University of Missouri, where he was promoted to the professorship of chemistry in 1907.

At Missouri he met not only Richard Moore, but Oscar M. Stewart, who had written the excellent surveys of radioactivity research for the *Physical Review,* in 1898 and 1900, and had left Cornell in 1901. Stewart apparently did not engage in any experimental work himself, but we may surmise that his interest in radioactivity persisted and influenced the several generations of physics students in his classes. Recognition of their teaching abilities at Missouri is seen in the naming of the present physics and chemistry laboratories after Stewart and Schlundt, respectively.[13]

Like Moore, Schlundt's primary interests were in the area of the radioactivity of waters and soils, and the chemical methods of extracting radium from minerals. In this latter connection, he established a research laboratory on campus, in the late 1920s, in which advanced students refined radioactive materials for the U.S. Bureau of Standards, the U.S. Radium Corporation, and other commercial firms.[14] Also in the 1920s, Schlundt was called upon as an expert witness during the famous trial in which luminous watch dial painters sued the U.S. Radium Corporation; these women "tipped" their brushes with their tongues and most of them ingested enough radium to poison themselves fatally.[15] Again like his colleague Moore, and so many other American scientists, Schlundt availed himself of an opportunity for advanced research abroad, and spent a sabbatical leave in 1921 at Cambridge University. There he worked in Rutherford's laboratory, measuring the rate of alpha particle emission from thorium C.

In their joint paper, Moore and Schlundt confirmed the Crookes and Becquerel methods for separating uranium X from uranium, and added several new and superior techniques.[16] Using Kahlbaum's best commercial grade of uranium nitrate,[17] they found that uranium is soluble in acetone, methyl and ethyl acetate, methyl and ethyl alcohol, and a few other hydrocarbons, while most uranium X precipitates upon the addition of some ferric hydroxide. Yet, even more interesting than this chemistry, was their claim that uranium X emits alpha rays as well as betas. Since Soddy had shown that uranium X would not affect an electrometer (which measures best the ionization produced by alpha particles), while it would affect a photographic plate, no one had believed it to emit anything but betas. But this new fact was quickly integrated into the decay scheme, for the twelve mass units between uranium and radium could now be explained in terms of three alpha particles from uranium, uranium X, and the parent of radium.

It is now known that this explanation is wrong, and that Soddy's work was accurate, in that neither of the two types of uranium X ($U X_1$ = Th-234, $U X_2$ = Pa-234) is an alpha emitter. What no doubt happened was that the soon-to-be-discovered ionium, the immediate parent of radium, being a thorium isotope, precipitated with the uranium X_1, and *its* alpha particle was detected. This means that only two alphas were actually accounted for between uranium and radium, i.e., from uranium and from the radium parent, which was yet unidentified, but whose existence was known. The third, for there are indeed three, was later found to be emitted from an unsuspected radioelement, uranium II (U-234), the daughter of uranium X_2.

Actinium Discredited

Almost half a year now passed before further results were reported on growing radium from actinium. Then, in June 1907, Rutherford announced that "*no appreciable growth of radium was observed over a period of eighty days,*" from an actinium solution purified in a different manner.[18] "If there were any growth at all, it was certainly less than one two-hundredth part of that normally to be expected." This was an important revelation, typical of Rutherford's unerring knack for seeing things correctly. "From these observations," he continued,

> I think we may safely conclude that, in the ordinary commercial preparations of actinium, there exists a new substance which is slowly transformed into radium. This immediate parent of radium is chemically quite distinct from actinium and radium and their known products, and is capable of complete separation from them.[19]

Of course, he added, if actinium was not the parent of radium, then its position in the decay series was as unsettled as before.

Nearly two years after Boltwood had written that certain of Soddy's results were without significance, and now that Boltwood's belief that actinium was the parent had been shown to be wrong, Soddy replied to the criticism.[20] He was particularly irked at the suggestion that he had unwittingly introduced radium salts into his solutions during the tests, although he had himself inferred the possibility of contamination. "Now such a criticism and such an imputation on the part of one investigator dealing with the work of another surely ought only to have been made if it was the only possible explanation of the discrepancy," he wrote. "As it was, to me at least, it was not even the most obvious explanation."[21] He reasoned that his uranium nitrate had been purified initially from radium by precipitating barium as a sulphate, while Boltwood had purified his uranium nitrate from radium by repeated crystallization. The one method removed only the radium from the solution, while the other removed *all* other substances from the uranium. Therefore, Soddy claimed that his procedure had not entirely removed the parent of radium, permitting him to observe some growth of this element. This view was supported by Rutherford in the published Silliman Lectures,[22] and is probably quite true. But the possibility of contamination still existed, and because his solution was not originally pure, Soddy had not really refuted Boltwood's claim that his results were without significance.

Yet, Soddy felt he could crow a bit because history had repeated itself. Boltwood had believed that actinium grew radium, but then, Soddy wrote, Rutherford showed that

> actinium *purified from radium in a different manner* yields no appreciable growth of radium. Is Boltwood's previous positive result then "without significance"? Surely not. But if Boltwood's result on the production of radium from actinium can be explained, as, of course, it can be explained, without charging him with introducing radium into his solution, so in the same way can mine with uranium.[23]

And Soddy was logically correct.

After this outburst of righteous indignation, he submitted new experimental results. During the past two years he had attempted to grow radium from a solution of uranium, *purified to remove all other products,* and wished to "entirely confirm and extend the results obtained by Boltwood,"[24] i.e., that no increase in radium could be detected.

The tone of Boltwood's reply was conciliatory, though its content unyielding.[25] He regretted that Soddy had taken his paper as such a serious "criticism" and "imputation," explaining that his remark that Soddy's results were without significance was an "unsuccessful effort at brevity." Yet, he was still "of the opinion that the experimental procedure which Mr. Soddy adopted was not suited to give conclusive results of either a positive or negative character." And, in defense of his own work, Boltwood observed that his finding "of the growth of radium in actinium preparations, even if it has served no other useful purpose, has certainly indicated where the immediate parent of radium is to be sought."[26]

Distasteful as controversy generally is to the participants, it can prove valuable when it highlights the problem to be solved. Unmistakably now, this problem was to separate the parent of radium from actinium and identify this radioelement. Would it be one of the known radioactive substances or, more likely, an entirely new one? Two months later, Boltwood gave the answer.

Ionium

Recalling his letter to the editor of *Nature*, of 15 November 1906, Boltwood explained why he had then written that actinium was the parent of radium.[27] His preparation had been a thorium compound which contained a radioactive body whose activity did not decrease over the course of several years. Since it was easily shown not to be radium, uranium, or polonium, and since André Debierne, in naming actinium in 1900, had said its chemical properties were largely similar to those of thorium, Boltwood assumed his radioelement was also actinium. As further proof, the emanation evolved from this material completely lost its activity in less than half a minute, which fit in well with the short half-life of actinium emanation (3.9 seconds).

But since Rutherford had found the radium parent to be distinct from actinium, Boltwood concluded that this mysterious radioelement was merely mixed with the actinium in his preparation. The 3.9 second half-life emanation was a true member of the actinium family, while the new product, he now showed, was an alpha particle emitter, chemically similar to thorium, which produced no emanation at all. Its characteristic properties were different from those of all known radioelements. "It is very likely," he wrote, "that this body is contained in Debierne's actinium preparations and in Giesel's 'emanium' compounds, especially in the former, and its presence may perhaps explain the confusion which has resulted from Debierne's earlier assertions that actinium accompanied thorium as opposed to Giesel's positive statements to the contrary."[28] And, almost as an afterthought, Boltwood added the information that these studies had enabled him to establish the half-value period of radium as approximately 1900 years.

Further conclusive evidence concerning the individuality of this new substance was almost immediately forthcoming through an examination of its alpha radiation.[29] By testing the distance at which a scintillating screen could be made to glow, the alpha ray range in air was found to be less than three centimeters, though the small quantities available made such determinations difficult. Yet, this short range was "sufficiently characteristic to serve as a definite means of identification."[30]

A weak beta emission was also reported, but here Boltwood was in error. The yet unresolved uranium X_1, which is a beta source, was probably mixed with his radium parent. Being also an isotope of thorium, it was, of course, chemically inseparable from it.

According to quantitative separations, the activity of the new element was about 0.8 times that of the radium with which it was in equilibrium, a figure Boltwood found "about the value to be expected . . . when the ranges of the α particles are taken into consideration."[31] Certainly, the 3.5 cm range for radium and less than 3 cm for the new element was a good confirmation, but Boltwood remained silent on another point. His earlier work had shown radium to be about 0.52 times as active as the uranium with which it was in equilibrium, and the radium parent about 0.8 times uranium. If, now, the activity of the parent was 0.8 times *radium,* this was only about half of that formerly expected. Yet, the order of magnitude was correct and the numbers would later fall into place.

As happens to all discoverers of elements, Boltwood had need for a name. Therefore, he wrote:

> The name "ionium" is proposed for this new substance, a name derived from the word "ion." This name is thought to be appropriate because of the ionising action possessed by this element in common with the other elements which emit α radiations.[32]

Though Rutherford had left McGill for the University of Manchester, Boltwood remained in correspondence with him and confided that

> it is a curious and interesting fact that ionium was the chief, if not the only, radioactive constituent of the radioactive substance that I separated from pitchblende in 1899 and which I had always supposed was actinium owing to Debierne's perfectly rotten statements in the matter. I think that De-

OTTO HAHN (1879–1968), BERTRAM BOLTWOOD (1870–1927),
AND ERNEST RUTHERFORD (1871–1937)

bierne has probably had the stuff in his hands for years and has not had the sense to identify it. In my own case I feel that with the insignificantly small amounts of material that I have been able to get to work with that there is some excuse for my having overlooked it.[33]

Rutherford replied with congratulations

on your success in getting one element to your credit. I feel a sort of paternal feeling with regard to it, having taken part in its inception . . . You deserve to get the last of the radioactive family. I never felt that Debierne deserved much credit for actinium—he couldn't miss it. As a matter of fact, if there had been a dozen elements with actinium, he had not enough radioactive sense to find it out.[34]

Though he apparently did not know it at the time, Boltwood had made his discovery of ionium none too soon, for his friend and colleague, Otto Hahn (1879–1968) was also hot on the trail.[35] Hahn was a German organic chemist who went to work in Ramsay's laboratory, in 1904, primarily to learn English as preparation for a commercial job.[36] Ramsay, disregarding Hahn's training, asked him to do some work on radioactivity, and so successful was the young man that he discovered radiothorium. Feeling that such abilities should not be wasted, Ramsay was instrumental in obtaining a position for Hahn as a specialist in radioactivity in Emil Fischer's Chemical Institute in Berlin. But before returning to Germany, Hahn decided to spend a year (1905–1906) learning more about radioactivity under Rutherford. It was during a visit to

New Haven, from Montreal, that Hahn met Boltwood, and the two later became good friends.[37]

In May 1907, Hahn inquired of Boltwood, "How is your work in actinium getting along? Don't you think, that the parent of radium might be *with* your actinium preparation and not your actinium itself?"[38] This was about a month before Rutherford announced that actinium grew no radium, and the then unconfirmed idea must have intrigued Hahn. He thereupon wrote to the Austrian government (which ran the mine) for a preparation from Joachimsthal pitchblende. But he was too late. Before the preparation arrived, Boltwood's letter announcing the new element (26 Sept. 1907) appeared in *Nature*.[39]

Within a short time of ionium's christening, Boltwood's colleague at Yale, Lynde Wheeler, who had a tendency to build such elaborate apparatus that he had no time to experiment with it,[40] did determine a range of 2.8 cm in air for the ionium alpha particle,[41] replacing Boltwood's estimate of less than three centimeters. Boltwood also soon found that the beta radiation came not from ionium but from uranium X, a fact made evident by the characteristic decay period.[42]

Regarding the name "ionium," Rutherford was less than happy, but could offer nothing better.[43] Yet, he would have preferred something suggestive of its place in the radioactive family. Norman R. Campbell, of Trinity College, Cambridge, objected on the same grounds.[44] He, however, proposed the name "uranium A," since he believed it a convention that the first decay product was called "X" and the following ones "A, B, C," etc. Rutherford replied that this "system is very excellent in theory, but . . . extremely difficult to carry out in practice . . . Besides uranium and thorium, twenty-four distinct radio-active substances are now known."[45] When it is generally agreed, he felt, that no new elements are likely to appear, the entire nomenclature might well be revised. As for the present, he did not care for the name "ionium," but neither did he care for Campbell's suggestion, for he pointed out that the suffix "A" was applied only to the first product of an emanation, and ionium's parent was not such a gas. Rutherford did half-heartedly suggest the name "paradium," which was indeed indicative of its position in the decay series, although the play on words made it possibly unsuitable. Boltwood would have none of this, and looked on in amused unconcern:

> If later we can succeed in getting the lady (Hahn calls it "die Mutter-substanz," so *it* must be a *she*) respectably tied up to uranium, then I will not have the least objection to her changing the name. But for the present my motto is "fest sitzen."[46]

8 Thorium Genetics

Whereas the history of uranium-radium genetics was to a rather large extent a story of domestic research, that of thorium genetics was more international. Yet, American chemists made no small contribution, and, again, Boltwood and McCoy were the leaders of this movement.

Boltwood's Thorium Research

Boltwood's work on thorium began in August 1904, as a natural consequence of his investigation of uranium-bearing minerals. He was at this time planning to study minerals containing small quantities of uranium, including some thorium ores, and wished, therefore, to know more about the active bodies associated with thorium. Apropos of this, he wrote to Rutherford:

> I am planning to carry out some work on which I should find your opinion most valuable. It is on the radio-active constituent of thorium, if there really is a special radio-active constituent. I am fairly familiar with your work on thorium, the thorium emanation and thorium X, but my knowledge is not sufficiently extensive to convince me that it is not possible for ordinary thorium to contain some special radio-active constituent. Please do not think that I make any reference to the work of Baskerville, but I am thinking of the apparent analytical complexity of the element, and the persistent assertions of Hofmann that he has obtained thorium salts of very low radio-activity from certain minerals.[1]

Rutherford encouraged Boltwood's project because he was "quite prepared to believe that its radioactivity may be due to an impurity," but for the present he took the position "that the published results to the contrary do not appear . . . conclusive."[2]

Boltwood's work on the activity of thorium and the chemical properties of thorium X continued throughout the rest of 1904 and early 1905. Speaking on the origin of radium before the New York Section of the American Chemical Society, he merely noted that he too had found uranium present in monazite.[3] But he used this fact as "a plausible explanation of the occurrence of helium in this mineral, without recourse to the unsubstantiated hypothesis of the formation of helium from thorium."[4] For some reason, Boltwood early made up his mind that radium could produce helium, but thorium could not, and he was long adamant in this view.

In April, Boltwood received the following letter from Rutherford:

> Strutt has just sent me an early proof of his paper.[5] As I know you are very anxious to see it, I forward it to you in *confidence* for a few days. I would like to have it back in a few days after you have soaked it in. What do you think of the thorium-helium argument? It doesn't appear to me very strong but of course thorium does produce He as it gives out α particles. After seeing Strutt's paper, I appreciate how good your results are. He fell through on monazite in great style.
>
> Strutt tells me that Ramsay has a substance (new element) which gives out Th emanation.... [Strutt] hazards that it may be Th X—I'll back 10 dollars it is. If it proves so, we shall have a good laugh on Ramsay. Ramsay says it is a substance like aluminium! He gets it from thorianite.[6]

Boltwood disagreed with Strutt's contention that helium is produced far more by thorium than by uranium, and even more with his experimental procedures. His reply was painfully blunt:

> I think that the thorium-helium part of the paper is all rot, both the methods and the conclusions. I notice that Sec. 5 was "Amended March 6" which was just three days after he had talked with J. J. [Thomson] and Bumstead at the Cavendish. What it was before that heaven only knows, for it is bad enough as it stands. If I felt any confidence in any of the figures given in the table ... I could perhaps rationally discuss his conclusions as to the occurrence of thorium and helium together, but as it is I don't find much basis for an argument either way.[7]

Boltwood also disagreed with Strutt's suggestion that thorium was the parent of uranium. Despite Baskerville's belief[8] in a thorium constituent of atomic weight 256, which would remove the difficulty of ordinary thorium (232) being able to produce an element of higher weight (uranium-238), Boltwood saw no compelling evidence of a connection between the two series.[9]

Radiothorium

With respect to Ramsay's announcement of a new element (discovered by Otto Hahn working in Ramsay's laboratory), Boltwood waited until he was able to read the report in *Nature*[10] before commenting: "I also am willing to bet that it is Th-X. Why doesn't Ramsay have one of his students rediscover radium? It offers lots of interesting p ossibilities!"[11] This was followed on 4 May 1905, with:

> After that synopsis of Hahn's paper "On a new element etc." in Nature I am positive that the substance he got was Th-X mixed with a little radium. If he made any experiments of the rate of decay he was thrown off the track by the activity of the radium rising, as the Th-X fell off. I have obtained exactly similar precipitates in working with other minerals. They give off thorium emanation when fresh and radium emanation when old. I am surprised that Ramsay has not yet named his "new element."[12]

And, still bursting with confidence, in the early autumn:

> Apropos of Ramsay's "new element which evolves thorium emanation" you will perhaps be interested to know that I have worked up 5 grams of thorianite, following the method described by Hahn, except of course, that I did not attempt to fractionally crystalize [*sic*] the barium-radium chloride, but simply treated the solution with ammonium hydroxide, and examined the precipitate thus obtained. This precipitate when fresh gave off large quantities of thorium emanation but its emanating power has steadily decreased and its activity now, after 7 days, is less than a third of what it was at the commencement. Miner, the chemist of the Welsbach Light Co., is working up a kilogram of thorianite for me, and is going to send me the insoluble residue left after fusing with sodium bisulphate. This should furnish enough material to throw a definite light on the nature of this new ? element. I am confident still that it is only a new compound of Th-X and stupidity....[13]

But not *all* the research on radioactivity done in Ramsay's laboratory was worthless, although Boltwood and Rutherford had good reason to regard the work of the leading British chemist with a jaundiced eye. In this case they were wrong, and Boltwood soon got the first intimation of his error:

> I have recently had an opportunity of pumping a young man named Gutmann, a former colleague of your Dr. Hahn, just fresh from Ramsay's laboratory and about to occupy a position at the feet of The Great Baskerville. Gutmann spent Sunday here in New Haven and I had a good opportunity of seeing him. From what I could get out of him, which was not very definite, there seems to be more reason than I had supposed for assuming

that Hahn may really have found something. You will have undoubtedly sized up the situation by this time, and I should be extremely interested to learn your conclusions. I suppose that it is not impossible that there may be an intermediate radio-active product between thorium and Th-X, and that thorium itself, like actinium may undergo a rayless change and be itself non-radio-active. If the intermediate product had a fairly long average life, then this would explain Hahn's results.[14]

By this time Hahn was working in Montreal and had convinced Rutherford of the existence of the new element. Hahn "thinks that he has separated a constituent between Th & Th X," the latter wrote, "pointing to the conclusion that Th itself . . . is non-active." And, Rutherford added maliciously, "Ramsay calls it radio-thorium for him and is apparently giving papers on Hahn's work to the Radiology Congress at Liège and to the French Academy. I wonder how much will be left to Hahn when these publications appear."[15] The name chosen for this distinguishable product, radiothorium, correctly suggests early recognition of its indistinguishable chemistry from thorium.

When, in the spring of 1906, Hahn planned a trip to New Haven to meet Boltwood, Rutherford warned his friend to "avoid the topic of Ramsay as if he were the devil himself . . . Hahn is naturally a Ramsayite," and there would be no sense in developing ill feelings.[16] Boltwood complied with this suggestion, but was astonished with the results:

> I had supposed that Hahn had come over to you for information and enlightenment, and from what you said I feared that you had neglected your first duty and had left him groping in darkness. But I took the advice in good grace and prepared to make you reap the whirlwind of your iniquity and to heap coals of fire on your head by taking advantage of any opportunity offered to laud and praise "the first chemist of Great Britain." . . . But all my well laid plans went astray and the best I could do was to feebly defend Ramsay against Hahn's frequent criticisms. "Great oaks from little acorns grow" and you must have dropped some of those little seeds quite unconsciously into Hahn's gardenpatch. At all events he is not the Ramsayite that you appear to believe and in all our conversation touching on the great master I was the only champion who spoke up for him. I hope that I did my duty.[17]

Rutherford was equally amazed, for he had "always avoided the subject of Ramsay's greatness. It would have been amusing," he added, "to hear you defending Ramsay—I am sorry I wasn't there."[18]

The happy sequel to this episode lay in the warm friendship that developed between Boltwood and Hahn. The latter frequently recalled with amusement Boltwood's assessment of radiothorium as "a new compound of Th-X and stupidity."

Other Chemical Work by Schlundt and Moore

Even before they published their paper on new methods for separating uranium X from uranium, Schlundt and Moore had successfully attacked the same problem for thorium.[19] Of course, others had been able to separate thorium X from thorium—most commonly by dissolution of thorium nitrate or thorium chloride in water and precipitation of the thorium by the addition of ammonia, leaving the thorium X in solution—but no one had yet been able to prove decisively that thorium was intrinsically inactive. In this, Schlundt and Moore also failed, but they did discover two new reagents, pyridine and fumaric acid, which precipitated the thorium and left the thorium X in solution. When tested, the filtrate residue possessed a higher activity than thorium X prepared by conventional means. The difference was due to the emanation decay products, since some are *in*soluble in ammonia and therefore precipitated with the thorium, while they are at least partially soluble in the new reagents. This radioactive matter, called by different writers the active deposit, the induced activity, the excited activity, the imparted activity, or a component of the secondary activity, was, therefore, complex. This was certainly not a new idea at this time, but the confirmation was welcomed. Decay curves plotted over the course of a month showed that it was thorium A (later renamed thorium B = lead-212) which separated from its daughter products and remained with the thorium X.

Boltwood's Theories

At about this time, Boltwood was working on the method of dating rocks, and had in his laboratory a sample of very old Ceylonese thorianite. It occurred to him that here was an opportunity to investigate thorium genetics in detail. His reasoning to Rutherford was as follows:

> Now taking the average age of thorianite as 500 million years, it is evident that since thorium predominates so largely in this mineral we ought to be able to determine from its composition what elements have been formed by the disintegration of the thorium. If the analyses of Dunstan and Blake are to be relied on at all, it would appear from them that, if the rate of decay of thorium as estimated by you is of at all the right order of magnitude, the only constituents of the mineral which are present in sufficient amount to be disintegration products of thorium are the rare earths, cerium predominating. Now the composition of thorianite itself does not exclude the suggestion of Strutt that uranium is a disintegration product of thorium, but a thorite containing 66% of ThO_3 which is found with the thorianite contains only 0.46% of UO_3, so that uranium is here excluded.

This thorite according to theory should contain about 0.08% of lead, and this amount might easily have been present and have escaped notice. It is an impressive fact that cerium is the predominating rare earth in this thorite, as it is in a number of others of which analyses are given in "Dana". If thorium breaks down into cerium, however, there is such a large difference in their atomic weights (92 units) that it would be expected that there would be some other substance formed also in considerable quantity. The difference (92) comes pretty close to the atomic weight of zirconium (90), and most of the analyses of Hillebrand indicate the presence of zirconium in the uraninites containing high proportions of thorium. It must be kept in mind that the analytical separation of thorium and zirconium offers very great difficulties, so that it is impossible to be sure from the published analyses that zirconium is not present in the theoretical proportions in the different minerals. The point is at all events well worth looking into, and I am going to make some analyses myself with this point in view, beginning with thorianite of which I now have a considerable quantity. It is also interesting that if it is assumed that thorium disintegrates according to the equation: $Th(232.5) = Ce(140) + Zr(90) + ?(2.5)$, that the amount of water shown by the analyses of the least hydrated minerals . . . is of about the right order of magnitude to warrant the assumption that the expelled alpha particle from the thorium family consists of hydrogen. You will see that a really good and reliable analysis of thorianite will throw some interesting light on this question also. I don't know how you will look on these speculations, but I feel sure that there is something at the bottom of them all, and they in any event are of great assistance to me in indicating *what* to look for and *where* to look for it.[20]

Boltwood did not often theorize and Rutherford, who may have wondered where thorium X, radiothorium, thorium emanation, and thorium active deposit would fit into such a proposed decay scheme, was particularly delighted with this letter, as his reply indicates:

You appear to have been diving deeply into the mysteries of matter and certainly manage to obtain very plausible results. The way the ages of the radioactive minerals work out is certainly striking and I am glad to find that a professional chemist when properly infected is quite as rash in theorizing as a physicist. I have been much amused at various [authors] the last six months, notably in Engineering Journals, who hold up their hands at the audacity of the imagination of the workers in radioactivity and sagely reflect how Newton would have sat down and worked out the whole subject and then given a theory. It never occurs to them that it would have wanted half a dozen Newtons to accomplish the experimental work in a lifetime and even these could not have put forward any more plausible theory than we work on today. These dam'd fools—whom I think must once have been chemists—(excuse me—no personal reference) haven't the faintest notion that the disintegration theory has as much evidence in support of it as the Kinetic Theory of Gases and a jolly sight more than the electromagnetic

theory which they all swallow as the eternal verities. Apart from this minia-
ture outburst, I quite agree with you that the only way to get any idea of
what are the products of the radio-elements is to examine carefully every
available mineral. If we don't find it that way, I think it is extremely impro-
bable we shall get any further at all. I feel *sure* helium is the α particle of Ra
and its products but it is going to be a terrible thing to prove *definitely* the
truth of this statement for I feel confident e/m will come out to be 5×10^3
instead of 2.5×10^3. It may conceivably be a hydrogen molecule—or half
atom of helium or helium atom with two charges and nothing but a pure
scientific nose can say with certainty that one is more probable than the
others. My nose (which may be prejudiced) leads me to avoid the H
molecule like the devil. It is too plebeian in character to be sired by such
blue-blooded stock like radium whose ancestors certainly existed before the
flood. However for plebeian thorium, hydrogen seems to be very well fitted. I really see no valid reason why *all* the active bodies should emit
Helium—actinium and radium certainly do—but H is the next most likely
material to be thrown off. However, it will be mighty difficult to prove.
Until you fix [?] me with more evidence, I shall still cleave to Helium if only
for Galileo's doctrine of simplicity.

You remind me of Japhet in search of a father for you seem bent on
finding a mother at any rate for the two waifs cerium and zirconium. Your
proofs seem to me too convincing to be true but all things are possible if
Thorium B breaks up into two fragments of about equal weights. I have
long thought such an effect must occur among some of the products and I
feel confident actinium owes its origin to some product which has two
distinct forms of equilibrium—the smaller percentage part yielding actinium. I thought I had a fair amount of scientific nerve but I am left far
behind in your last essay. This search for a father is becoming positively
indecent. Why not accept the chemist's view of separate creation and rest
happy?[21]

The provoking suggestion that atoms may fission (not discovered until
1938 by Otto Hahn and Fritz Strassmann) was not commented upon, but
Boltwood felt compelled to defend his profession:

Your very interesting and amusing letter of the 5th afforded me much
pleasure and entertainment, but really, you know, I do think that you are a
bit hard on the chemists, whose chief fault is a defect in early education, and
whose attitude toward recent advances in science has been vastly more
liberal than anybody could have expected from their past behavior. And
then too, there are some physicists who seem to be quite down to the
chemical standard and I don't see any excuses to be offered in their case.
Apropos to this, have you seen the review of Duncan's "New Knowledge" by
your friend and colleague, the Hon. R. J., in a recent number of the
"Speaker". Well, I thought that Ramsay laid it on pretty thick, but Strutt has
quite outdone him in praising that literary and scientific abomination. And
Strutt is certainly no chemist!!

As to my attempts to trace the lineal decendents of thorium, my efforts may be misdirected but they are none the less earnest. I am beginning to believe that thorium may be the mother of that most abominable family of rare-earth elements, and if I can lay the crime at her door I shall make efforts to have her apprehended as an immoral person guilty of lascivious carriage. In point of respectability your radium family will be a Sunday school compared with the thorium children, whose (chemical) behavior is simply outrageous. It is absolutely demoralizing to have anything to do with them.[22]

Boltwood was inclining toward error not only in believing thorium the parent of the rare earths, but also in accumulating evidence to show that the thorium alpha particle was hydrogen. Water found in thorianite, he felt, supported this latter view. Analysis of thorianite gave a much lower percentage of rare earths than he had expected, but this could be explained if Rutherford's calculated rate of disintegration for thorium were too high. He was "sure that some, if not all, of these substances [cerium, lanthanum, didymium] are the thorium disintegration products, because they are the *only* constantly present constituents of thorium minerals."[23]

But Boltwood was, at this time, deeply involved in research on the relative activities of the uranium products, and his work on thorium was relegated to infrequent interludes. During the remainder of his research career he returned to the subject on but two significant occasions.

Specific Activity and the Discovery of Mesothorium

On the first instance Boltwood presented evidence of the constant specific activity of thorium in equilibrium with its decay products.[24] Most remarkably, this paper was followed, in the same issue of the *American Journal of Science,* by *two other* papers, by H. M. Dadourian (1878–1974), and by McCoy and Ross, proving the same thing. This was simultaneous discovery with a vengeance!

Since the specific activity of thorium minerals was a constant, Boltwood felt certain that Hahn's radiothorium was a disintegration product of ordinary thorium. Interesting and unexpected results were obtained when thorium *salts* were tested, however. Four of the eight samples examined were oxides prepared from commercially obtained thorium salts, and their activities averaged less than half that of the other oxides prepared directly from minerals. These last four, further, had activities comparable to those of the natural minerals tested earlier. Boltwood could only conclude that the commercial processes, for example of the Welsbach Company, of Gloucester City, New Jersey, somehow separated radiothorium from thorium far more efficiently than even Hahn's method.

Not long after, this conclusion was revised, for Hahn had by then discovered a new product, mesothorium, between thorium and radiothorium.[25] It was this product which was being separated from thorium. Originally, Hahn had believed radiothorium possessed a very long half-life, but further work in Rutherford's laboratory forced him to change his mind. "I myself get only very disagreeable decay curves of the radiothorium," he wrote Boltwood, "which do not agree very satisfactorily together, but it seems to me that the period to be half transformed is somewhat about 1.5–2 years. . . . I feel pressed to state this decay before somebody else will find it out and correct my former statement."[26]

Boltwood, in his final contribution to thorium genetics, and Dadourian had, however, come to the conclusion that "if radiothorium was a product formed directly from thorium it was obvious that its period of decay (recovery) could not be less than half-value in about six years and might be somewhat longer."[27] This view was privately communicated to Hahn in April 1906, and was the immediate reason[28] for the addition, in Hahn's letter quoted above, of an inclination to state "the possibility of the existence of another rayless product between thorium and radiothorium." With a relatively long half-life it would account for the longer time observed by Boltwood, assuming his preparation contained the new radioelement while Hahn's did not.

Hahn soon left Montreal, returning to Germany. In Fischer's laboratory in Berlin, he was allowed to examine the many thorium ores, salts, and residues of Knöfler and Company, on the condition that any commercially valuable product or processes discovered would be kept secret.[29] By testing the activity of materials prepared months and years earlier, he was able to conclude that something with a half-value period of about six years was indeed separated from the thorium. Even then, however, he was "not yet sure whether mesothorium is a simple product; there might be two."[30]

Whereas Hahn almost gained priority over Boltwood in the discovery of ionium, Boltwood almost beat Hahn to the discovery of mesothorium. Had Boltwood not been so occupied with work on the specific activity of the uranium series and on the parent of radium, he would have had more time to examine the thorium family. "I have been very much interested," he wrote Rutherford, "in Hahn's last paper on 'a new intermediate product in thorium' and his conclusions are not only certainly correct but I should have reached the same views myself in a week or so if his paper had not arrived just at the critical moment."[31] Boltwood was in the process of remeasuring some old thorium oxide films which had been prepared over a year before and was finding a noticeable decrease in the activity of all of them. No doubt, he reasoned, the radiothorium which remained with the thorium when the meso-thorium was removed, was decaying and not being replenished. In his August 1907 paper, therefore, Boltwood was able to add confirma-tory evidence to Hahn's discovery of mesothorium.[32]

Even more, he wrote, "a further point which appears to be worthy of notice in passing is the similarity in chemical behavior shown by thorium and radiothorium on the one hand, and by thorium-X and mesothorium on the other."[33] Not publicly announced was a further revelation:

> I have also got a lot of data about mesothorium, which as a matter of fact has the same chemical properties as radium and can be separated from minerals in the same manner. But I shall hold up this for some time yet in order to give Hahn a chance, for I feel that his priority in the matter should be recognized.[34]

Thus, the concept of the existence of radioelements having different radioactive properties (and names), but similar chemical properties, was being clearly formulated in the minds of a handful of chemists in 1907. It was based on evidence long available (e.g., radiolead), and on new data, but it was still to be several years before a theory of isotopy would be proposed.

H. M. Dadourian

We may return now to that impressive occasion on which three papers describing the constant specific activity of the thorium series appeared in the June 1906 issue of the *American Journal of Science*. The second au-thor, Haroutune M. Dadourian, was born in 1878 in the town of Everek, "on the south side of Mt. Arjias, the highest peak in the Taurus Mountains, in Asia Minor."[35] In 1900, he came to the United States and entered Yale's Sheffield Scientific School, receiving his bachelor's degree in electrical engineering three years later. In another three years he was awarded the Ph.D. degree in physics, having done his thesis research on radioactivity under Henry Bumstead. Between 1906 and 1917, he taught physics at Yale, then worked on sound ranging for the Signal Corps during World War I, and in 1919 joined the staff of Trinity College, in Hartford, Connecticut, where he was soon appointed to the chair of mathematics and natural philosophy.

In his 1906 paper, which was first read before a meeting of the American Physical Society, on 24 February, Dadourian showed that the constant specific activity of thorium from minerals was about twice that

HAROUTUNE M. DADOURIAN (1878–1974)

from commercially prepared salts.[36] Boltwood had, of course, stated the same thing just a few pages before, but what made this work valuable was the different method used to get the same results. Dadourian was a physicist who had earlier performed some research on the radioactivity of underground air; in general, such work on the radioactivity of airs, soils and waters was based upon a study of the emanation present, or the emanation's active deposit on a negatively charged body. Bumstead had taught his student the techniques and this familiarity led Dadourian to test thorium's activity in this manner.

He was fortunate in having Boltwood's private laboratory not far off, for the latter supplied him with certain preparations and their chemical analyses, and no doubt was a valuable source of information. Soon, Dadourian also was able to show that the thorium minerals exhibited twice as much activity as the salts from commercial sources. As the most logical explanation, he suggested that the latter's low specific thorium activity is caused by the removal of part of the radiothorium in the chemical preparation. To confirm this, he examined some thorium nitrate which was prepared by Boltwood in a manner different from the commercial method, and obtained an activity equivalent to that of the *minerals.*

Thus, he had confirmed Boltwood's findings by different techniques, and, supposedly, the matter was now well-explained: Thorium, with a very long half-life, decayed into radiothorium, which in turn transformed more quickly to thorium X, which similarly produced the thorium emanation. If, during commercial production, some of the radiothorium was removed, it would not be replenished in any observable period of time, and the commercial salts would therefore indicate a lower specific activity.

But there were two unanswered (even unmentioned) points. These were the lack of explanation as to why only a *part* of the radiothorium present was separated, and if a part, why the seemingly *constant* fraction, as shown by the fairly regular specific activities of the various salts. Within the year, however, these questions were answered by Hahn's discovery of mesothorium. This decay product was being entirely removed in the commercial processes, while its daughter, radiothorium, remained with the thorium. In time, the radiothorium decayed and was not replenished, there being no mesothorium in the salt. Thus, the salts tested, being perhaps a few years old, had a very small amount of fresh mesothorium in them (newly produced from the thorium), a little radiothorium and its products not yet decayed, and nearly the original amount of thorium. The minerals, on the other hand, had all these in full equilibrium amounts. For Dadourian, this meant that he measured the activity from a maximum amount of emanation in the minerals, and from a small quantity of emanation (evolved by the radiothorium through the thorium X) in the salts. Regarding the "constant" fraction mentioned above, this was illusory, for the amount of emanation in the commercial salts was steadily decreasing as the radiothorium became depleted.

McCoy and Ross

Besides the specific activity of the thorium series, McCoy and Ross had hoped to present, in that notable issue of the *American Journal of Science,* information on the activity of thorium completely free from its products. However, the complete separation "has proved to be so difficult," they wrote, "that we shall no longer delay reporting on the results already obtained."[37] The removal of radiothorium, in particular, was a trying problem. The work described was, in effect, the logical sequel to McCoy's investigation of the specific activities of the uranium family, and was carried out in a similar manner. In fact, since all the thorium minerals tested contained some uranium, the earlier results were necessary to be able to know the latter's contribution.

Using the same unit of activity, i.e., that due to one square centimeter of a thick film of U_3O_8, they found the average activity of one gram of thorium from a mineral, together with its decay products, was 953. As determined earlier, the specific activity of the uranium series, from a mineral source, was 3280, while that of uranium in pure compounds was 791. It was impossible for them to determine this figure for pure thorium compounds, of course, since they could not prepare the element free of radiothorium. Much was left, therefore, for future investigations. In conclusion they wrote:

> It is possibly still a question whether thorium entirely freed from radio-thorium, ThX, etc., will produce rays capable of ionizing gases; but that it is undergoing transformation, rayless or otherwise, which gives rise to active products, seems certain from the fact that the portion of the radio-activity due to thorium, of any mineral, is directly proportional to the thorium content of that mineral. We believe our experiments also show clearly that the activity of thorium compounds is not due to bodies accidentally retained by thorium (as radium frequently is by barium sulphate), but that the radio-thorium, ThX, etc., are disintegration products of thorium.[38]

In December 1907, McCoy and Ross published another paper on thorium, admitting the necessity, as in the case of uranium, of raising the value of the thorium series' specific activity (from 953 to 1009), because the original electroscope used had been too small, preventing complete ionization by the alpha particles.[39]

Additionally, Hahn's discovery of mesothorium cleared up the various growth and decay curves for them, as well as others, and they realized that the best (if not the only) way to prepare radiothorium was to chemically separate its parent, mesothorium, and allow it to grow, free from the grandparent, thorium. Even more, they now had circumstantial evidence of an intrinsic activity in thorium. When mesothorium, thorium X, and the other products were removed from thorium and radiothorium, the latter mixture's initial activity was very low. One month later, its activity reached a maximum as the radiothorium produced an equilibrium amount of thorium X. The ratio of maximum to minimum activities was found to be greater for samples relatively rich in radiothorium, showing conclusively that thorium itself is active. It should be noted, however, that Hahn published this view first, and that McCoy and Ross, while emphasizing that this was their conclusion as much as a year and a half before, did not wish to claim priority.

Another contribution reported at this time concerned the period of mesothorium. Based upon rough ideas of the specific activity of thorium and of its products, separately, and using the standard decay equations, they were able to calculate an approximate half-life of 5.5 years for mesothorium. Using this, they next computed the time for a preparation of thorium and radiothorium, from which mesothorium had been removed, to reach minimum value.[40] This was found to be about 4.5 years after preparation. Tests upon several samples of thoria revealed activities that compared well with the values calculated, thereby confirming the initial approximation of mesothorium's period. Once again this illustrates the circular reasoning used so often to fill out knowledge of the radioelements.

Of the greatest ultimate importance, however, was their report that various chemical procedures, including precipitations repeated *one hundred* times, failed to isolate radiothorium. "The direct separation of radiothorium from thorium by chemical processes is remarkably difficult, if not impossible,"[41] they concluded. Not until Soddy in 1911 declared such bodies chemically *identical,* and Kasimir Fajans and Soddy in 1913 explained their locations in the periodic table, were stronger cases presented for the existence of what were later called isotopes. But such ideas took time to gestate; more examples of chemical "inseparability" were needed before the anomaly cried for resolution. McCoy and Ross initiated no new research program for this problem. Probably they regarded it as analogous to especially difficult separations among the rare earths, and a way eventually would be found. Yet, whether or not they recognized the significance of their chemical curiosity, it was certainly the extent and quality of this work in 1907 upon which the ideas leading to isotopy were based. This contribution, among his several others, was publicly recognized when McCoy was awarded the Willard Gibbs Medal of the American Chemical Society in 1937.[42]

9 Radioactivity in Medicine

The scientific discoveries concerning radioactivity, and most especially radium, were so striking that even the layman developed an intense interest in the subject. The manifestation of this interest, the "radium craze," has been described in an earlier chapter. Fanning the flames of such curiosity to even higher levels was the increasing number of reports that radium exhibited curative powers over a range of distinguished diseases. While the prime purpose of this book is to describe the physical and chemical developments in radioactivity, and there only the main lines of "pure" research, it will not be amiss to sketch, ever so briefly, some of the activities in related fields. Concern for medical applications did, in fact, compete with the more academic interests of the physicists and chemists—because, for example, these people suffered from radiation burns, because they were called upon for their expertise, because they had similar needs for reproducible standards, and, perhaps most importantly, because the medical use of radioactive materials stimulated demand for the radioelements which resulted in creation of an important industry to which the scientists were also beholden.

Early Findings

Since the radiations from radioactive substances were initially compared so closely with the x-rays discovered only a few months before them, it was inevitable that their applications would also be contrasted. As the x-rays' most popular property was the ability to penetrate opaque materials and produce clear shadow photographs, it was learned with disappointment that uranium rays yielded very poor pictures indeed. While the outlines of assorted pins, keys, coins, and chains, taken through a wooden box, might be acceptable to some, the extremely long exposure (often on the order of days), coupled with the weakness of the radioactive salts generally available, suggested no practical future for such a use. And if radioactivity photographs were impractical for inanimate objects, there was no likelihood of application in medical diagnosis, where the patient might be unable to remain still for extended periods.

Yet, diagnosis is but one aspect of medical practice; therapy is another. And in this latter area radioactive materials were to see extensive application. During the last years of the nineteenth century the physiological effects of such radiations as x-rays and Finsen rays (ultraviolet light) were widely recorded and numerous claims of cures advanced. Because of the weak preparations available, similar effects due to radioactivity were not as quickly seen. The earliest observation that a burn could be produced on the skin by these rays, reported in 1900 by Walkhoff in Germany,[1] seems to have excited little interest, although it did induce his countryman, Friedrich Giesel, to deliberately expose his forearm for two hours to a capsule of radium.[2] Unlike a fire burn, there was no immediate pain; this developed later and, as was subsequently learned, often took as long as a few weeks to appear. This lack of immediate reaction was to prove a serious problem in establishing radiation safety procedures. Giesel also experienced other effects which were soon to become common: the burned area turned red, the wound opened, and the scab which formed took an unusually long time to heal.

It was, however, in France, not Germany, that the most significant steps occurred, and this because the Curies possessed the most purified radioactive substances in the world. In April 1901 Becquerel borrowed from them a tube of radium, of the then enormous activity of 800,000 times that of an equal amount of uranium, and carried it for six hours in his vest pocket while attending a conference.[3] The redness which appeared nine days later was followed by the other stages described above, and even two years after the event the area appeared whiter than the neighboring skin. At about the same time Pierre Curie, fascinated by Giesel's experience, deliberately experimented upon himself, with the expected results.[4]

These discoveries initiated a series of biological investigations which extends to the present time. In the early years the action of radiation was tested upon bacteria, plant life, lower forms of animal life, and tissues from some higher forms.[5] It was found, for example, that isolated bacteria were easily destroyed, but when they were *in vivo* the tissues might succumb first. In other areas, plant seeds could be made to lose their power of germination, and abnormalities could be produced in certain organisms.

Of greater immediate importance were the medical applications.

Besnier, the dermatologist to whom Becquerel showed his burn, noticed the similarity to an x-ray dermatitis and suggested the therapeutic use of radium. The Curies thereupon made a sample of the radioactive salt available to Danlos at the Hôpital St. Louis in Paris, who reported success in a limited variety of cases.[6] Others became intrigued with the subject also, and medical investigation of radiation proceeded on a wide front during these years. While the greatest attention was given to the subject in France and Germany, the American developments paralleled those abroad and may be described as representative. Moreover, there was so much simultaneous experimentation without controls, with radioactive sources of unknown intensity, that assignment of priority for various contributions is hazardous, and a general survey is to be preferred.

American Beginnings

An eminent radiologist has described the stages of x-ray thereapy development as optimistic, pessimistic, and realistic.[7] If a certain amount of overlap in the periodization is permitted, the same may be said of radium therapy. In the first stage the new "wonder treatment" was credited with the ability to cure or help every affliction of mankind, even sexual impotency.[8] When this was seen simply to be not true, general pessimism set in. The "professionals" then established scientific standards and methods which allowed the treatment to be applied to maximum advantage in the areas where success could be achieved. The years 1905 and 1910 may be taken as very imprecise markers of the transition from one stage to the next.

In all the periods information about medical applications of radiation reached the public in newspaper and magazine articles, editorials, and letters, besides the normal vehicle of medical journal reports. A measure of the early interest generated is seen in the rise of published articles: one or two in 1901, ten in 1902, twenty-five in 1903, fifty in 1904, etc.[9] Despite this evidence of concern for radium, the first decade was plagued by the small amounts available and the lack of a method for standardizing preparations. Another early feature was the collaboration between technically minded individuals and physicians. Prior even to the European thoughts of therapy, William Rollins, the Boston dentist and x-ray protection specialist mentioned in chapter three, gave to Dr. Francis H. Williams in 1900, 500 milligrams of radium chloride of strength about 1000. "The first person," Williams later recalled,

> so far as I am aware, to appreciate that radium salts would probably be of service in the treatment of certain diseases was Dr. William Rollins, of

Boston, who has done so much to promote the use of the x-rays.... Dr. Rollins put into my hands a metal box with an aluminum front, containing some chloride of radium, with the suggestion that I use it for therapeutic purposes. I did so employ it, but at that time the radium to be had was weak, and I did not obtain definite results....

I used this radium in rodent ulcer and lupus, but cannot give you a definite date as to my first use of it. It was too weak to compete with x-rays, although after some hours' application to the normal skin it caused a redness of the size of the capsule. In the summer of 1903 I went abroad and obtained 100 mgm. of pure radium bromide, and a large amount of radium of less radioactivity. On my return, I used the pure radium bromide with satisfactory results and started a radium clinic at the Boston City Hospital.[10]

Although the results were inconclusive in this first test, the cause was correctly seen to be the weakness of the radioactive source. Subsequent work involved salts of greater strength. Drs. Walter B. and Carroll Chase, in 1903, treated a woman with a preparation of strength 7000. An interesting sidelight is that the patient's husband purchased the tube of radium salt.[11] Apparently physicians attempted to avoid such large capital outlays themselves, until radium's widespread use was confirmed.

The innovative Rollins continued to advocate tests using radium in those areas where x-rays had been proven beneficial: lupus, superficial cancer, and skin diseases. His technical mind turned to means of application:

Radio-active substances can be used in sealed capsules held against the body by adhesive plaster, or they can be made to cover larger areas by mixing them with rubber or celluloid to form moisture-proof plasters. These plasters may be still further protected by being coated on the sides nearest the body by aluminum foil and on the opposite sides by lead foil. They could be kept in stock by the yard by druggists and given to patients by prescription, with proper directions as to the length of application. They could be worn at night. Their use would prevent the poor from making such frequent visits to a physician as are now required when x-light obtained from a vacuum tube is used. This is a matter of some importance, as the present treatment takes many sittings which require time and cost money.[12]

The amount of radium in such a device would necessarily be small, to avoid both dangerous burns and high expense. Implied here was the concept that long exposure to a weak source was as good as or better than a single massive dose, a concept which was repeatedly argued over the next decades and which assumed a nationalistic flavor. In this paper of January 1902, Rollins told also of the radium capsule he had given to Williams in 1900, and offered another to any Boston physician willing to give it a fair trial. Despite the growing interest in the subject, he commented sadly in 1904 that no applications had been received.[13]

As was to be expected, external irradiation of superficial maladies was first attempted. Robert Abbe of New York, who started working with radium in 1902, and who was able to obtain a sample of 300,000 strength from the Curies by cabling an early order, found that the ordinary wart could be made to disappear. He also achieved notable success in shrinking a giant tumor in a patient's jaw, which surgically would have required removal of at least half the lower jaw. In this case the results were visibly remarkable, though the disease was not entirely eliminated.[14] Radiation therapy was seen to be especially efficacious when the alternative of surgery would have left obvious deformity of the tongue, jaw, eyelids, or face.

Thorium, despite its lower activity, also was used for therapeutic purposes because its cost was correspondingly lower than that of radium. The practitioners of this new field thereupon replaced the name "radium therapy" with "Curie therapy."[15] Samuel Tracy, of New York City, was in 1904 among the early users of thorium, which he made into a paste or ointment and applied to chronic skin diseases, especially those of a parasitic origin.[16] Tracy also made use of the active deposit left by radium emanation, in the form of a spray and on dressings to cover skin afflictions.[17]

Subsurface Treatment

Human diseases are not, of course, only superficial, and thought was given to treatment of deep-seated disorders. These may be attacked in three ways: from the surface, by surgically inserting radioactive materials into the malignant tissues, and through the body's cavities, and all were, indeed, pursued. The physicians who early chose to use radium had little to guide them but guesswork. Assuming that any skin change was undesirable, their frequent charge to the therapist was "Give my patient a good dose of radiation but don't damage the skin."[18] However, this was more easily requested than achieved. The alpha rays were known to be stopped by the walls of the radium container and only slowly did they learn that the beta rays caused skin irritation while the more penetrating gamma rays produced the effects desired. Gradually also they learned to filter out the unwanted radiation, to calculate proper source-to-target distances, and, in the second and third decades of the century, when radium "packs" or "bombs" of a few grams were acquired by some hospitals, to "crossfire" at the malignancies, thereby giving the intervening healthy tissues a much lower dosage.

Alexander Graham Bell, inventor of the telephone, in 1903 advocated another method of attacking subsurface cancers:

> I understand ... that the Röntgen rays, and the rays emitted by radium, have been found to have a marked curative effect upon external cancers, but that the effects upon deep-seated cancers have not thus far proved satisfactory.
>
> It has occurred to me that one reason for the unsatisfactory nature of these latter experiments arises from the fact that the rays have been applied externally, thus having to pass through healthy tissues of various depths in order to reach the cancerous matter.
>
> The Crookes' tube, from which the Röntgen rays are emitted, is of course too bulky to be admitted into the middle of a mass of cancer, but there is no reason why a tiny fragment of radium sealed up in a fine glass tube should not be inserted into the very heart of the cancer, thus acting directly upon the diseased material. Would it not be worth while making experiments along this line?[19]

Abbe was one of the pioneers in this field, experimenting upon a cancerous breast before amputation in 1904,[20] and therapeutically inserting a tube or needle filled with radium into a goiter the following year.[21] He found that when radium was placed on the skin above a subsurface area of cancer, the cells suffered a softening and devitalization. But, much more effectively, when the radiation source was surgically placed within the cancer, an area of complete destruction of cells occurred, though only within about a quarter inch radius. Some, perhaps teleologically inclined, at first believed that radium rays exercised a *selective* action on cancerous cells. In time, however, it was understood that they act indiscriminately; it is the various cells which react differently. Those cells which multiply most rapidly are most sensitive to radiation, and cancer cells are of this type. Thus, it is possible to irradiate a patient sufficiently to kill the diseased tissue, yet leave the healthy neighboring tissue alive. Upon this difference in susceptibility rests the success of the treatment, and this difference varies greatly in diseases and in patients.[22]

The use of radium in the natural cavities of the body began at about the same time as surgical implantation, and both were seen to have advantages of accessibility over the more widely used x-rays. William James Morton, in 1903, enclosed his source in a small celluloid tube, which allowed more of the radiation to pass than his glass container, and treated cancer of the throat and the uterus from within.[23] When success was reported, a number of applicators were designed and soon marketed. Max Einhorn, of New York, and J. A. Storck, of New Orleans, for example, in 1904 devised essentially similar capsules for the stomach, to be fastened to the end of a string and swallowed, and capsules for the esophagus, rectum, and vagina, which were to be attached to the end of a hard rod. Knots were tied in the string of the stomach apparatus to ascertain the radium's position, and the free end was looped over the patient's ear until time for removal.[24] In the same year, Joseph G. Beck applied radium to nose, throat, and ear diseases.[25]

Einhorn tested a range of materials in an effort to avoid using fragile glass. Hard rubber and celluloid proved better than aluminum, ivory, or wood, though none of these substances passed as many of the radium rays as the thin glass. He treated nine cases of cancer of the esophagus, reporting only slight improvement in six of them. Yet, this early effort was the basis for some enthusiasm, for the radium rendered the stricture more pervious, a situation which allowed some patients to eat comfortably—some even just to eat.[26]

Radium salts were also administered by mouth and by injection. Large quantities could not be consumed in these manners because of cost and the long, active life of radium, but mineral waters, many of which were found to contain traces of radioactivity, and injected soluble salts were often prescribed, since they could be naturally eliminated in time.[27] This was because these latter frequently contained only radium's daughter products which seemed to pass readily through the body, while a fraction of the radium itself would have remained permanently in the bones since its chemistry is similar to that of calcium.[28] Although the full impact of this became apparent only in the 1920s, as a result of the famous radium poisoning cases of women in New Jersey who painted luminous watch dials,[29] medically prescribed consumption of this element appears to have essentially stopped a decade earlier for purely economic reasons.

But costly or not, this "one shot" use of radium could only have ended because a substitute had been found. As early as 1903, Frederick Soddy, by then in London, called attention to the possible use by tuberculosis patients of radium emanation and thorium emanation.[30] Being gaseous, they could be inhaled and thus exert their beneficial influence directly upon the lungs. With decay periods of about four days and about one minute, respectively, and with active deposits also of relatively short half-lives, any material retained in the body would soon be rendered harmless. If the precise dosage was difficult to determine, at least a maximum could be calculated.

Soon emanation treatments became a fad and were undertaken for a wide range of complaints. Since the gas was most easily applied by inhalation, small spray devices and masks were early devised.[31] Then "inhalatoria" and "emanatoria" were constructed. These were small rooms in which several people could sit for a number of hours and enjoy the benefits of emanation, sometimes "spiked" with oxygen.[32] Emanation baths, which also were popular for respiratory diseases, may be classified as an inhalation method, since very little is absorbed by the skin. Aside from the hydrotherapeutic gains, this treatment was not as desirable as straight inhalation because the dosage could not be determined as accurately. Water impregnated with emanation was given by mouth and by injection, although the latter method seems little used. When thus taken internally, it is excreted more slowly than when inhaled, allowing more time for an active deposit to form in the body. One of the reasons behind the public enthusiasm for emanation may have been the widespread belief that uranium mine workers never suffered from rheumatism or gout. While the validity of this belief is questionable, successful treatment of these illnesses was reported.[33] It is ironic that, as the emanation craze reached the height of popularity (in the years immediately preceding World War I), the first evidence was produced showing uranium miners' major occupational *malady* was cancer of the lung. Yet, it was more than a decade later before emanation (now called radon) in the mine air was suggested as the carcinogenic agent.[34]

Professionalization

The inhalation and ingestion of emanation were not, however, the only reason for the decreased internal consumption of radium. Ever more significantly, the surgical implantation of emanation containers marked this notable step in the professionalization of radium therapy. Following a 1908 suggestion by William Duane, then working in Madame Curie's laboratory, Henry H. Janeway, of Memorial Hospital in New York, developed the technique of filling glass capillary tubes and later glass "seeds" with emanation. When it was evident that the beta rays which penetrated the container walls were causing local inflammation around the insertion, Gioacchino Failla, the engineer-physicist whom Janeway brought to the hospital in 1915, suggested pure gold capillary tubing. This filtered the beta rays, leaving only gamma rays to attack the malignancies. By this time it was understood that these highly penetrating rays came not from the disintegration of emanation, but from the decay products radium B and C.[35] Also by this time most hospitals using emanation kept their radium in solution, locked in safes, and pumped off the gas periodically as it formed, much as a cow is milked daily.

There were, however, pessimists about the applications of radioactive bodies. Some early researchers could not even kill bacteria or create a red, burned area on human skin.[36] Another reported that, except for some relief from pain, not one of his twenty-two cases showed improvement from radium treatment.[37] In such cases it is likely that the salts used were far weaker than believed, the techniques inferior, and the diseases not the most amenable to this treatment. There were also some physicians, not directly involved in the subject, who found it easy to restrain their enthusiasm over the uses and possibilities of radioactivity. In a conservative summary of the facts known by 1904, William A. Pusey told the annual meeting of the American Medical Association that "while the public press is teeming with vague and sensational accounts of the use of radium in disease, actual tangible facts are surprisingly few in the authentic literature of the subject." Pusey, who had a vested interest in x-rays, further remarked that the use of inhaled emanation, recently suggested by Soddy, was a "bizarre idea."[38]

Despite such doubts, research and application increased. The great majority of papers written on the subject included case history reports or statistical summaries that showed impressive successes, while at the same time expressing an ever-increasing rigor about what was and was not possible. While critical authors continued to attack what they felt to be the weak points, by about 1910 the field was beginning to mature. England, in 1909, announced plans to build a Radium Institute for therapeutic purposes, which was opened two years later. In France, where a few medical laboratories already existed, the University of Paris and the Pasteur Institute decided to found a similar institute, whose opening was delayed until after World War I. Germany, Austria and other countries had such establishments before hostilities began.[39] Several Columbia University physicists (including George Pegram), New York physicians (including Robert Abbe), and other gentlemen of that city in 1909 incorporated the Radium Institute of America.[40] They seem to have failed in their goal of establishing radium clinics, but even without their instrumentality such departments were created in numerous hospitals in this country during the second decade of this century.

In 1911, N. S. Finzi, of St. Bartholomew's Hospital in London, became the first to use large quantities of radium (over 600 milligrams), on a millionaire patient.[41] This is an indication of the fact that increasing quantities of radium were becoming available. Yet supply never seemed adequate for demand, and demand increased by notable jumps, as after the 1913 Gynecological Congress at Halle, Germany, where astonishing results were reported using massive doses of gamma rays, from heavily shielded mesothorium ("German radium"), in cases of cancer of the uterus. In the ensuing "radium fever" cities vied with each other to raise funds by public subscription and governments also contributed toward the purchase of this rare element for local hospitals.[42] Soon, medical establishments all over the western world were acquiring radium sources, some even as much as several grams. More and more, radium therapy was taken out of the hands of the individual practitioner and institutionalized. Men and women trained in science, not medicine, were employed by hospitals to look after their radium supplies. More and more, precise knowledge of source strength and distribution of the radiation within tissues was demanded. The definition of the "curie" of activity at the 1910 Congress of Radiology and Electricity in Brussels, and the subsequent preparation of an international standard by Madame Curie, were as much desired by the medical specialists as by the physicists and chemists.

By 1921, there were two institutions in the United States which owned over four grams of radium, four with at least a gram, and seven with more than half a gram. The total amount of radium in use was estimated at between thirty-five and forty grams, while the number of specialists employing radium therapy was four to five hundred.[43]

From this level of professionalization, begun about 1910, the major trends of emanation container implants and beam or "telecurie" irradiation developed, while that of intracavitary use was maintained. After about 1920, radioactivity entered diagnostic medicine with the tracer techniques devised by Hevesy and Paneth. With the discovery of artificial radioactivity in 1934, by Marie Curie's daughter, Irène, and her husband, Frédéric Joliot, the enthusiasm for radium began to abate, for then a variety of sources, with a choice of energies and with half-lives shorter than radium's (hence, safer when used internally) could be produced artificially. Ever more, this ability to select from a large "catalog" of radioelements is a feature of modern medicine, brought about in reactors by the release of neutrons during nuclear fission. Artificially produced radioelements (cobalt-60, cesium-137, gold-198, etc.) and high energy x-rays now substantially share with radium the applications in beam therapy and implantation.

The magic element radium, though it can kill as well as cure, and though it has not been the final answer to cancer, nevertheless holds a secure place in the history of medicine. Beyond its real therapeutic benefits across a wide range of diseases, it brought physicians and physicists into contact, resulting in a desirable gain in precision in the healing arts. The standards, procedures, and experience acquired in these early years, moreover, proved valuable after World War II, as human populations are faced with greater exposures to radiation.[44]

10 The Radium Business

European Origins

Radium was first found in the uranium ore called pitchblende, mined from the famous deposits at Joachimsthal, Bohemia (now Jáchymov in north-western Czechoslovakia). Originally known as a silver-producing region, Joachimsthal gained additional fame when the uranium extracted there proved increasingly important as a coloring agent in the national glass and ceramic industries towards the end of the nineteenth century. Mendeleev's formulation of the periodic table of elements in 1869 had drawn attention to the atomic weight of uranium, the highest then known, and for a number of years armor plate and cannon of a heavy uranium-steel alloy were contemplated. Such applications, however, consumed relatively small quantities of uranium. Becquerel's discovery in 1896 of the radiation emitted by this element once again caused interest in it, but scientific investigation could never stimulate major production. The other uses were minor and total output was modest indeed.[1] Not until the medical demand for radium arose did uranium mining assume the status of a significant industry, and then, ironically, the uranium was mostly discarded as residue.

With a chemical separations process presumably devised by their colleague Gustave Bémont, Marie and Pierre Curie extracted and purified the first minute quantities of radium from tons of pitchblende (actually the residue from which the uranium had been removed). This well-known story is outlined by Marie in the biography of her husband,[2] and is told with greater feeling and imagination in the biography *Madame Curie* by their daughter Eve.[3] Once the process was proven satisfactory, and it would appear even before a substantial commercial demand existed for the new element, a benevolent industrialist named

Armet de Lisle performed the gross separations in his factory, where the Curies supervised final purification. In keeping with their oft-stated disinclination to make money from their science, the Curies' gain from this arrangement was possession of the most highly purified radium.[4] In Germany also there were small factories producing microscopic quantities of radium for the scientific and medical research communities. E. de Haën of List, near Hanover, and the quinine factory of Buchler in Braunschweig (Brunswick) were prominent during the first few years of this century. Friedrich Giesel, the chemist at Buchler, not only furnished higher purity radium to his colleagues at a time the Curies retained all material greater than a certain activity, but performed notable research in radioactivity himself.[5]

From the discovery in 1898 of thorium's activity, independently by Gerhard C. Schmidt and Marie Curie, this element too was regarded with great interest. But here, in contrast to uranium, there was a sizable industry. Though electric arc lighting originated decades earlier and Edison had invented the carbon filament lamp in 1879 in his quest for a suitable incandescent light, home electric illumination was not generally available until the twentieth century. Not until the 1920s, however, did the demand for radios and household appliances spur the final electrification of urban areas. In the gap between the candles and oil lamps of olden times and the future promise of electricity—indeed, giving the latter strong battle—was the gas mantle. Having recognized that a fabric bag, impregnated with a mixture of thorium containing about one per cent ceria, produced strong illumination in a gas flame, Carl Auer von Welsbach about 1890 created a lucrative new worldwide industry.[6] It was estimated, for example, in 1912 that 400 million incandescent gas mantles were consumed annually.[7] At the same time Welsbach unconsciously aided those who shortly would search for new radioelements in thorium materials by causing large supplies to be available. Yet, because of radium's enormously greater activity, uranium production clearly was of most central industrial concern; Hahn's mesothorium ("German radium"), produced from thorium ores, was to be of considerably lesser interest.

American Origins

The first small pitchblende deposits located in the United States were found in 1871 in Gilpin County, Colorado.[8] Another source of uranium (though insignificant commercially) was the plentiful monazite sand of North and South Carolina, mined primarily for its thorium content. But the mineral destined to be of greatest significance was not formally discovered until 1897. Then, a French chemist named Charles Poulot, acting for a syndicate in his country, created excitement in Denver by offering good prices for all the uranium that could be produced.[9] Cut off from German-Austrian sources by the prevailing political alliances, already at odds with England over Egypt and Morocco, and no doubt sensing the naval construction race begun in earnest the following year, the French sought supplies of uranium for hardening gun metal and armor plate.

Much of the ore Poulot purchased had only shortly before been recognized to contain uranium. Unlike the hard, blue-black pitchblende, it was a bright canary yellow, soft, powdery mineral found scattered in the sandstone, limestone, and clay of Paradox Valley, Montrose County, Colorado. Its uranium content, generally below 2 percent, the circumstance that it does not occur in veins, and the difficulty of concentration made it of marginal economic value, but since usable vanadium was also a constituent, mines and treatment facilities were given encouragement to expand.[10] A sample was sent by Poulot to the University of Paris chemists Charles Friedel and E. Cumenge, who recognized it as a new mineral. *Not* in honor of Sadi Carnot, whose law of conservation of energy the radiation from uranium was supposed by the Curies to violate, but after Adolphe Carnot, an eminent French mining engineer-chemist, did Friedel and Cumenge name the mineral "carnotite."[11]

This carnotite, a complex potassium uranium vanadate, was mined and milled only for its uranium and vanadium content up to 1904. By this time the radium craze affected even miners, who carried photographic plates with them to test the radioactivity of suspected rocks. Carnotite deposits were found in various locations across the Colorado plateau of southwestern Colorado and southeastern Utah. Apparently the first to conceive of radium extraction from these vast mineral beds was Stephen T. Lockwood of Buffalo, New York, a young Princetonian who later was appointed a United States Attorney by his former professor, Woodrow Wilson. As early as 1900, Lockwood brought back carnotite samples from Richardson, Utah. The plan slowly matured in his mind, and two years later he purchased 500 pounds of hand-picked, high grade ore, on which he tested various extraction techniques. Lockwood, no chemist, at first secured technical help from Alexander Phillips on the Princeton faculty, who had a passing interest in the subject. But Lockwood required more expert assistance. Boltwood had not yet presented strong evidence of the relationship between radium and uranium, whatever the nature of its mineral, and some prominent voices, among them Sir William Ramsay, claimed that only pitchblende contained the new, wonder element. Yet Lockwood was able to obtain a clear radiograph of a key and a Chinese coin from the residue of his carnotite after he had extracted the uranium and vanadium. Convinced that he had radium in his ore, yet uncertain how to remove it, he boldly wrote to the world's leading authorities, the Curies.[12]

Pierre and Marie Curie had published a number of articles in which their separations techniques were outlined. Even more, Marie's doctoral dissertation of 1903 contained detailed information of the chemical processes, and its English translation was serialized later that year in the *Chemical News*. These revelations, however, seem not to have precluded the possibility of patenting the process, had they wished, for specifics of acid concentration, temperatures, the time for each stage, etc., were not yet common knowledge. The decision—to patent or not—appears to have been taken early in 1903, on receipt of Lockwood's inquiry. In her "official" history, Eve Curie introduces the matter as an offhand decision, then reconstructs some dialogue between her parents attaching great significance to the question, and ends with the interpretation adopted by later generations: "They had chosen forever between poverty and fortune." Marie is supposed to have firmly said that to patent would be "contrary to the scientific spirit," largely, but not entirely a truism, for scientists have been known to patent their discoveries without apparent criticism. Equally appropriate in this case would have been consideration of the disapprobation expected in France for patenting aspects of dissertation research. Pierre, adopting a role of devil's advocate, urged her to think of their children, of a life of comfort, of the fine laboratory they could purchase, for with the use of radium against disease there promised a fine commercial future. "This patent would represent a great deal of money, a fortune."[13]

But in early 1903, were there really such clear visions of a fabulous industry, especially to such ivory-tower introverts as the Curies? Radium was indeed being touted as a wonder cure, but such claims still were widely regarded with suspicion. Deposits of high percentage uranium ore were extremely scarce and it was uncertain if the more common low grade supplies could profitably be treated. Not until the medical applications created a large and sustained demand for radium, about 1905 at the earliest, was there much thought of a significant industry in this metal and the high prices it would bring. Before this time radium was known to be expensive to produce because of its scarcity, the quantity of chemicals required in its treatment, and the manpower needed, but efficient industrial production could be expected to lower the costs, even if it did not in fact lower prices (which were set at what the market could bear). Before 1905 it was impossible to predict that any significant quantity of radium ultimately would be produced. Thus it is extremely unlikely that the Curies did foresee and consciously reject the fortune in royalties that was theirs for an inexpensive patent application. Not for humanity, not for the sufferers of cancer, not for a masochistic love of poverty did they freely publish their information; they simply did not see great financial gain, but quite properly saw the ethical overtones.

Consequently, Lockwood's letter was answered with some specific suggestions.[14] The problem, as it would be with almost every ore, lay in the series of initial treatments to separate milligrams of radium-containing compounds from tons of other materials. Each mineral posed its own special difficulties and many competent chemists would fail to establish successful commercial processes for concentration, conversion into chlorides or bromides, and final purification. Buoyed by Pierre Curie's response and by his own tests on the carnotite, Lockwood in May 1903 incorporated the Rare Metals Reduction Company. Before the end of the year the first plant in America for the separation of radium was under construction, near what is now Lackawanna, New York. Early enthusiasm then turned to discouragement. The processes used failed to extract sufficient radium to show a profit, the low grade ores further depressed the output, and a poor market for the uranium and vanadium "by-products" caused prospectors to look for other minerals more in demand. By early 1908, Lockwood simply could not obtain enough carnotite to keep his plant running; "the production of radium was not to be contemplated as a commercial money-making project," and no Lockwood radium ever appeared on the market.[15]

Rise of an Industry

Coincidental with the collapse of his faltering enterprise Lockwood was approached by Thomas F. Walsh, a millionaire who had made his fortune in Colorado gold. Walsh was dumfounded that so desirable a commodity as radium was in such short supply, to the extent that uranium shipments could not be furnished with confidence by any ore dealers. Taking Lockwood as his radium adviser, Walsh subsidized certain research on carnotite at the Colorado School of Mines, investigated reported deposits of this ore, and in general endeavored to stimulate prospecting. His untimely death early in 1910 came just before he acted upon plans to acquire claims, concentrate the ore to a high grade in Colorado, and then ship it to the Rare Metals Reduction Company for further refinement.[16] Once again, Lockwood had failed in his desire to be associated with the production of American radium.

Thus, by the end of this century's first decade, the moderate amounts of radium produced worldwide had not only been prepared abroad but had come largely from foreign ores. Uranium deposits of varying extent and richness were found in England, Portugal, Norway, and in Saxony, on the other side of the Erz Gebirge from Joachimsthal. Yet the Bohemian mines continued to yield the greater amounts and, despite periodic reports of exhaustion, remained active for many years. An embargo, placed towards the end of 1903 by the Austrian Government on the export of uranium and its residues,[17] led their northern neighbors to the manufacture of "German radium," following Hahn's discovery of mesothorium. The French, however, seem not to have been affected by the ban, perhaps because of the cordial relationship established between the Curies and the Vienna Academy of Sciences, which accepted supervision of this resource from the government. As mentioned in chapter two, a contemporary observer interpreted the embargo as a protection of the state glass industry,[18] but more likely it reflected official displeasure that much of this material intended for scientific purposes found its way into commercial channels.[19]

Into this somewhat controlled and constrained European radium industry, which had produced less than an estimated ten grams total,[20] came a flood of American carnotite beginning about 1911. The main supplier was Thomas F. V. Curran, a New York dealer in ores and concentrates. The shipment of carnotite abroad increased so rapidly, in fact, that within a few years the major portion of the world's supply of purified radium was American in origin.[21] But even before public outcry arose over the circumstance that hospitals in the United States wishing radium often had to purchase the refined Colorado product overseas, Joseph M. Flannery moved to dominate the American market.

Though he had founded the American Vanadium Company in 1906, Flannery seems not to have considered carnotite as a source for this metal, at least before 1911.[22] When he did turn to carnotite it was for a far more precious constituent indeed. In rapid succession he purchased numerous mining claims, acquiring nearly one hundred by 1913.[23] He also incorporated the Standard Chemical Company in 1910, to whose plant near Pittsburgh carnotite concentrate was sent via a transportation network that included burro, horse-drawn wagon, truck, narrow gauge railroad, and standard railroad. The basic chemical process was devised by Boltwood, acting as a consultant for Flannery, and initially supervised by a young Yale graduate named Rowland Bosworth.[24] Boltwood did not limit his activities to the Standard Chemical Company, however, and was involved in numerous dealings with prospectors, miners, industrialists, and financiers. He not only tested mineral samples, but also devised various commercial separations processes, advised on standards of activity, and sought to channel the investments of wealthy Yale alumni such as Alfred I. du Pont.

His protégé Bosworth found the Standard plant undergoing a chaotic birth during 1911, with rapidly changing personnel and indecisive and contradictory management decisions. After about half a year he left, despairing that radium would ever be produced there.[25] Others were skeptical too, for the disbelief persisted that carnotite contained any radium.[26] Furthermore, Flannery had gained something of an unsavory reputation as the purveyor of patent medicines containing vanadium during his earlier activities with the American Vanadium Company. Would now this same person misdirect scarce radium away from reputable medical uses and into nostrums? Early advertisements were not reassuring, for radium compresses, muds, drinking solutions, ampoules, ointment, and bath waters were offered for sale in the pages of his monthly house organ, *Radium*, which posed as a scientific periodical. And finally, even before his carnotite was refined, it appeared that Flannery had contracted to sell most of his production abroad, to those "sinister" groups popularly known as the "Radium Trust."[27]

Whatever his shortcomings and the legitimacy of others' doubts, Flannery was determined to market radium, and market it he did, successfully, honestly, and in quantity. He hired as chief chemist Charles Viol, a talented student of Herbert McCoy's, who held this position until Standard's demise a decade or so later. Viol perfected the techniques for treating carnotite, the first American radium being obtained in January, 1914. In the months before war erupted in Europe 7 1/2 grams of radium element (generally in the form of crystalline radium bromide, $RaBr_2 \cdot 2H_2O$) were produced, and by the end of 1918, Standard claimed to have prepared 37 grams of radium out of a total American yield of about 56 grams.[28] (A gram has been pointed out to be the amount of salt commonly used at dinner.)

There were others who entered the radium business, as the numbers above suggest, some for profit and some for humanitarian reasons coupled with fears of fraud by Flannery. Among the former were the American Rare Metals Company, which operated a mill on the old Poulot property in Paradox Valley, and the Radium Company of America, with mines near Green River, Utah, and its plant at Sellersville, Pennsylvania.[29] Prominent in the latter category were Howard A. Kelly of Baltimore and James Douglas of New York. Kelly, a noted gynecologist and one of the famed four physicians (along with Osler, Halstead, and Welch) on the Johns Hopkins medical faculty, became seriously interested in radium around 1910 and began acquiring considerable quantities of the rare material from manufacturers in Paris and Germany. Since Boltwood, recently returned from a year in Rutherford's Manchester laboratory, was widely regarded as America's leading worker in radioactivity, Kelly asked him to test his purchases. An international standard had not yet been prepared, and there was too much discrepancy between quantity ordered and quantity delivered to place

great confidence in the manufacturers.[30] When he learned of the carnotite resources in the United States, Kelly had the requisite missionary
zeal to urge creation of a domestic industry.

Douglas more than shared Kelly's enthusiasm and energy. First
trained for the ministry, then for a career in medicine, he was permanently sidetracked into science and engineering by a fascination with
chemistry, geology, and mineralogy. These interests merged in an early
position he held as superintendent of a chemical company specializing in
copper, and remained in the forefront as he moved through copper
mining interests, ultimately to the presidency of the giant Phelps-Dodge
Corporation. His daughter's death of cancer in 1910 made him aware of
the inadequacy of medical treatment, which was primarily surgical. He
had taken her abroad for specialist care and had been shocked to learn
that the one promising cure, radium, was in such short supply that it was
rented from a central Banc de Radium in Paris. One outcome of his deep
concern was support of the General Memorial Hospital for the Treatment of Cancer and Allied Diseases in New York; some $600,000 even-
tually was given, much of it directed to creation of a laboratory and
purchase of radiological equipment.[31]

Another result was a personal commitment on Douglas' part that
the supply of radium should be increased. Apparently, as a first step he
employed Rowland Bosworth, who was no longer with Flannery, in an
effort to learn something of the problems of carnotite chemistry himself
and, hopefully, to have Bosworth devise a commercially economic process.[32] Douglas also became acquainted with Kelly, possibly through
Boltwood to whom he too sent samples for testing.[33] In the autumn of
1913 the two men formed the National Radium Institute, whose purpose
was both to obtain radium in quantity and to investigate its therapeutic
properties. By this time Kelly had about a gram but, with visions of the
suffering he was convinced he could alleviate, desired four or five grams
more.[34]

Governmental Action

The National Radium Institute (about which more shortly) operated in
partnership with the Federal Bureau of Mines. This was not, however,
the only interest in radium shown by this government agency, created
only a few years earlier and seemingly intent on making an impression
on the public's mind. In the fall of 1912 Richard B. Moore, chief of the
Bureau's Denver laboratory made the first careful survey of the
Colorado and Utah carnotite deposits and to a surprised world announced that America possessed by far the largest resources known of

radium-bearing ore.[35] When it suddenly dawned upon those concerned with radium that this American ore was going abroad for processing, and was only at great expense returned across the Atlantic, there seemed sufficient justification for governmental intervention. And when an already suspicious public heard authorities such as Moore charge that radium prices exceeded the cost of production by four times, action could not fail to be demanded.[36]

There was precedent for control: the mines in Austria were under government supervision; in England, Frederick Soddy urged nationalization of radium,[37] and in America such natural resources as oil (for example, Teapot Dome) were conserved (and uranium after World War II). Amid newspaper reports of the newly formed European Radium, Limited, whose stated, if unlikely, plan was to monopolize the world's radium,[38] and moved by sincere humanitarian as well as political considerations, Interior Secretary Franklin K. Lane, in the closing days of 1913, publicly advocated protection of the national radium resources. In a letter to Representative Martin D. Foster of Illinois, chairman of the House Committee on Mines and Mining, and a concerned physician as well, he pleaded for speedy legislation, before speculators could gobble up the claims. In fact, Lane considered the matter so urgent that he recommended a joint resolution of Congress, in order to bypass lengthy hearings. Under the bill he proposed all radium-bearing land in the public domain would be withdrawn; existing claims, considered private property, would be untouched. The government would then mine and treat the ores and distribute the product to the country's hospitals.[39]

Immediate support came from Dr. George Otis Smith, director of the Geological Survey. Such action is necessary, he said, "if America is to have first claim on the benefit of its own resources."[40] Approval was voiced also by Howard Kelly, who lectured on radium in Secretary Lane's office before fifty members of the Senate and House Committees on Mines and Mining and on Public Lands—the men who would consider the bill submitted by Representative Foster.[41] The deeply religious Kelly, who looked upon radium as fulfilling biblical prophecies,[42] also lobbied on a more subtle level. Robert G. Bremner, a young Congressman from New Jersey and a personal friend of President Woodrow Wilson, lay ill with cancer in Kelly's private Baltimore hospital, where he was being treated with one hundred thousand dollars worth of radium. Over the course of several weeks, until his death in February, 1914, numerous newspaper reports of his condition appeared, replete with praise of radium and of Kelly.[43] Kelly lamented that he had insufficient radium for Bremner;[44] Flannery charged that too rapid application of radium killed the Congressman,[45] and the Maryland medical society sought to investigate whether Kelly was violating medical ethics in his outspoken acclaim of radium.[46] Whatever the merits of the turmoil, radium was kept before the public.

Immediate condemnation of the proposed legislation was heard from two Colorado Congressmen, and the groundswell of protest from the mining community of that state appeared well-orchestrated, if unconvincing. While agreeable to a ban on exports and on monopolizing of radium, they feared the further withdrawal of any Colorado land, as this would "deplete" the state's mineral resources.[47] More straightforward were visions of the industry coming to a standstill, with loss of work for a potential 5,000 miners, as if the government planned not to work the carnotite deposits.[48] A protest strike was suggested by O. Barlow Willmarth, president of the Colorado Carnotite Company, which claimed to control the deposits from which nearly half the world's supply of radium had come. The garrulous Willmarth then undercut his case by adding that every pound of his output in the past three years had been sent to Paris, London, and Berlin, all American purchase offers being rejected. He justified this behavior by his obligation to get the best price for his ore.[49]

Joseph Flannery, just marketing the first radium produced by his Standard Chemical Company, argued that the government should not compete with private industry. This might not be a significant problem for the radium itself, since foreseeable demand would exceed supply, but difficulties would probably arise in the disposal of the by-product uranium and vanadium, which any efficient business must try to do. Moreover, there is radioactivity, however minute, in nearly every mineral. If the government now wants 2 percent carnotite, what will it take in the future? where will it stop?[50] Flannery was every bit as good a publicist as Kelly, and in the Congressional hearings (not vanishingly short, as desired by Lane) casually remarked about some monumental philanthropy to which he was privy. In response to a committeeman s assertion that it was necessary for the government to furnish the radium, Flannery disclosed that a financier, later claimed to be Henry Phipps of Pittsburgh, was ready to donate $15 million for the erection of twenty special hospitals across the country, and their endowment with five grams of radium each. Since the Bureau of Mines estimated that only thirty grams of purified radium then existed in the world, this proposal for one hundred grams was headline material. Phipps, aged seventy-four, was a partner of Andrew Carnegie and second largest holder of Carnegie Steel Company stock.[51] Since Phipps never gave this money, the story may have been a ruse; if so, it was effective. A few days later the *New York Times* editorialized against the legislation, saying that the reported philanthropy would likely make governmental intervention unnecessary.[52]

But there were other forces at work that eventually made governmental intervention impossible. In his Washington testimony Flannery admitted that he had recently purchased five more claims in Colorado, but denied giving his agents instructions to "buy up everything."[53] However, it shortly became evident that, if not Standard, then others were quietly filing claims to public land containing carnotite deposits, in an effort to circumvent the feared legislation. Soon, it reached noticeable proportions, and was compared to a remembered gold rush.[54] By the end of March 1914, even proponents of the bill admitted that there were few known deposits left to protect. And for good measure, the Senate committee helped to kill the proposal by voting crippling amendments.[55]

The National Radium Institute

Not all government interests, however, were doomed to failure. Aware of work being done by the Austrian and British Radium Institutes, Charles L. Parsons, Chief of the Division of Mineral Technology of the Bureau of Mines, hoped for the establishment of a domestic counterpart. It was apparently at his suggestion that Kelly and Douglas formed the National Radium Institute, in October 1913, and Parsons even accompanied Kelly on a western trip to inspect some carnotite mines in Colorado. With the Institute providing funds for the acquisition and treatment of the raw materials in a newly constructed plant, the Bureau agreed to furnish the necessary scientific and technical skill. The Denver laboratory under Moore was already investigating radium extraction and purification, so the Bureau could now extend its competence in the field to the commercial scale. Their cooperative agreement, certified legal by analogy with Department of Agriculture support of farmers, provided that any radium, beyond seven grams of anhydrous radium bromide produced from any one thousand tons of carnotite ore, was to become federal property and be used for research or in government hospitals. The first seven grams were to be divided between the General Memorial Hospital in New York and Kelly's hospital in Baltimore.[56]

Twenty-seven claims in Paradox Valley were leased from the Crucible Steel Mining and Milling Company, which was still interested in the property's uranium and vanadium.[57] Alfred I. du Pont turned over to the government his pitchblende and carnotite holdings as a contribution toward the establishment of a cancer cure. Like Douglas, du Pont was clearly a major industrialist, and like his fellow corporate businessman held no fear of creeping socialism—at least as concerned the wonder element. "Government ownership of railroads and telegraph lines," said

Mr. du Pont, "is nothing compared with the necessity of taking over the radium producing deposits in this country."[58] As we have already seen, such legislation failed, yet the country was increasingly being supplied with purified radium.

Kelly and Douglas each contributed $75,000 for establishment of the Denver plant, which by early 1915 produced its first radium, and by the end of that year claimed an output of five grams. Most interesting to those concerned about the reported 75,000 cancer deaths in the United States each year was the financial news. Suspicions of enormous profits in the radium business were confirmed when the Bureau of Mines announced that its production costs were $36,000 per gram, while commercial selling prices (including distribution costs and profits) ranged about $120,000 to $160,000 per gram.[59] A spokesman for the industry challenged the government's figures, arguing that they did not include invested capital, overhead, research expenses, salaries, etc.,[60] but such charges seem unsubstantiated.[61] Another controversial point was the Bureau's assertion that future supplies of radium ore would not support the current production rate; thus, lowered production costs did not ensure reduced prices.[62] Standard's Charles Viol had a running battle with the Bureau's Richard Moore, who commonly made such predictions, saying that his company's holdings alone contained more than anticipated future demands.[63]

With the summertime help of Herman Schlundt, and the continuous employment of Samuel C. Lind and others, Moore developed and applied a new nitric acid leach technique for extracting the radium fraction from the carnotite ore. The final purification followed the fractional crystallization procedures set down by Bémont and the Curies. One gram of radium required at least one hundred tons of ore, and usually several times that, plus water, twice the ore's weight in chemicals, and about 150 days processing time.[64] When the National Radium Institute-Bureau of Mines association ended in 1917 (for reasons unspecified), about 8.5 grams of radium had been produced.[65] With hindsight, the effort bears some resemblance to the World War II Manhattan Project, for, in both, government direction was exerted to separate microscopic quantities of constituents from tons of uranium.

Uses for Radium

World War I broke out just as American producers, especially Standard Chemical Company, were beginning to market radium. Because a substantial portion of their sales were planned to foreign customers, they immediately curtailed production. This decrease in domestic output,

coupled with the total loss of European suppliers, generated fears of a
"radium famine" in America.[66] Soon, however, domestic sellers and
buyers seem to have made contact, and the mines and plants worked at
something approaching capacity. Indeed, radium production in the
United States grew during the war years, particularly as new applications
were devised.

That the radiations exerted germicidal powers was generally conceded. Bacteria had been killed in laboratory tests, and suggested applications of radium therefore included incorporation in mouthwashes,
toothpowders and pastes, as well as the destruction of fungi which attacked drying codfish. Curiously, however, if smaller amounts of radium
were used, bacteria growth was stimulated, not killed. This suggested use
in cheese manufacturing, where the process time might be shortened.[67]
Cell growth also was stimulated, a property which was seized upon by the
bottlers of drinking water, who claimed that the secretory and excretory
organs would be more active in throwing off the body's waste products.[68]
And farmers were destined to benefit from the wonder element, for
traces of radium were believed to aid plant growth, and fertilizers bearing such names as "Nirama" and "Liquid Sunshine" were sold during the
century's second decade, if not earlier. It was hoped that intensive farm-
ing would be encouraged, at the same time that radium factories would
find a use for their residues, which still contained minute amounts of
radium since their processes were not 100 percent efficient.[69] None of
these applications proved effective in practice, largely because of variable or insufficient intensity of radiation, inapplicability of the idea, or
misunderstanding of the reactions involved. Nevertheless, radium was
regarded as potentially promising in diverse areas and, indeed, some
uses ultimately proved of distinct value. Though it is unlikely that codfish ever were treated with radium, other sources of radiation have been
applied to numerous foodstuffs since World War II to protect them
from spoilage. Radium has also been used in the nondestructive testing
of metal castings for flaws, being more easily handled than x-rays, and to
dissipate hazardous static electricity through the ionization produced by
its alpha radiation.

But these are valid applications of later decades, and for every success there were several failures, often of striking interest, such as the
claim in 1914 that radium was a cure for insanity, and a government
warning in the mid-1920s that numerous hair tonics, bath compounds,
suppositories, tissue creams, tonic tablets, face powders, ointments,
mouthwashes, opiates, ophthalmic solutions, healing pads, and other
preparations, claimed to contain radium, were frauds. A glass rod,
coated on one end with a yellow substance and designed to be hung over
one's bed, where it would disperse "all thoughts and worry about work

and troubles and bring contentment, satisfaction and body comfort that soon results in peaceful, restful sleep," also was considered to have a "highly exaggerated therapeutic claim."[70]

The single most important serious use of radium, which rivaled and even for a time exceeded medical demands, was in the manufacture of self-luminous paint. Early visions of radium lamps, for home, factory, and automobile, had proven unrealistic; too much of the expensive, rare element would be required, while the intensity of the illumination was inferior to that obtained from gas mantles and electric lamps. In a paint, however, the radium could be mixed to extreme dilution, such that costs were not excessive, yet the radiation remained intense enough to produce luminescence in the paint's zinc sulphide crystals.

In the 1870s, Balmain's patented luminous paint (using calcium sulphide) had appeared in Europe, where it was applied to clock and watch dials and hands. By about 1900, as prior exposure to light was necessary to generate phosphorescence, it was used infrequently. With the availability of radium, however, the paint could contain its own phosphorescence "generator." As early as 1903, the Tiffany gem expert, George F. Kunz, prepared such a mixture and painted the numerals and hands of a watch.[71] He seems not to have perfected the material or technique, though, as it remained for World War I to serve as the catalyst bringing this product to popularity. Before the war self-luminous paint was a novelty; during and after that time its use was widespread.

Paint production received its greatest impetus from the demand for timepieces that could be read in the dark. One gets the impression that every soldier saw frontline combat, where striking a light could give away his position to the enemy, for it seems that whole armies in Europe were equipped with radium-painted wristwatches. Even before the United States entered the war the Ingersoll Watch Company was reported negotiating for enough radium to manufacture one million watches—one year's anticipated sales.[72] Once America became a belligerent, demand for radium paint increased. By 1920, it was estimated that over four million clocks and watches had been painted with radium, though less than one-third of an ounce of the element was used (one ounce equals 28.35 grams).[73] Part of the fascination surrounding this product was that so costly a material as radium could be sold on a two dollar watch or a twenty-five cent keyhole locator. Moreover, the radium did not wear out within human lifetimes (though the zinc sulphide crystals' efficiency was impaired under prolonged alpha particle bombardment).

Clocks and watches were not the only items painted with radium. Aircraft instruments, ship compasses, and gunsights were among the military applications, where night vision might be temporarily ruined if electric illumination were used. No need to fear mechanical breakdown was another significant benefit here. Peacetime uses, actual and proposed, included luminous door plates, street signs, mine signs, telegraph dials, light switches, fire exits, automobile instruments, poison bottle indicators, bedroom slipper buttons, theater seat numbers, fish bait, and glowing eyes for toy dolls and animals.[74] While mesothorium was substituted for radium in some instances, in order to conserve the longer-lived radioelement, more and more radium did go into luminous paint. In 1917, one knowledgeable observer estimated that the radium output was equally divided between medical uses and paint; the following year, with America geared to a wartime economy, about 95 percent of radium production went into paint. Some physicians, moved by patriotism, even sold their radium supplies to paint manufacturers, a practice fairly common in Europe. Others condemned the waste of this valuable resource in a largely nonrecoverable application.[75]

The Post-World War I Period

Radium production continued to increase after the war, with medical applications regaining a significant portion of the output. While the story of the 1920s and 1930s is beyond the scope of this volume, it will not be amiss to indicate the direction taken by the radium industry.

By 1920, it was estimated that about five ounces (*ca.* 140 grams) of radium had been purified in the world.[76] Three years later more than that amount was in use in the United States alone.[77] Despite oft-repeated warnings that the Joachimsthal mines were near exhaustion, and that the American carnotite beds would soon decline in productivity, there seems to have been little fear of a contracting supply. Perhaps the public chose to believe those who had more faith in the Colorado deposits. Or, perhaps, after years of periodic rumors about great discoveries in such exotic places as San Salvador, Madagascar, and New Zealand, the public considered that sooner or later one of these stories would contain more fact than fiction. Indeed, that is precisely what occurred, in the heart of darkest Africa: the Belgian Congo.

Near the copper-mining center of Elizabethville, prospectors of the large Belgian corporation, Union Minière de Haut Katanga, found uranium deposits of great extent and richness. The year was 1913, however, and the war prevented immediate exploitation of this ore. Because of the distinct possibility of a German victory, Union Minière kept the discovery secret. Even after the Armistice secrecy was maintained, until the largest plant in the world designed for radium-ore treatment was completed and ready for operation, at Oolen, Belgium. Their announcement in 1922 stunned the American manufacturers. Most of the mines in Colorado and Utah closed immediately; the rest followed before long. When Belgian radium appeared on the market the price dropped from one hundred twenty thousand to seventy thousand dollars per gram, since not only were the Congo deposits richer in uranium (as high as 60 percent uranium oxide[78]) and more easily mined, but labor costs were considerably less. In this situation the American industrialists could do nothing but accept the Belgian offer to market their product.[79]

By the end of 1924, in just two and a half years of operation, Union Minière had produced about 110 grams of radium. But their output of four grams per month exceeded demand and they were forced to limit the amount of ore being processed or accumulate reserves.[80] Whatever their choice—the company was remarkably secretive—they held a virtual monopoly on the world's manufacturing of radium for a decade. This domination was broken when, in 1930, rich pitchblende deposits were discovered at Great Bear Lake in Canada, within sight of the Arctic Circle. Extracted by Eldorado Gold Mines, Ltd., the ore was first processed in 1933, at a refinery in Port Hope, Ontario, some 4,000 shipping miles away. Despite the long transportation requirements, waterways frozen closed nine months of the year, higher labor costs, and difficult working conditions, these and other mines established in northern Canada prospered and took a substantial amount of business from the Belgians because considerable quantities of silver and copper were also removed from the ore. Radium prices now dropped to twenty thousand to twenty-five thousand dollars per gram.[81]

As the 1930s drew to a close, million volt x-ray tubes, neutron beams, artificially radioactive elements, and charged particles accelerated in the cyclotron and other machines, began to supplant radium in certain areas of medical treatment. Following World War II, reactor-produced radioactive nuclides contributed to an even greater eclipse of radium therapy, though the uranium industry for a while was greatly stimulated—this time for the uranium content of the ore. The scientists, whose activities form the major thread of this book, were not overly important in the "radium business," yet not without significance. Boltwood's brief work for the Standard Chemical Company, including his recommendation of Bosworth to Flannery; his longtime friendship with H. S. Miner, chief chemist at the Welsbach Light Company, Gloucester City, New Jersey, from whom he obtained various thorium products; his contacts with Kelly; and his voluminous correspondence with prospectors, mine owners, ore dealers, industrialists, and investors certainly stimulated the radioactive materials industry to some degree. McCoy, besides training Viol, the key technical person at Standard, went further than Boltwood by entering the business himself. However, if his Carnotite Reduction Company, in Chicago, produced radium at all, it made little impression on the American market.[82] Schlundt and particularly Moore, among these research-oriented academics, were most closely associated with the substantial production of purified radium. A handful of scientists, a handful of companies, a cupful of radium in the world, yet enormous public interest in its application.

11 Chemical Maturity
Radium Data

If we may say that radioactivity in America reached maturity with the discovery by Boltwood in 1907 of a new radioelement, ionium, then this level of attainment lasted but several years, declining with the advent of World War I and the coincident involvement in other activities of the leading figures in this field, Boltwood and McCoy. Yet, even before the war the "center of gravity" of radiochemistry research had shifted to Europe where the group displacement laws and concept of isotopy placed the capstone on this subject. By the end of that decade the finishing touches had been added, the questions this science asked had been answered, and radiochemistry was suicidally successful. The physical side of radioactivity persisted a little longer, but from the second decade of the century was itself evolving into both atomic and nuclear physics.

To a large extent the areas pursued in radioactivity in America remained the same in maturity as in adolescence. Decay series holes were filled, half-lives better determined, various constants more accurately measured—essentially a quantitative improvement that led to qualitative insights. The few new topics pursued in the United States included atomic weight determinations, which confirmed the existence of isotopes, to be discussed in chapter fourteen, and an upsurge in physical investigations, examined in later chapters. In this chapter and the next we review the increasingly sophisticated and accurate work which maintained America's high standing in radiochemistry.

During the period under discussion, of approximately 1908–1920, perhaps the most spectacular scientific event connected with radium was its preparation in 1910 in elemental form. Marie Curie, whose Nobel Prize in chemistry the following year seems largely based upon this accomplishment, with André Debierne obtained the pure metal by the electrolysis of a mercury amalgam.[1] Yet this work, based upon a comparable procedure for barium, whose chemical properties, as we know from the earliest preparation of radium compounds, are similar to those of radium, appears to have been little more than a *tour de force*. Of greater value were the data acquired about the "wonder element's" radioactive properties.

The Half-Life of Radium

Probably the most significant constant in radioactivity studies was the half-life of radium, the time required for half its amount to transform into emanation. As the only highly active radioelement that could be prepared in pure and reasonably large quantities, with a half-life long enough to consider those quantities constant for most practical purposes, radium was regarded as the standard substance in this field. Its physical and chemical properties, such as radiation range, spectrum, and atomic weight all were carefully determined, but its activity was the real yardstick against which other radioelements were measured. So important did this need for standardization become that Madame Curie was prevailed upon in 1910 to prepare an international standard for comparison purposes, as will be discussed in chapter seventeen. The value of such precision was apparent not only throughout the community of radioactivity workers, but in medicine and in geology as well, where calculations of the age of rocks increasingly hinged upon radioactivity techniques. Though the half-life was too long to be measured directly, it was fortunate that such an important constant could be indirectly determined by several more or less independent methods. Their eventual agreement, moreover, gave increased confidence in various other radioactivity constants used in related calculations. By this common sort of feedback mechanism the science became ever more precise.

A generally accepted corollary of the 1902–1903 Rutherford-Soddy transformation theory was that each atom of radium emits one alpha particle as it decays. A count of these alphas ejected each second from one gram of radium would yield the fraction of the mass decaying, i.e., the decay constant, and from that the element's lifetime. The problems encountered centered on imprecision in the counting results and inaccurate knowledge of the number of atoms in a gram-atom of radium (Avogadro's number). Rutherford attacked the problem early, and in his 1904 book *Radio-Activity* was able to describe three methods of determining the number and energy of alpha particles emitted. These were based upon comparison with the number of beta emissions, as measured by their accumulated charge, and upon two different approaches to the ionization produced by the alphas. He reported that approximately 10^{11} alpha particles are projected each second, or, considering that there are four alpha decays in radium and its products, one gram of radium element alone emits 2.5×10^{10} alphas per second.[2] Other experimental evidence showed that one cubic centimeter of hydrogen at standard temperature and pressure contains about 3.6×10^{19} molecules (a figure over 30 percent too large which, when multiplied by 22.4 litres, would give a value of 8.06×10^{23} for Avogadro's number instead of the now accepted 6.02×10^{23}). Using the Curie atomic weight determination of radium as 225, this meant that one gram of radium contains 3.6×10^{21} atoms. With 2.5×10^{10} of these transforming each second, the decay constant λ was calculated to be about 7×10^{-12} per second or 2.2×10^{-4} per year. This converted to a half-life for radium of about 3000 years.[3]

That the order of magnitude was correct was confirmed by other data, such as the heating effect of radium due to the collision of its alpha particles with neighboring atoms. Such measurements had been made from 1903 onward, principally by Pierre Curie and his associate Albert Laborde, and yielded estimates of the kinetic energy evolved per time unit. From Rutherford's experiments of 1903, showing the deflection of alpha particles in electric and magnetic fields, their mass and energy were calculated, and these combined with the heating effect data gave the number of particles expelled from radium each second as 5×10^{10}. Using the same figure above for the number of atoms in a gram of radium, Rutherford calculated a half-life now of 1600 years.[4]

Yet another technique involved measurement, by Ramsay and Soddy, of the volume of emanation released by one gram of radium in a specified time unit. Using the figure mentioned above for the number of molecules in a cubic centimeter of a gas, and assuming the emanation molecule is monatomic, it was but a simple step to the number of radium atoms breaking up each second and thence to a half-life of 1050 years.[5] A variation of this method was the calculation that if one gram of hydrogen occupies a volume of 11.2 liters, it would fill 22.4 liters if its molecules were monatomic. Then a gram of radium, assuming it to be obtainable as a monatomic gas, would occupy a volume of $(22.4)/(225$ atomic weight$) = 0.1$ liter $= 10^5$ cubic millimeters. Ramsay and Soddy had shown that one gram of radium produced 1.3 cubic millimeters of emanation in 5.3 days. By Avogadro's law (equal volumes of gases, under the same conditions, contain equal numbers of molecules), therefore, the ratio of the number of emanation atoms to radium atoms is 1.3:100,000. If one atom of emanation is produced from one atom of radium, in a year's time the ratio, i.e., the decay constant, would be 90:100,000. This converted to a half-life just under 800 years,[6] which suggests that their volume of emanation was about twice as large as it should have been.

Rutherford's particular specialty was work with the alpha particles, those "jolly little beggars," whose charge and charge-to-mass ratio he measured with increasing accuracy in the middle of the century's first decade. Initially assuming each alpha to carry a single electronic charge, and measuring the total charge carried by the rays from a known quantity of radium, he calculated a half-life of about 1300 years.[7] A new determination of the charge-to-mass ratio, which indicated the likelihood of the alpha particle being a helium atom, with a double electronic charge, then made him double radium's half-life to 2600 years.[8] In 1908, partly to avoid the need to assume the magnitude of the alpha's charge, Rutherford and his assistant at Manchester, Hans Geiger, developed an electrical method of counting individual alpha particles,[9] and then made a refined measurement of the total charge from a gram of radium. This yielded further certainty of the alpha's identity with helium, a better value of the electronic charge e (since the number of alpha particles n, multipled by $2e$, equals the total charge Q carried by the alphas), and also a more precise value of Avogadro's number (which equals the Faraday, long known from electrolysis, divided by e). Then, adapting a procedure developed by Boltwood, they derived a half-life for radium of 1760 years.[10]

It is apparent that these values were taken as approximations, for though they all were of the same general size, they varied too much to regard them as results of decisive precision. The uncertainties lay in important assumptions and/or measurements, such as the number of atoms present in a cubic centimeter of a gas, the number of disintegrations per second, the charge on the alpha particle, the volume of microscopic quantities of gases, etc. Some techniques were little more than "back-of-the-envelope" calculations; others involved great experimental skill or innovation. The American contributions now served to place the half-life between narrower limits, towards the value accepted at present.

As a chemist and as the discoverer of ionium, Boltwood understood that the rate of radium's production compared with its equilibrium amount would give exceedingly accurate information about radium's constant of change. In 1908 he carefully measured the amount of radium present in his samples. Then he separated the ionium from these minerals, sealed the ionium solutions in glass bulbs, and from time to time determined, by the presence of its emanation, the amount of new radium formed. By the laws of radioactive decay, at equilibrium the amount that forms equals the amount that decays. The radium in Boltwood's solutions was not, of course, in equilibrium with the ionium, but this did not matter. For determination of the amount of radium grown, the important equilibrium was that between radium and emanation, which, because of the latter's relatively short half-life of less than four days, was reached after about two months.[11] The method is advantageous in that being a ratio, it does not depend on the purity of any radium standard used. Rutherford and Geiger's value of 1760 years, by contrast, being directly dependent on the purity of the radium salt used as a standard in their counting experiments, had to be revised to 1690 years upon adoption of the international radium standard in 1912.

But Boltwood had reservations about the completeness of his own ionium separations, for the starting materials were relatively impure and not entirely soluble in any one reagent. Moreover, uranium and thorium have strikingly similar chemical properties (ionium at this time was said to "resemble" thorium strongly), which would further affect their separation. Half-lives obtained of 3100 and 2400 years, he felt, were too high—a good example of expectations influencing the acceptance of data. In this final test, therefore, he began with a very pure uraninite from North Carolina and subjected it to lengthy treatment, which presumably removed more of the ionium. The growth of radium in this solution, over the course of 147 days, gave a half-life of 1990 years, which Boltwood was convinced was the most reliable value. It was published after Rutherford's 2600 year figure and shortly before Rutherford and Geiger's 1760 year announcement.[12] Since Boltwood's result was higher than that of his Manchester colleagues, there was a suspicion that perhaps, even in his final experiment, not all the ionium had separated from the mineral. Work by Bruno Keetman in Berlin, yielding 1800 years,[13] and by Stefan Meyer in Vienna, giving 1730 years,[14] was more in accordance with Rutherford's own determinations.

Rutherford's fertile laboratory in 1911 produced yet another means of closing in on radium's half-life. Years earlier he had wondered aloud whether any relationship existed between decay constants of radioelements and the ranges of their alpha particles. His concern, then, however, had been with the energy and velocity of emission of the alphas.[15] Now, Geiger and J. M. Nuttall attacked the problem using a novel method to measure range, which experimentally enabled them to determine this value for some weak radiations, and found that the logarithm of the decay constant plotted against the logarithm of the range yielded a reasonably straight line for the members of a decay series. Not only were they able to extrapolate ionium's half-life to a little below a million years, but radium's period was judged to be around 2000 years. As evidence for a rough approximation, the data and technique were satisfactory. But since the measurable half-lives were not precisely linear on the graph, the line drawn through them offered no *definitive* values for the long-lived radioelements.[16] Boltwood's method still promised the greatest precision.

Upon his return to Yale from a European holiday in the summer of 1913, Boltwood was confounded to learn that he had a research student waiting on his doorstep. Even more perplexing to the confirmed bachelor, and a source of merriment to his friends, the student was female. Ellen Gleditsch, a very able Norwegian chemist who had done research for five years in Marie Curie's laboratory, had won a fellowship from the American Scandinavian Foundation and wished to work in New Haven.[17] Boltwood suggested that she determine more accurately the half-life of radium using his method, and though it was not a collaborative effort, he kept in close touch with her progress.[18]

At this time Boltwood's own value of 1990 years and the Rutherford-Geiger determination of 1760 years (corrected against the International Standard to 1690 years) were the leading "contenders" for general acceptance. The unreasonably large difference between them motivated Gleditsch's investigation, for the English value might be in error due to its reliance upon the purity of a standard, while the American method was subject to incomplete chemical separation, though better in principle.

First Gleditsch tested Boltwood's best ionium solution of 1908, and confirmed that radium was being produced at the same constant rate as he had then reported. This underscored the reliability of the technique. Gleditsch's chief task, therefore, was to be certain that all ionium was separated from the minerals to be tested. Her general procedure was to precipitate and filter out all the rare earths contained in a mineral solution. To the remaining filtrate she added a solution of rare earths which contained no radium or ionium and precipitated and filtered this, as above. Once again the process was repeated, providing, in all, three precipitates. The first two were combined and dissolved. Presumably and hopefully, this "main solution" contained all the ionium. The third precipitate was similarly dissolved and constituted the "test solution," in which there should be no growth of radium. A range of minerals and a variety of purification techniques yielded two tests in particular in which the ionium separation was assured as complete. With the amount of radium in the original mineral and that grown in the "main solution" determined by the presence of emanation, Gleditsch in 1916 published values for the half-life of radium of 1642 and 1674 years, in close agreement with the 1690 years of Rutherford and Geiger.[19] As he had suspected, Boltwood's longer period was due to incomplete ionium separation.[20] This most important constant of radioactivity received further revision in later decades, to the presently accepted 1620 years, but the work of Gleditsch had developed assurance that future changes would be small.[21]

Workers in radioactivity viewed with increasing confidence the accuracy of the physical constants they used in their calculation since, as mentioned before, different methods of determining radium's period yielded concordant values. In a symbiotic relationship the half-life of radium supported, and was supported by, the accepted number of alpha particles emitted per second per gram of radium (3.57×10^{10}), the ionic charge (4.65×10^{-10} electrostatic units), the number of atoms in one gram of hydrogen (Avogadro's number, 6.2×10^{23}), the mass of the hydrogen atom (1.61×10^{-24} gram), the number of molecules in one cubic centimeter of any gas at standard temperature and pressure (2.72×10^{19}), the volume of the radium emanation in equilibrium with one gram of radium (0.62 cubic mm), and the rate of production of helium per year per gram of radium (163 cubic mm).[22] Many of these values could be derived from the others; often, however, they could be generated independently, outside of radioactivity studies. Most would be further refined in subsequent years, but the changes then would be relatively minor.

Helium from Radium

When Boltwood accepted Rutherford's invitation to spend a year in Manchester, they immediately planned their work. Rutherford characteristically enjoyed short vacations during which he liked to drop from his mind the cares of the laboratory, a trait seemingly shared by Boltwood. Because, without attention, his large supply of radium would produce an explosive mixture of about forty cubic centimeters of hydrogen and oxygen each week, through dissociation of the water of its solution, Rutherford proposed that they precipitate and dry the radium upon Boltwood's arrival in the summer, leaving it until the school year began in October to produce helium. In that way they would be "in position to have a holiday with the full knowledge that a research is in progress notwithstanding."[23]

Helium had been discovered in the mineral cleveite by Sir William Ramsay in 1895, and it was soon recognized to be present only where uranium and thorium were found. Its association with these heavy elements, often in minerals impervious to the passage of water and gases, was at first puzzling. Then, in 1902, Rutherford and Soddy suggested that helium was a product of radioactive disintegration, a belief confirmed the next year by Ramsay and Soddy, who detected helium's spectral lines "growing" in the gases of a radium compound. But experimental proof of this association raised the need to define helium's precise role in the decay process. Work by Rutherford on the deflection of alpha particles in electric and magnetic fields indicated this particle to be either a charged hydrogen molecule or a charged helium atom, the latter being more likely because of the noted presence of helium. On the basis of the imprecise data then available, Rutherford calculated in 1903 that between 20 and 200 cubic millimeters of helium should be produced per year by one gram of radium in equilibrium.[24]

It was, however, exceedingly difficult to resolve the identity of the alpha particle with certainty, despite the increasingly accurate velocity and charge-to-mass data collected by Rutherford in 1906 and by Rutherford and Geiger in 1908. In the latter year the two men in Manchester directly counted the number of alphas emitted and calculated the charge on each to be about twice that taken as the unit charge carried by a hydrogen atom.[25] For most, this indirect evidence would have sufficed for acceptance of the alpha particle as a double charged helium atom instead of a singly charged hydrogen molecule. For Rutherford it was tantalizingly close to the absolute proof he desired. This came later in 1908 when, with the aid of the talented spectroscopist Thomas Royds, and the extraordinary glassblower Otto Baumbach, who constructed glass apparatus so thin that alpha particles could be fired right through it, these rays conclusively yielded the helium spectrum.[26]

With this identity finally established, and with the alpha particle such a major manifestation of radioactivity, information about the rate of helium production became of even greater significance. Sir James Dewar, famous for his low temperature work at the Royal Institution in London, and keenly interested in helium because he had long tried to liquefy it, in 1908 made the first direct determination of its generation. Other gases present in his radium salt were removed by exposure to coconut charcoal cooled by liquid air, and any hydrogen present was accounted for by further cooling to liquid hydrogen temperatures. Dewar measured a rate of production of 135 cubic millimeters of helium per year from one gram of radium in equilibrium (corrected later to 182), a figure revised by additional, longer-duration experiments in 1910 to 169 cubic millimeters.[27] These values differed sufficiently from the 158 cubic millimeters calculated by Rutherford and Geiger in 1908, based on their alpha particle counting experiments, to warrant further attention.[28]

Boltwood and Rutherford felt they could resolve the discrepancy because they had at their disposal 183 milligrams of radium element loaned to Rutherford by the Vienna Academy of Sciences. With this large quantity, measurements would be more significant. By contrast,

Dewar's values were based on only 70 milligrams of radium chloride. Boltwood treated the radium solution to remove radio-lead and polonium which had accumulated through decay, leaving radium as essentially the only active material in quantity. The dried salt was then placed in an exhausted and sealed glass tube, in which new helium was allowed to collect. The amount of radium was determined not by the old method of measuring its emanation, which being a gas like helium would have made testing awkward, but by the gamma ray technique which was replacing it in precision use. In this method the gamma ray activity of the salt was compared against that of the Manchester laboratory's 3.69 milligram radium bromide standard in a thick-walled lead electroscope, which filtered out any alpha and beta radiation, and the quantity of radium thereby calculated to an accuracy of about 1 percent.

Boltwood's superb design and manipulative skill are nowhere more evident than in this complicated experiment which had to be performed within a sealed maze of glassware, presumably constructed by the Manchester wizard Baumbach. The problem of measuring a minute volume of helium was complicated by knowledge that hydrogen was not completely removed from it, and by recognition that some of the alpha particles embedded themselves in the glass walls of the vessels. Control experiments were performed to assure that all helium was removed and that the apparatus was correctly calibrated. This was the first joint experiment performed by Boltwood and Rutherford, and the American's foresight made a lasting impression upon his host.

Not long after their return from the 1909 summer holiday, they presented a preliminary report to the Manchester Literary and Philosophical Society, a provincial scientific group of distinguished history, saying that one gram of radium produced 163 cubic millimeters of helium per year.[29] Subsequent experiments of slightly modified procedure, longer duration, and hence greater accuracy, yielded a final figure of 156 cubic millimeters per year for one gram of radium in equilibrium with its first disintegration products (emanation, radium A, radium C), in striking agreement with the calculated 158 cubic millimeters. By way of "spin-off," to use the NASA term for related, derived benefits (not "fallout"), they also determined that emanation and polonium produce helium. This was not, of course, unexpected. Debierne, for example, had earlier shown this for actinium preparations, but it was evidence confirming the belief that all alpha particles emitted from radioelements were identical in their chemical nature. Though both Rutherford and Boltwood had at one time or another speculated that the alpha particle emitted from radium was different from that emitted, for example, from thorium,[30] they were willing now to conform to Isaac Newton's centuries-old third Rule of Reasoning in Philosophy: "The qualities of bodies . . . which are found to belong to all bodies within the reach of our experiments, are to be esteemed the universal qualities of all bodies whatsoever."[31] In acknowledgment of the valuable loan of the Austrian radium, the paper was published first in the *Sitzungsberichte* of the Vienna Academy of Sciences, and only half a year later in the English World's foremost physical journal, the *Philosophical Magazine.*[32]

With experiments and theory so well in agreement, Boltwood and Rutherford felt justified in claiming as proven not only the assumptions upon which the calculations were based—that the alpha particle in flight is a doubly charged helium atom and that helium is monatomic—but the "essential correctness of the atomic theory of matter." From the direct counting of the number of helium atoms expelled per second by one gram of radium, and from the direct measurement of the volume of helium produced, they calculated 2.69×10^{19} atoms of helium in one cubic centimeter of that gas at standard temperature and pressure. This figure, of course, is but a different way of expressing the all-important Avogadro's number.

The Radium: Uranium Ratio

The quantity of radium found in one gram of uranium in equilibrium had been determined by Rutherford and Boltwood in 1906 as 3.8×10^{-7} gram.[33] The derivation of this value, described in chapter five, had been plagued with difficulties, primarily the circumstance that some of the radium in Rutherford's standard solution had adhered to the walls of the glass container, making the portion sent to Boltwood weaker than believed. With that problem recognized, and solved by the introduction of some acid to keep the radium in solution, they hoped that this number, in *principle* a constant according to the transformation theory of radioactivity, would in *practice* remain unchanged. That this was not to be so was a reflection of the uncertain state of much chemical knowledge early in this century.

In his later work on the relative activities of radioelements, to be discussed in the next chapter, Boltwood's methods often depended on the quantitative separation of constituents from complex minerals. In this period prior to the group displacement laws and the concept of isotopy, information about the chemical properties of many radioelements was meager or in error, or both. Consequently, the techniques he devised frequently included tests to confirm that separations had, indeed, been complete. Ellen Gleditsch's need to ascertain that all the ionium was isolated from her starting materials is in this tradition of experimentation. The uncertainty of chemical behavior extended beyond the specific properties of the radioelements, however. Uranium was long known to exist in the form of several oxides, including UO_2 and U_3O_8. The textbook wisdom of analytical chemistry, initially followed by Boltwood, asserted that *moderate* heating of UO_2 in oxygen converted it entirely to U_3O_8, while the reverse reaction would occur in an atmosphere of hydrogen. Discordant results apparently made Boltwood question this procedure and he was led to discover that the reactions require very *high* temperatures to go to completion. This underscored the need for much precise investigation of chemical practice—the uranium-oxygen system was not systematically studied until the 1920s, and later received even more attention in the Manhattan Project[34]—and in this case caused Boltwood in 1908 to recalculate the proportion of uranium in the mineral used for his radium determination. Instead of 68.2 percent uranium, based upon the assumption that his uranium oxide was entirely U_3O_8, the new value was 75.8 percent uranium. This required a revision of the amount of radium associated with one gram of uranium in a mineral from 3.8×10^{-7} gram to 3.4×10^{-7} gram.[35] The new value was widely adopted for several years, becoming known as the Rutherford and Boltwood standard, only to be replaced by the figure of 3.23×10^{-7} gram around 1914, when comparison with the new International Radium Standard showed there was slightly less radium in their solution than believed. This change in turn forced the revision of several other standards of radioactivity—e.g., half-life of radium, production of helium per gram of radium per year, volume of emanation from one gram of radium in equilibrium, number of alpha particles expelled per second per gram of radium, charge carried by alpha particle per second from one gram of radium—which had been based on the Rutherford-Boltwood Standard.[36]

Though the precise value of the radium-uranium ratio varied in the fashion just described, this was due to accuracy of measurement, not fundamental change in their relative amounts. However, into this atmosphere of certitude that the ratio was a constant in minerals which had reached an equilibrium state of decay, there arose unsettling reports of intrinsic variability. The matter was serious since the transformation theory of radioactivity required a fixed ratio, because all primordial uranium was assumed to have formed about the same time and to have decayed at an unalterable rate. McCoy and Ross, for example, had concluded in 1907 that the specific activity of uranium was strictly constant.[37] If the variability of the radium-uranium ratio now were proven, the assumptions underlying the transformation theory, and perhaps the explanation of radioactive decay itself, would be threatened.

Jacques Danne in Paris had shown in 1905 that radium was found in a uranium-*free* mineral from Issy-l'Évêque.[38] With a sample provided through the courtesy of Robert Millikan and Paul Langevin, McCoy and his student Ross a few years later proved the radium to be concentrated on the surface of the mineral, confirming Danne's supposition that the radium had been transported there by percolating water from a distant uranium source.[39] Less easily explained was the work of Ellen Gleditsch, then in Marie Curie's laboratory, who separated the two constituents from different minerals and found their ratio to range from a low of 2.85×10^{-7} gram in French autunite to a high of 4.19×10^{-7} gram in Ceylonese thorianite, with the accepted value about midway between them.[40] Subsequent work in France, England, and Germany confirmed such variability, but other investigation suggested once again that loose formation of the minerals and percolating ground water were responsible, as well as the likelihood that some of the minerals were of relatively recent geological age and hence not in radioactive equilibrium. Still, certain old, massive, unweathered crystals of Portuguese autunite seemed to require another explanation.[41] John Joly, the eminent geologist from Trinity College, Dublin, studying inexplicable characteristics of pleochroic haloes, offered the view that uranium's transformation period had changed over the ages, but this violation of the concepts of radioactivity was not widely accepted.[42]

In effect, what followed was that evidence was never offered disproving these findings of disparity between different minerals. Belief in the transformation theory of radioactivity was so strong that these discordant experimental results just *had* to be wrong. The explanation *had* to involve minerals formed too recently, and/or minerals that had experienced some sort of chemical or physical alteration. Thus, in the mid-1910s, several German chemists who rederived the radium-uranium ratio largely limited their investigations to varieties of pitchblende, reporting results close to Rutherford's value, though presumably more accurate.[43] They simply worked under the assumption that the ratio was constant and restricted their samples to "trustworthy," i.e., old and unaltered, minerals. That they were correct is of less interest than the insight obtained about the nature of scientific progress. Belief in a theory can supersede apparently contradictory evidence.

By 1920, while this supposition of constancy remained unchanged, the need to reconfirm the ratio's value became manifest. The motivation arose in the American radium industry, which was concerned about giving proper value for money. Pitchblende was commonly used to standardize the electroscopes with which the radium content of ores was determined, but the gamma ray method, which did not depend upon the radium-uranium ratio, was used to test the separated product. Hence, there could be sizable accumulated discrepancies between input and output unless both tests measured the real, not apparent, amount of radium. Samuel Lind and L. D. Roberts, U. S. Bureau of Mines chemists, were able to profit by the previous work performed, such as the accurate determination of radium's atomic weight by Marie Curie and by Otto Hönigschmid, recognition of partial precipitation in some radium-barium solutions which occluded emanation, greater understanding of the difficulties in the correct analytical determination of uranium, and so on, which made their own work more precise. Their value raised the radium-uranium ratio to 3.40×10^{-7} gram, in good agreement, with some European results obtained five years earlier.[44] Radium's position and properties were now substantially settled; interest shifted increasingly to scientific and commercial applications of the element.

12 Chemical Maturity
Other Radioelements

In the public mind radioactivity may have been synonymous with radium, but the term, of course, embraced other radioelements and their interrelationships as well. Simultaneously with the study of radium, further information about these other bodies was acquired during the period now under examination, roughly 1908–1920. To a remarkable degree, this information fulfilled the existing needs. Certainly, there were lacunae that were not bridged until the 1920s or later, but enough data were collected, a comprehensive interpretation proposed, and verification obtained, such that the system could be regarded as complete. The group displacement laws and concept of isotopy, to be discussed in the next chapters, placed the numerous radioelements in the periodic table, giving them all chemical identity, while other work assured that all (or almost all) members of each decay series were accounted for. It was a time of problem solving, of organization, of placing constituents into proper boxes; it was also a time in which detailed information about specific radioelements and series relationships became ever more precise. Because of the trees the forest became visible.

Ionium

Boltwood's first major contribution to radioactivity studies had been proof of the constant ratio between radium and uranium in old minerals, circumstantial evidence of their genetic connection. More direct proof of this relationship came with his discovery in 1907 of the parent of radium, based upon evidence of the production of measurable quantities of radium from ionium. This was, of course, only half of the picture; that ionium grew from uranium had also to be demonstrated to forge the complete linkage between uranium and radium. Because the long periods of these radioelements meant that growth proceeded at a slow rate, this would be a time-consuming project.

Tending first to more immediate ionium problems, Boltwood tied together some loose ends in the literature. Debierne's "actinium" was largely ionium, and Giesel's "emanium" largely actinium, he explained, thereby resolving that dispute, while the unusually high activity of thorium separated from uranium minerals, as found by Hofmann and Zerban, was due to the inclusion of ionium, whose chemical behavior was "similar" to that of thorium.

Boltwood's first sample of ionium had been both small and weak; hence the uncertainty of some of his quantitative data. In 1908, he therefore prepared a more active source by dissolving carnotite (a thorium-free mineral) in hydrochloric acid, adding several grams of rare earths as carriers, and then separating the earths. This fraction was next treated repeatedly in sodium thiosulphate, like ammonia a standard reagent for precipitating thorium, the material settling in this case, however, being ionium (and uranium X, another isotope of thorium). New measurements of its alpha particle range were made by Lynde Wheeler and T. S. Taylor, his colleagues at Yale, who independently obtained a value of 2.8 centimeters in air. While the preparation was not strong enough to settle the question of a beta ray activity, Boltwood inclined to his previous (erroneous) belief that ionium did indeed emit betas. On the question of relative activities, he simply refined his earlier result, reporting that ionium was 0.76 times as active as radium, in good agreement with the ratio of the ranges of their alpha particles (0.8), but he remained silent about the poor fit with the relative activities of the rest of the uranium series.[1]

Not its chemical nature nor the range of its radiation, but its half-life was the major fingerprint of a radioelement. Since the activity of ionium did not change measurably in several months of observation, direct determination was impossible. But the radioactive decay laws, which specified that the product of decay constant and number of atoms was the same for each radioelement in a decay series in equilibrium, would permit calculation of half-lives longer than could be measured by laboratory experiment. Boltwood had found a constant rate of radium production in an ionium solution over the course of 500 days. From this he estimated that the time required for half a given quantity of ionium to transform itself was at least 25 years. Other work had shown him that the amount of radium formed in a solution containing 48 grams of purified uranium, observed over two and a half years, was less than 10^{-11} gram.

This indicated a far greater period, at least as long as that of radium, assuming no other products came between ionium and the uranium X from which it was believed descended. A lifetime approximately of the same order of magnitude as that of radium meant that similar quantities would exist in minerals in equilibrium, and gave promise that macroscopic amounts of ionium might one day be separated. Since half-life is essentially the inverse of decay constant (fraction of atoms decaying per unit time), and, as noted above, if this could be *estimated* only, then the key to *accurate* determination of half-life, by the decay law ratios, was knowledge of the number of atoms present. This gave added significance to efforts to detect and measure the growth of ionium in uranium. In early 1908, Boltwood could only report that he had begun such work.[2]

It was left largely to Frederick Soddy, however, to complete the task. Boltwood, as described earlier, soon spent his year overseas with Ernest Rutherford, and on his return to the United States completed little more research. First from Glasgow, then Aberdeen, and finally Oxford, Soddy published a series of papers detailing the relationship between radium and uranium. He wisely chose to measure the growth of radium directly, rather than the increase of ionium, because the former's presence was so much more easily detectable through its gaseous daughter product, emanation. Aside from a false indication of radium growth in 1908, it was not until 1915 that unmistakeable amounts of radium were detected and the growth shown to be proceeding in accordance with theory, and not until 1931 that Soddy finally concluded his refinement of ionium's half-life, determined as 73,500 years (now considered to be 8×10^4 years).[3]

A detailed description of Soddy's work is beyond the limits chosen for this monograph. The story of ionium, however, bears upon other topics in radioactivity worth noting. In itself ionium was of great experimental interest, for it possessed the quality that made polonium preparations valuable, namely, it was a pure alpha ray emitter, while it was even more desirable than polonium in having an effectively constant activity. These properties, for example, made a sample prepared by Boltwood in 1909–1910, while he was in Rutherford's laboratory, useful to James Chadwick in 1942, when he was engaged in war work.[4]

Ionium was also important in advancing knowledge of the uranium decay series. It was not the last member to be discovered—uranium X, for example, was later shown to consist of two different radioelements, while two different types of uranium itself (isotopes) were also shown to exist—but the growth of ionium in accordance with theory proved that no further products of long half-life lay between uranium and radium. Ionium figured significantly in the steps leading to the concept of isotopy, as well, once its distinctiveness from actinium and similarity to thorium were recognized.

Relative Activities of the Uranium Series

The alpha particle activities of specific radioelements in the different decay series, relative to the activity of the entire series or of other individual radioelements, seemed to be fairly well in hand by 1907. As described in chapters six and seven, experimental values were theoretically interpreted by breathtaking feats of number juggling. The agreement, however, appeared better than it was, for actinium's relationship to uranium remained unresolved, the thorium series was largely neglected, and the uranium-radium series had yet to make room for new alpha-emitters ionium and uranium II. Refinement of the activity measurements of constituents in a series could only highlight such remaining uncertainties.

In his 1904 text, *Radio-Activity*, Rutherford had assumed that all alpha particles from all sources contributed equal amounts to the total activity.[5] This view was modified when Bragg and Kleeman found the activity proportional to the alpha's range, and confirmed by the Boltwood work on relative activities discussed earlier. Another modification came from the realization that spontaneous loss of emanation from the mineral and interception of the alpha particles by the apparatus before they reached the end of their range had to be considered. On Boltwood's advice of this nature, McCoy and Ross revised their uranium mineral-uranium ratio, from 4.15 to 4.54, which was a bit closer to Boltwood's value of 5.3.[6] In April 1908, two years after this suggestion to McCoy, Boltwood lowered his own value to 4.69. Because he had found textbook procedures for the preparation of UO_2 and U_3O_8 to be inaccurate, as discussed in chapter eleven, he determined directly both the amount of uranium in minerals and its fraction of the total activity. The latter (4.69) compared well with the sum of the individually measured activities in the uranium series (4.64), which by this time included ionium, although the list still contained the actinium products (for lack of knowing where else to place them), and omitted the yet unknown uranium II.[7]

By 1920, when uranium II was known to exist but actinium's position remained uncertain, Boltwood and a student revised the ratio by a mere 1 percent, to 4.73.[8] Another early Boltwood value, the relative activity of the alpha-emitting products of radium to radium itself, 4.65, which he found in 1906, was revised by McCoy and his student, Edwin D. Leman, in 1915, to 4.11. Significantly this work confirmed the proportionality of ionization to the two-thirds power of the alpha's range instead of to the first power.[9]

Boltwood in 1908 also offered a new measurement of the radium-uranium activity ratio, which he found now to be 0.45.[10] This was a change from his earlier value of 0.52,[11] but the revision was insignificant beside the fact that radium's activity remained about half that of uranium's, while their alpha particles' ranges were about equal.[12] The most likely explanation was that "two distinct α ray changes may exist in ordinary uranium," though it was hesitantly advanced, for where would two uraniums fit into the decay series or, alternatively, how could one radioelement double its radiation output?[13] This difficulty apparently bothered him more in April 1908 than it had in January 1907, when he referred somewhat favorably to the report by Moore and Schlundt that uranium's daughter, uranium X, emitted alpha as well as beta particles.[14] As related in chapter seven, they had detected an alpha activity in uranium X, which heretofore was thought to emit only betas. Though the activity in question was actually due to ionium, which is isotopic to uranium X_1, and therefore was separated with it, at the time it was felt that the twelve atomic mass units between uranium and radium could be explained by alpha particles (mass = 4) from uranium, uranium X, and radium's parent.

Based on his publications of January 1907 and April 1908, Boltwood often was given credit for suggesting that uranium emits two alphas. In fact, priority goes to McCoy and Ross who presented the idea at the Chicago meeting of the American Physical Society, on 1 December 1906, and who felt obliged to claim their precedence some six years later. They too had looked at the relative activities of radium and uranium and concluded that either "each atom of uranium produces, upon disintegration, two α particles, or that there is an α ray product, not yet isolated, between U and UX."[15] In the spring and summer of 1908 Boltwood tried hard to split uranium into two constituents. But his "prolonged and elaborate chemical treatment" failed to change the activity even 1 percent, "and if there are really two substances present," he wrote Rutherford, "they are more intimately related (chemically) than are any two of the rare earths which have been separated from one another."[16] Possibly at Boltwood's request, for he used materials furnished by the Yale chemist, Hans Geiger then applied the relatively new technique of counting scintillations produced by alpha particles bombarding a zinc sulphide screen to examine the radiation from uranium. His early results seemed to indicate two uranium scintillations for each of radium.[17] He soon was joined by Rutherford in the counting, in which both had become expert, and the results were verified.[18] The two ranges, however, were difficult to determine due to the weakness of the thin films used as sources. Not until 1912 did Geiger and Nuttall succeed in measuring these values, 2.5 and 2.9 centimeters.[19] Other work in Rutherford's laboratory had shown that the scintillations from uranium did not appear as doubles, and this lack of simultaneity meant that there were two successive components.[20] That uranium was complex and its components chemically inseparable was now clearly recognized. Indeed, the situation was compared to the chemical similarity of thorium and ionium, and radium and meso-thorium.[21] Just where the two uraniums, UI and UII, would fit into the decay series would soon be resolved by the group displacement laws.

The work just described was stimulated by Boltwood's measurement of the radium-uranium activity ratio at equilibrium as 0.45. This figure was not immediately challenged, but would be in time. At the end of 1907 McCoy and Ross had revised the value for the activity of a gram of uranium element relative to that of a square centimeter of uranium oxide (U_3O_8), from 791 to 796.[22] This became known as the "McCoy number," and thick films of oxide prepared according to his directions were a common standard of radioactivity for smaller activities, being listed in the *International Critical Tables*. Using McCoy's number and other widely accepted values, Boltwood's friend at the Vienna Radium Institute, Stefan Meyer, and Fritz Paneth in 1912 made the following type of calculation:[23] The ionization current from one square centimeter of U_3O_8 is 5.79×10^{-13} ampere. The current from one gram of uranium, therefore, is $(796) (5.79 \times 10^{-13}$ ampere$) = 4.61 \times 10^{-10}$ ampere. One gram of radium gives a current of 8.06×10^{-4} ampere. But at equilibrium one gram of uranium exists with 3.2×10^{-7} gram of radium, which means the equilibrium amount of radium produces a current of $(8.06 \times 10^{-4}$ ampere$) (3.2 \times 10^{-7}) = 2.58 \times 10^{-10}$ ampere. The relative activity of radium to uranium, thus, should be $(2.58 \times 10^{-10})/(4.61 \times 10^{-10}) = 0.56$. Since this differed significantly from Boltwood's finding, they themselves tested the activity of a U_3O_8 film, measuring a current close to that which McCoy had obtained. Again using the McCoy number, they calculated Ra:U = 0.57.

Boltwood's reply was delayed eight years, both because he had drifted away from research activity and because the student with whom he ultimately published served in the Canadian forces during World War I and consequently put off converting his dissertation into a journal article. Just as Boltwood earlier had assigned to Ellen Gleditsch continuation of one of his old problems, presumably he did the same here to J. H. L. Johnstone. Together they reported new experiments which yielded Ra:U = 0.49, but the reply must surely have been penned by Boltwood. Its critical tone, in fact, is reminicent of his earlier comment about Soddy, for he wrote "Very little, if any, weight can . . . be attached to [Meyer and Paneth's] determination of the value of the ratio."[24] He did not even deign to elaborate, beyond commenting that the Viennese method used to purify the uranium salt was totally inadequate. As for their own experiments, they claimed 0.49 was the best value, more accurate certainly than Meyer and Paneth's 0.57, and better also than the ratio of the two-thirds power of the alpha particles' ranges, which too gave 0.57—another approach which heretofore had been accepted even by Boltwood as valid and to which Meyer and Paneth pointed in confirmation of their own work. The Yale chemists in turn suggested that the measured ranges might be in error or, more likely, that a series of decay products, such as actinium and its descendants, splits off from the main uranium series. The problem was that whether the proposed branching occurred at uranium I, uranium II, or radium, the amount of material passing into each branch required for the radium-uranium ratio conflicted with relative activity measurements made on the actinium family. Yet another contradiction of the branching at radium lay in the agreement for that element's half-life by Gleditsch and Rutherford and Geiger. The latter two had based their value on alpha emission by radium C in equilibrium with radium, whereas Gleditsch depended on the growth of radium from ionium. Were there a branch at radium, the two methods, involving radium's parent and descendent, would not have agreed. Moreover, assuming that each series leading off from the branch began with an alpha decay, as proposed for example by Soddy and John Cranston in England,[25] would not the first products be identical?

When Johnstone and Boltwood wrote their paper in 1920 these questions were unanswerable. This seems, in fact, to have been another controversy that died without complete resolution, for subsequently Boltwood was busy designing laboratories and serving in administrative capacities, while Meyer tried hard to hold his Radium Institute together in the postwar chaos. The existence and location of branching and actinium research, clearly, were areas that remained alive past 1920.

Actinium

Actinium, discovered by the Frenchman André Debierne in 1899, had long been a puzzle. It was always found associated with uranium in constant ratio to it, and therefore considered one of its decay products, but where it fit into the decay series could not be determined. Because the radioelements *following* radium were better understood at an early date, Rutherford, for example, in his 1904 Bakerian Lecture suggested actinium's position was somewhere in that unexplored region between uranium and radium.[26] Boltwood, whose experiments established the constancy of the actinium-uranium ratio, as he had done earlier for radium and uranium, inclined to agree with Rutherford, although he was disturbed by the unexpectedly small amounts of actinium found.[27] In an effort to resolve this problem, and because actinium was seen to have several decay products of its own, together they advanced the first suggestion in 1905 that "actinium is not a direct product of uranium in the same sense as is radium"—it forms a separate, side chain.[28] This was a hesitant step toward the concept of branching, i.e., one radioelement serving as the parent of two different daughter products. For a few years little more was heard of the subject, largely because Boltwood seemed to be placing actinium back in the main line of the uranium series, as the parent of radium. However, when Rutherford in the summer of 1907 disproved the actinium connection and Boltwood subsequently isolated ionium[29] interest in branching was renewed.

In a detailed paper on the activity of uranium, published at the end of that year, McCoy and Ross explicitly discussed branching.[30] The topography of the uranium series prior to radium was still largely uncharted. Max Levin, a former student of Rutherford's then in Germany, had tried unsuccessfully to separate from uranium any radioelement other than uranium (and actinium, radium, and their products),[31] yet the work of McCoy and Ross had suggested the probability of an alpha-emitting product between uranium and uranium X, though possibly it would be inseparable from uranium. Given this uncertainty of sequence and completeness, they arbitrarily placed the branch at uranium X since there was some evidence that that product decayed by both alpha and beta emissions while uranium was believed to emit only alphas.

$$U \xrightarrow{\alpha} U' \text{ (hypothesized)} \xrightarrow{\alpha} UX \underset{\alpha}{\overset{\beta}{\lessgtr}} \begin{matrix} (?) \to Ra \to \text{etc.} \\ Ac \to \text{etc.} \end{matrix}$$

Through the emission of an alpha particle uranium X was supposed to produce actinium and thence its products; a much faster beta decay led to an inactive product (Boltwood's recent discovery of ionium was not incorporated here) which in turn produced radium. But even more arbitrary than the location of the branch seems to be their assignation of the products of each type of decay. Though in the same paper they invoked the necessity for three alpha transformations prior to radium, in order to satisfy atomic weight differences, only two were incorporated into their scheme. The alpha from uranium X did not lead to the radium family. Probably they were led into this corner because uranium X's beta activity was known to be far greater than any alpha activity it might possess, and based on relative activities actinium had to be placed in the weaker branch. The simple alternative expedient (in fact true) of having radium's immediate parent emit alphas was precluded by calling it inactive. Yet, despite these conceptual difficulties and the lack of real evidence for their suppositions, the possibility of branching was rooting firmly in the soil of this field of science. This was not idle activity, for without such cultivation there would likely have been more resistance to such ideas, ideas which after all called for a departure from the ingrained scientific belief in nature's consistent behavior. In branching a radioelement apparently had a *choice* of its mode of decay.

But with such weak activity as actinium and its products exhibited, how could one untangle their puzzle? During his unsuccessful efforts in the summer of 1908 to split uranium into two radioactive constituents, Boltwood had worked up about a kilogram of what he believed was the purest uranium salt (nitrate) ever prepared. He carefully sealed it in a bottle, free from radioactive contamination, but vented his frustration to Rutherford about how to proceed.

> I am considering the possibility of excavating a sepulchre and publicly entombing this uranium with the hope that some scientist of future generation may examine it and solve the mystery of the birth of actinium. I do not see any direct method of tackling this problem, but I should like to get my hands on a moderate amount of a good, strong preparation of actinium and see if I can't get it to grow radium or something. I hate that notion of an illegitimate branch of the uranium family like the very devil.[32]

Would it be possible, he wrote to his friend in Manchester, for Rutherford to use his influence to obtain some very active actinium for him from the Austrian Radium Commission? Less than a year later Boltwood had his actinium—in Rutherford's laboratory, where he visited in 1909–1910. Actually, he had at his disposal certain "actinium residues," separated from five hundred kilograms of pitchblende residues which the Royal Society of London had purchased from the Vienna Academy of Sciences and then had treated for radium at the Nogent-sur-Marne works of the French industrialist Armet de Lisle. Several chemists, including Boltwood, had noted that the precipitation of actinium by ammonia was often incomplete. He therefore developed a new technique, involving fractional recrystallization, boiling in acids, and further separation by precipitation, that yielded a small but highly active product. However, here he stopped, his time in Manchester at an end, and he did not return again to the subject when back in the States.[33]

Like Boltwood, Soddy in Glasgow was interested in actinium, not the least because of its one-time supposed parentage of radium. But Soddy, far more than Boltwood, was a serious and accomplished theorist, as well as a fine experimental chemist. In a 1909 paper entitled "Multiple atomic disintegration," he reasoned that there was nothing known that would disallow the possibility that a radioelement might decay by two or more modes, each proceeding simultaneously and independently.[34] The law of probability followed in disintegration by an atom would apply whether that atom decayed in only one way or by multiple modes. The decay constant λ would be the sum of the separate constants. But since these separate constants could not be directly determined (only the overall period being detectable), proof of the number of modes and their relative importance would have to be established indirectly, as by the ratio of the amounts of the final, inactive products of each chain. A case in point would be the end products of the radium and actinium series, assuming that these series did not somewhere converge.

Going a step further, Soddy took from Boltwood's relative activity data for uranium minerals the values of radium = 0.45 and actinium plus products = 0.28, argued that the four or five alpha particles emitted from the latter group gave actinium alone a value of about 0.06, and concluded that radium is roughly seven times as active as actinium. If the alpha particles, as a first approximation, are considered to impart equal ionization, then of every eight atoms of uranium seven go to form radium and one to produce actinium.[35] Branching now was beginning to undergo numerical analysis.

In 1911, Kasimir Fajans and G. N. Antonoff, both chemists visiting in Rutherford's laboratory from abroad, and Ernest Marsden and T. Barratt, at Manchester and East London College, respectively, all developed experimental evidence of branching. Fajans examined radium C, already known for its complexity, and detected a product called radium C_2 (now C''),[36] Marsden and Barratt found an analogous thorium C_2 (now C'),[37] while Antonoff isolated uranium Y, a product of uranium chemically "very similar" to uranium X, but having different radioactive properties.[38] Since uranium Y existed in only a few percent of the equilibrium amount expected if it were in the main line of uranium descent, it was looked upon as a branch product of uranium. Even more, because actinium's origin had long been felt to require branching, it was clearly possible that uranium Y was somehow associated with it.

Marsden and a former Yale student of Bumstead's then spending a year in Manchester, Perry B. Perkins, in 1914 found a branch at actinium C analogous to those at radium C and thorium C,[39] while McCoy and Leman filled out other actinium data by comparing alpha particle ranges with relative activities of the members of this series.[40] The close agreement between experiment and theory established more strongly the usefulness of the two-thirds power of the range calculation, as well as that of another fundamental assumption, that each exploding atom emits at most one alpha particle. Several Americans, including Perkins, McCoy, and Alois Kovarik, contributed increasingly accurate measurements of the half-lives of actinium series members,[41] although actinium's own period remained uncertain until after World War I. The only early evidence bearing on the matter was presented in 1911 by Marie Curie, who noted an erratic decrease in the activity of an old sample, which suggested the surprisingly short half-life of about twenty years, a value which proved quite accurate (now determined as 21.8 years).[42] However, for the most important question concerning actinium, namely, the origin of its decay series, further insights now awaited the ideas of group displacement and isotopy.

Thorium

There was in the thorium family a little less uncertainty than in the actinium group, and a lot less interest. Despite a suggestion tossed off by Soddy in 1909, and disregarded by skeptics, that, like actinium, thorium might be a branch product of uranium through that atom's multiple disintegration,[43] its genetic relationships seemed fairly clear. Even the question of its intrinsic activity generated little further concern; the simultaneous publication in 1906 by Boltwood, Dadourian, and McCoy and Ross convinced most that thorium was indeed active in its own right.[44] The case was not similar to that of actinium, whose beta rays were so weak that they were long undetected, causing the radioelement to be classified as undergoing a rayless transformation, and, hence, by one definition, not radioactive. There *was* activity in thorium preparations, and though some felt that it might belong entirely to the radiothorium from which thorium could not be separated, this view was not dominant. Regardless whether the definition of "radioactive" was taken to be "emitting ionizing radiation" or "transforming into a daughter product", thorium was accepted as a radioelement.

What, then, remained uncertain about it? More precise information was desired about thorium's half-life, end product, and, as with the other series, about any unknown radioelements, the relative activities of all constituents, and possible branching locations.

Such precision took a step forward with the publication in 1909 of G. C. Ashman's work on thorium's activity.[45] His professor at Chicago, McCoy, earlier had urged the view that thorium was intrinsically radioactive, and with Ross had estimated the limits for the specific activity of thorium dioxide as 100 and 130 (out of a total activity of 1009 for thorium and all products in a mineral, the unit being the activity of one square centimeter of a thick film of U_3O_8).[46] Otto Hahn, too, when back in Germany, had shown that samples rich in radiothorium possessed a greater ratio of maximum to minimum activity, indicating that thorium must produce *some* alpha ionization.[47] The problem in obtaining quantitative information was one of separating pure radioelements. To remove from thorium and radiothorium the thorium X, emanation, and thorium A, B, and C, with which it occurs, the standard technique of precipitation with ammonia was unsatisfactory. Because the separation was not complete in a single precipitation, over a dozen were required. This took several days during which activity measurements were difficult, and involved the introduction of large quantities of silica and other impurities from the reagents and glass vessels. Precipitation of thorium by fumaric acid, as devised by Schlundt and Moore in 1905, had the advantage of faster separation and easier activity readings, but required a rather expensive reagent and posed some other chemical problems. Since McCoy and Ross had earlier used meta-nitrobenzoic acid successfully in their work with thorium, Ashman now tried it specifically as the reagent for this separation. With but a few precipitations, requiring only several hours time, he obtained thorium dioxide (including radiothorium) whose purity was attested to by its immediate increase in activity. Had some other radioelements remained in the material the activity would have decreased for a while as they decayed but were not being replenished by equilibrium amounts of their parents.

Using the same electroscope employed by McCoy and Ross, Ashman measured the activity of several films made with the oxide, over the course of several weeks until they reached a maximum. Extrapolation of the plot of these data gave him a minimum activity at time zero. By also including samples that McCoy and Ross had earlier precipitated one hundred times (with ammonia) and forty times (with hydrogen peroxide), which meant that radiothorium's parent, mesothorium, had been removed, Ashman was able to measure the activity of thorium known to be poor in radiothorium. This variety gave him four equations with four unknowns, and permitted him to solve for the intrinsic activity of thorium. His value was 104 for thorium dioxide and 119 for thorium itself, or over 11 percent of the total equilibrium activity of the thorium series. Radiothorium contributed 20 percent, while thorium X and subsequent products furnished the other 69 percent. Thorium's activity clearly was too large to be due to ionium or actinium from any uranium that might have been associated with the thorium mineral.

In 1913, McCoy and Viol, the latter by this time having received his doctoral degree and been appointed an assistant in chemistry at Chicago, further codified knowledge of thorium.[48] Not long before Soddy had published his monograph entitled *The Chemistry of the Radio-Elements*.[49] While comprehensive, it was essentially a review of the existing literature. McCoy and Viol went further and both more systematically and personally confirmed all the information they reported. Their paper was submitted before the group displacement laws and concept of isotopy were conceived by Fajans and Soddy, so even though they (and Soddy earlier) "escalated" their comparison of mesothorium I and thorium X, and radiothorium and thorium, from "chemically similar" to "chemically identical" they could not yet call them varieties of the same elements. Nor could they determine any chemical properties for the short-lived thorium A (half-life 0.145 seconds; now accepted as 0.16 seconds), which, like an analogous actinium A, had been discovered recently in Rutherford's laboratory,[50] nor for thorium D (now C″, with half-life of 3.1 minutes), discovered by Otto Hahn and Lise Meitner.[51] What they did do was give an extensive survey of existing wet chemical knowledge, all the more useful for the quantitative information included.

Then they went beyond this "analytical chemistry data sheet" to apply it. These chemical separations were a means to an end, the end being an understanding of the thorium series as a whole. As before, relative activity studies offered the insight. In the 1905 edition of his book *Radio-Activity,* Rutherford showed that the activity of thorium B + C + D (effectively that of C since the others are beta emitters) was 0.44 times the activity of thorium X + emanation + (then unknown) A.[52] May Leslie, in Marie Curie's laboratory, in 1911 expanded on Ashman's work, described above, and found the activity ratio of thorium B + C + D to that of X + Em + A to be between 0.33 and 0.42.[53] McCoy and Viol, recognizing that it was of crucial importance for the thorium X they started with to be initially free from all other radioelements and feeling that they had the chemical techniques now to do it, decided to remeasure this ratio. Since mesothorium I would normally follow thorium X in any separation, they first removed mesothorium from thorium and radiothorium, allowed radiothorium to grow anew in the mesothorium, and then separated the two. Finally, the radiothorium was used as a source for thorium X, the former being precipitated with aluminum hydroxide by means of ammonia. Several precipitations with mercuric sulphide freed the thorium X from the active deposit members B, C, and D. Activity measurements taken in an electroscope over the course of several days, when inserted into the radioactivity decay equations, yielded a ratio of 0.427, in very good agreement with the earlier Rutherford-Soddy value. The calculation's accuracy hinged also upon how well the periods were known of the radioelements, but McCoy and Viol had by that time values very close to those accepted today.

By a similar separations process, McCoy and Viol prepared pure radiothorium whose activity they measured over the course of several months, as its decay products grew in equilibrium amounts. From this they calculated a ratio of the activity of radiothorium products to radiothorium itself of 5.23. As in the preceding work, the ratio's accuracy depended upon the values of periods used; that for radiothorium was taken as 2.02 years (now 1.91 years), which introduced a slight error. They had not carried out experiments long enough to determine this half-life themselves, but relied upon the value given by Gian Alberto Blanc in Rome.[54]

Aware that so much depended on knowledge of the periods of thorium's products, they tested several of them after preparing pure samples. Thorium X, whose half-life was measured during the experiments described above, was found to decay to half its initial activity in 3.64 days, in perfect agreement with one of the earliest determinations by Friedrich von Lerch in 1905.[55] A second mesothorium, called mesothorium II, found by Hahn in 1908,[56] yielded a half-life of 6.13 hours, a slight reduction from the German's value. Its beta ray activity was measured after filtering off any alpha radiation from decay products by a thin sheet of aluminum. Thorium B, with a 10.6 hour period which confirmed von Lerch's earlier finding,[57] and thorium C, whose 60.8 minute half-life was higher than von Lerch's 60.4 minutes,[58] were collected upon a charged platinum plate exposed to thorium emanation.

About the same time as McCoy pursued this work with Viol he was calculating periods for thorium itself, and uranium as well.[59] Earlier methods had depended on counting scintillations, using the decay series relationship $N_1\lambda_1 = N_2\lambda_2$, where relative amounts of two series members and period of one were known, and the less accurate Geiger-Nuttall relationship of ranges and decay constants. This work, largely by Rutherford and his circle, had yielded half-lives for uranium somewhere between 4.8×10^9 and 7×10^9 years, and for thorium between 1.31×10^{10} and 3×10^{10} years. McCoy employed a variation of the other procedures, calculating with alpha particle ranges and with ionization currents per unit weight, derived from his own experiments on specific activities of thick films, and reported periods for uranium of 5.04×10^9 years and for thorium of 1.78×10^{10} years (now accepted as 4.51×10^9 years and 1.39×10^{10} years, respectively). In themselves they were not overly accurate, due essentially to the greater complexity of his technique, but they did support the contemporary inclination to place far greater confidence in the lower value of each spread of periods. There was no doubt, also, that thorium was the longest-lived radioelement.

Illustrating still further the incredible interdependency of one calculation upon another, McCoy's half-life determinations relied on the validity of the relationship between alpha ionization and ranges. With Viol, his work on relative activities furnished strong evidence that the ionization produced by an alpha particle was indeed proportional to the two-thirds power of its range.[60] This in turn depended for completeness upon the finding that 65 percent of thorium C branches into thorium C_2 and the remainder into thorium D, far different from the other known branches where the lion's share followed one of the trails.[61] It depended also upon McCoy and Viol's redetermination of thorium X's range as 4.1 centimeters instead of Hahn's value of 5.7 centimeters.[62] With this readjustment, activity-range calculations corresponded with measurements and left them feeling that "the uncertainty . . . in regard to the thorium series has at last been removed."[63] No other alpha emitter lay hidden in the thorium series and each exploding atom was felt to expel only one alpha—except thorium C, where alpha activity had been detected in each line of the branch. Here too the group displacement laws soon would iron out the last errors of interpretation, showing that thorium C decayed not by two alphas but by both alpha and beta emission, the daughter of the latter process being the then-unknown alpha source. McCoy and his students contributed still further to a quantitative understanding of thorium and its products during the century's second decade, but he and Viol were essentially correct in 1913: the uncertainty had been removed.[64]

In just one area was there major continued puzzlement. What was the final disintegration product of the thorium series? Boltwood's passing suggestion that thorium was the mother of the rare-earth elements was not seriously considered. Lead was widely accepted as the end of the uranium chain, but discounted for that of thorium. Since the ratio of lead to uranium in many minerals seemed unaffected by the presence of thorium, it was logical to believe thorium produced no lead. On the other hand, six alpha emissions meant the end product was about twenty-four mass units below thorium (232.5), or 208.5. Bismuth, of atomic weight 208.0, was the closest element, but mineralogical analyses showed it to be lacking in sufficient quantity. The picture was further confused by the possibility that each of the lines from the branch at thorium C gave a different end product.[65] Again, it was the group displacement laws that brought some order to this chaos, although in this case the order was not complete.

Clearly, this chemical work was a precision specialty, plagued by numerous pitfalls. Many ores of the same element were involved, each requiring unique processing because of its different constituents. Usually associated with the radioelements were the rare-earth elements, whose separations were notoriously difficult. Part of the problem was that separations often were not complete, and repeated precipitations or crystallizations were required to obtain a pure substance. Confusing the matter further was the circumstance that sometimes a substance would precipitate in a given reagent, but redissolve in an excess of that reagent. With radioelements continually growing and decaying, their quantities could easily change during the course of a lengthy chemical operation, and the purity of a substance could be affected. And in addition to their transitory nature, the quantities of radioelements commonly handled were exceedingly small. No wonder that British Association Section B (Chemistry) president, A. Smithells, in 1907 called this subject the "chemistry of phantoms."[66]

Phantoms they may have been to some, but real, elementary forms of matter, with individual characteristics, they were to those whose work has been described in these last two chapters. Exploration of the personalities of, and relationships between, the few dozen radioelements discovered, and fitting new pieces of information into the existing outlines of knowledge were their tasks. The simple problems had been solved long before; mature, incremental understanding was more the pattern now. Yet, one last great contribution was near to hand. The group displacement laws would serve to organize and explain much of the disparate data, providing an understanding of radioactivity much as the transformation theory had done a decade earlier.

13 Toward the Group Displacement Laws

The significance of radium has already been discussed in some detail, yet one other characteristic deserves emphasis. By the turn of the century the accepted criteria for calling a substance a new element were its isolation, its unique spectrum, and its measured atomic weight. Uranium and thorium had been regarded as elements long before their radioactivity was discovered. Proof of radium's elemental nature was accepted when Marie Curie determined its atomic weight in 1902 and this weight fit reasonably in the vacant alkaline earth box of the periodic table, below barium. But during the next decade and more, not a single other radioactive substance fulfilled the criteria for designation as a chemical element.

The problem, to a certain degree, was that spectroscopic evidence was not always clear cut. As we recognize today, an element may yield somewhat different spectra when taken from the neutral atom and from the ion. A greater difficulty, however, was that of isolating the substance in sufficient quantity for atomic weight determinations. Virtually all the radioelements found had half-lives shorter than that of radium, which meant that their equilibrium amounts were correspondingly smaller. This meant that evidence of chemical behavior from the classical reactions of chemistry was meager, leaving the placement of a radioelement in the proper column of the periodic table to indirect means. Thus, atomic weights were left to arithmetical calculation (by the number of alpha disintegrations between the unknown and uranium, radium, or thorium), rather than chemical experimentation. Moreover, the diminishing activity (and therefore quantity) of most substances separated raised the fundamental question whether they could be real elements. Were not the atoms of elements *permanent* bits of matter?

The Curies were sufficiently disturbed by the loss in time of polonium's activity to effectively abandon claiming it as an element. Only when Willy Marckwald seemed to be appropriating it for himself under the name "radiotellurium" did Marie hasten to its defense.[1] By 1913 there were over thirty radioelements known, but no more than twelve boxes in the periodic table in which to house them. The general consensus among radiochemists seems to have been that the objects of their study, possessing precise chemical properties, were certifiable elements. How to arrange them in the periodic table, however, was a monumental puzzle. Could radioelements with different physical properties (e.g., half-lives and radiation emitted) be placed in the same box? If so, did this mean that all the known elements were mixtures of bodies different in some ways? Did the radioelements belong in a separate region of the table, as did the rare earths? If so, how could their remarkable chemical similarity to some stable elements be explained? These questions were resolved in the masterful group displacement laws and concept of isotopy of 1913, which, along with the transformation theory, were the grand, unifying principles of radioactivity. The American contributions to this problem were not pervasive, but because Boltwood and McCoy were among the earliest of the radiochemists their work almost inevitably raised the issues and pointed in the direction of an explanation. Subsequent advances were virtually all made abroad, except that final confirmation of the new ideas relied heavily on the precision of the Harvard atomic weight expert, Theodore W. Richards.

Chemical Similarity

When the Curies discovered radium they were able to isolate it by following the separations techniques for barium, and then by using the method of fractional crystallization of the compounds of the two elements. The physicochemical methods were straightforward; it was the minute quantity involved that was relatively unique. In its chemistry, at least, radium behaved like any other respectable element; the trace amounts, however, raised fears that, in certain processes, for example precipitation, the element might be entrained or adsorbed, rather than chemically combined. Of the other radioelements encountered early in the history of the subject, uranium and thorium were not chemically controversial, while polonium's chemistry was wrapped in opposing statements, and few chemists seemed to care about actinium.

Perhaps the first radioelement found whose chemical similarity to another appeared even closer than the type of relationship between radium and barium was radiothorium in its behavior with thorium. Following Hahn's discovery in 1905,[2] Boltwood, Dadourian, and McCoy and Ross the next year showed by specific activity measurements that radiothorium must be a disintegration product of thorium. Despite this genetic connection, however, they observed that commercial salts of thorium were markedly depleted in radiothorium.[3] Somehow the industrial process effected a separation which they were unable to repeat in their laboratories. Hahn solved the mystery by postulating and then detecting a product between thorium and radiothorium, which he named mesothorium.[4] In the commercial treatment, the mesothorium was removed from the thorium and radiothorium, which remained together. Since the radiothorium then was without its parent in the latter fraction, it was not being replenished there and its activity decreased. It was, however, regenerated in the mesothorium fraction. Such an interpretation joined the concept of a decay series with the idea that some radioelements within that series exhibited similar chemical behavior. Before long chemical similarity was recognized also between radioelements from different series, such as radium, mesothorium, and thorium X.[5]

But it is one thing to recognize that isolation of different radioelements requires the same chemical procedures, or that certain radioelements in the same series can not be separated unless there exists a separable intermediate product. It is quite another level of sophistication to test deliberately the nature of this inseparability. McCoy and Ross in 1907 reported upon several different treatments to which they subjected samples of thorium nitrate, precipitating the thorium repeatedly with such compounds as ammonia, oxalic acid, sodium thiosulphate, and potassium chromate, but even after one hundred precipitations no radiothorium could be said to have separated. As already quoted in chapter eight, they concluded that "the direct separation of radiothorium from thorium by chemical processes is remarkably difficult, if not impossible."[6]

Soddy, with clearer vision than most, and writing annual reviews of radioactivity for the Chemical Society, emphasized that the method of preparing the thorium salt was really of minor importance; a variety of techniques would work satisfactorily. The significant factor in isolating daughter products was the age, the amount of time allowed for regeneration and decay.[7] This became apparent after Antonoff in 1911 claimed his discovery and separation of uranium Y (seen to be similar to uranium X) was due to his chemical process, while it actually depended upon sufficient time elapsing between successive operations.[8]

There were other cases of chemical similarity that, increasingly, became recognized as a not uncommon phenomenon in radioactivity.

Boltwood's ionium "was separated with thorium [i.e., used as a carrier], with which it remained very persistently,"[9] so persistently that when he returned some thorium to the Welsbach Company, whose chemist frequently furnished him with samples, he sent back "alas also the ionium."[10] Furthermore, if the mineral with which Boltwood worked happened to contain both thorium and uranium, then uranium X also was inseparable from the thorium and ionium.[11]

Bruno Keetman, a student of Marckwald's in Berlin, confirmed in 1909 the closeness of ionium and thorium when neither precipitation, nor crystallization, nor sublimation could alter their relative amounts in a sample. His work was notable in that his material contained about 1 percent of ionium. Previous separations attempts had been made wherein one of the "similar" radioelements was present in less than weighable quantity, allowing the criticism that the trace amounts were merely adsorbed by their similar counterparts. After Keetman's exhaustive experience, this chemical closeness had to be regarded as something fundamental.[12]

His work emphasized the incompleteness of the known chemical reactions and therefore the difficulty of devising separations techniques: standard precipitating reagents for thorium, such as sodium thiosulfate and oxalic acid in strongly acid solution, were not at all quantitative for small amounts of thorium. Obviously, thorium as a carrier could have been added, to bring down all the thorium and ionium in solution, but sources rich in ionium could not be prepared in this manner. Why, he wondered, do these elements exhibit such strong chemical similarity? Traditional wisdom asserted that elements were defined by their atomic weights; after all, the successful periodic table was so ordered. By Keetman's calculation, both uranium X and ionium had atomic weights of 230.7, quite close to thorium's measured 232.5.[13] To him this indicated that they probably all belonged in the same box of the periodic table, a striking anticipation of the concept of isotopy.[14] Yet, significant advance that this was, it must be recognized that progress is generally a process of accretion, rather than continual bolts out of the blue. Keetman's training and thinking were in the group of rare earths, where he had recognized the great difficulty of separating such elements as neodymium and praseodymium, whose atomic weights are close. Indeed, most of the chemists who pursued radioactivity were analytical chemists, nurtured in this baffling rare earth group. They were both helped and hindered in their radioactivity thinking by their analogies between the two species of elements.

As the century's first decade closed, lead and radiolead[15] and uranium I and II[16] could be added to the list of paired radioelements. Otto Hahn found that his mesothorium consisted of two products,[17] though he refrained from announcing their chemical identification in deference to the commercial interests of his patron, Knöfler and Company. Marckwald had no such reservations, and in 1910 publicly identified mesothorium I as similar to radium.[18] Indeed, this substance soon was marketed as "German radium." Within a short time Soddy made the same discovery, though he carried its implications further.[19]

Thrown away now was the equivocal word "similarity"; in its place Soddy used chemical "identity." After submitting a mixture of radium and mesothorium I to extensive fractional crystallization, without altering the relative proportions whatever, he overcame his natural reluctance to this step.[20] According to his insight, bodies with different atomic weights, different half-lives, and different radiations could yet be completely chemically identical. Nor was this possibility limited to the radioelements. Perhaps *all* the elements were mixtures of chemically inseparable varieties, a proposal that could explain measured departures from whole number atomic weights. In all but name, isotopes had been described.

Additional Keys to the Puzzle

It may be tautological to say that great unifying principles absorb and explain considerable bodies of evidence, yet it is worth emphasizing that the group displacement laws and concept of isotopy were based on a wide range of investigations. Consider the following variety, which added information and force to the results of conventional wet chemistry.

When Rutherford first detected the mysterious thorium product that was affected by drafts of air, he called it emanation in a deliberate attempt to avoid the more precise words gas or vapor. Radium too produced a corresponding substance. Soon, by liquefying emanation at low temperature, with Soddy, and by estimating its molecular weight, with Harriet Brooks, he convinced himself that it was not only a gas, but—due to Soddy's inability to make it combine chemically—a member of the novel family of inert gases. They investigated the weight in 1901 by comparing its rate of diffusion down a long cylinder with those for carbon dioxide and ether, and found it to be between 40 and 100.[21] In the following several years other approximations were reported from a few European laboratories and by Bumstead and Wheeler at Yale, who compared the diffusion rates of emanation and carbon dioxide through a porous plate. Their value for emanation's molecular weight was near 180.[22]

Since formidable amounts of the gas would be necessary to tip a balance, diffusion studies, though indirect, were more practicable. Such studies were more congenial to physicists, even though the goal was to place emanation in the chemists' periodic table. In this tradition, and probably as his doctoral dissertation research under Bumstead (with considerable assistance from Boltwood), Perry B. Perkins (b. 1877) in 1908 made further measurements on radium emanation. (Of course, the other emanations had half-lives too short for diffusion.) In view of the fact that its immediate parent, radium, at that time had a measured atomic weight of 226.5, and was separated from emanation by only an alpha particle, emanation's weight must be near this value. Since this gas combined with no other elements, it probably was monatomic. Previous experiments had used comparison gases that were both greatly different from emanation in weight and of complex molecular structure, conditions that made the calculations (Graham's Law) unreliable. Perkins therefore chose mercury vapor for his comparisons, because it is monatomic and has a molecular (or atomic) weight of 200. Though the experiments had to be conducted at high temperatures, to vaporize the mercury, his results yielded a molecular weight of 235, higher than that of radium and thereby viewed as suffering from experimental error, but definitely in the right region of the periodic table.[23] From this time onward emanation seems to have been regarded as firmly placed among the elements, little thought being given until 1913 that the emanations of radium, thorium, and actinium might have different atomic weights.

At the University of Upsala, D. Strömholm and The Svedberg applied concepts of isomorphism to the radioelements. Since bodies of similar chemical composition were regarded as having the same form, the chemical nature of an unknown substance could be ascertained by observing the form in which it crystallized. In 1909 they reported that salts of thorium X crystallize with lead and barium salts, and no others. Earlier studies of its diffusion coefficient and ionic mobility had characterized thorium X as monovalent, but the two Swedish chemists correctly concluded that it is an alkaline-earth element. They similarly called actinium X an alkaline earth, and could find no isomorphic differences between thorium X, actinium X, and radium. More important than this experimental evidence, however, were their theoretical insights. They specifically tried to fit these few radioelements into the periodic table, and called attention not only to their chemical nonseparability but to the three parallel and, except for actinium's possible connection to uranium, independent radioactive series. These series bore a number of similarities, each, for example, having an emanation followed by a group of active-deposit radioelements. As more was learned about the chemical nature of the radioelements (such as thorium X and actinium X), many corresponding products in the different series were seen to appear chemically identical. Strömholm and Svedberg wondered in print whether the three genetic series extended down the periodic table among the stable elements, each of whose atomic weight would be an average of its components.[24] Their work, which helped Soddy to gather the courage to use the word "identical" in describing some radioelements, was done at about the same time as Keetman's. Both they and Keetman saw the need to place more than one substance in a given box of the periodic table, a concept later formalized with the term "isotope," but the Swedes seem to have gone further in beginning to think about the transitions between the boxes.

Again, such thoughts did not originate in thin air; to an extent this advance had been foreshadowed by earlier work. The disintegration theory of radioactivity, showing genetic connections between series members, of course, was a major foundation. More specifically, however, Friedrich von Lerch, in Vienna, postulated electrochemical relationships between the radioelements. As early as 1903, at the time George Pegram in New York was electrolyzing thorium solutions, Lerch too separated some components of thorium active deposit. By dipping plates of different metals into the solution, various activities would be electrochemically deposited upon the plates, depending in each instance upon the type of metal. For example, a nickel plate dipped into a solution of thorium's active deposit yielded a coating of thorium C in a pure state. The C product could also be deposited on the cathode by electrolysis in an acid solution.[25] These were not only useful means of isolating products to determine their half-lives and radiations, but, because many known metals were arranged in a hierarchy of electrochemical potentials, allowed some radioelements to be similarly ranked.

Von Lerch, and independently Richard Lucas, in Leipzig, saw that radium C was more easily deposited electrochemically than radium B, radium E more readily than radium D, and radium F more than radium E. They generalized these relationships into a "rule" that the product of radioactive decay is electrochemically nobler than its parent.[26] But as more radioelements were tested, discrepancies appeared. Lerch himself, with E. von Wartburg, in 1909 showed that thorium C was more easily deposited than thorium D, instead of the reverse.[27] Georg von Hevesy, a Hungarian working in Rutherford's Manchester laboratory, in 1912 and 1913 published extensively on the electrochemical behavior of radioelements. The relationship was not experimentally simple, he pointed out, but depended upon the neutrality of the solution and the potential of the electrode. With care to these variables, Hevesy showed that the B members of the three series possessed the same electrochemical properties; similarly the C members.[28] He also indicated that radium A was more, not less, noble than radium B, though it was less noble than radium C.[29] The Lucas-Lerch rule, while a stimulus to thought about general relationships, was incorrect.

Hevesy went on to determine the valency of a number of radioelements. The customary procedures were stoichiometrical, an examination of the compounds formed by the element under consideration, and electrochemical, a comparison of the amount of current and the quantity of metal passing into, or deposited from solution in electrolysis. But with the microscopic amounts of radioelements, neither method was suitable. Hevesy, therefore, proved the applicability of a new approach. He calculated the number of charges borne by the negative ion, its valency, from its diffusion velocity in the presence of a large excess of the positive ion. As a check on this method, he also found the diffusion constant of a negative ion by measuring its mobility in an electric field. Along with values previously determined, he was thus able to list valencies for twenty-three radioelements, which, when combined with mobility data, offered a means of positioning radioelements in the periodic table. Hevesy, having largely laid to rest the Lucas-Lerch rule, was quick to note that in a number of alpha-ray transformations the valency of the daughter atom was two units smaller; for many other alpha decays information was lacking. Expulsion of a beta particle appeared to shift the valency in the opposite direction, though information about beta emitters was even more incomplete, and in the German publication of his paper he was forced to indicate valence changes of 0, +1, and +2. To Hevesy, these apparent rules, despite the remaining questions, were more than fortuitous. Valency change and particle emission must be connected.[30]

Hevesy's work was published just as a graduate student in Chicago was perfecting a novel technique to accomplish the same results. McCoy had given Leonard Loeb (1891–1978) the task of determining the valencies of radium A, B, C, and D, in order to position them in the periodic table. Yet the technique of studying the diffusion rates of ions in solution by placing a movable barrier at different points was unappealing to Loeb. His father, the eminent physiologist Jacques Loeb, had shown that agar gel was ideal for "immobilizing" an aqueous solution, while in no way restricting the migration of ions. Leonard cast cylinders of agar gel, placed the radium active deposit on a plate at the base of the cylinder, and then sliced the cylinder after a period of diffusion. He then evaporated the water and placed the residue in an electroscope for a prolonged count. Still, though he did it the "hard way," Hevesy had published first, and this single American effort during the "homestretch" to the group displacement laws came to a premature end.[31]

The spectroscope was yet another tool used to penetrate the mysteries of the radioelements, in this case the arc spectrum of ionium. During the 1909–1910 academic year that he had spent in Manchester, Boltwood had separated a preparation of thorium and ionium oxides from the Royal Society's residues. The proportion of ionium was undetermined, but considered to be at least 10 percent, given the measured alpha particle emission rate and taking ionium's period to be not less than 100,000 years. This estimate of the half-life was reasonable, assuming there were no other long-lived substances between uranium X and ionium; its magnitude was supported by the independent Geiger-Nuttall relationship. Rutherford, therefore, expected the preparation to yield a strong ionium spectrum as well as all the thorium lines.[32]

For this investigation he chose Alexander S. Russell, a brilliant radiochemist who had come from Soddy's Glasgow laboratory (the only person to work with both Rutherford and Soddy), and R. Rossi, a talented spectroscopist. Though it was well known that radium's strongest line could be detected amidst those of barium, with concentrations as low as a few parts per million, they could not find a single ionium line in the thorium spectrum. One possible conclusion was that there was indeed an undiscovered radioelement lying between uranium X and ionium, vitiating most calculations of the latter's period, and lowering the sample's estimated ionium content as well. Another explanation was that ionium simply had no arc spectrum in the region examined, yet this would have made it unique among all solids of high atomic weight. A final possibility was that thorium and ionium, already called identical chemically by Soddy, now in 1912 could be said to have identical spectra. Such elements would be distinguished only by their atomic weights and their radioactive properties. Russell and Rossi inclined toward the first of these three interpretations, perhaps because hypothesizing unknowns had a long and honorable tradition in radioactivity. But within months Russell himself, as well as others, would recognize the merit of the third view.[33]

Radioelements and the Periodic Table

In 1911, Frederick Soddy published a thin but important volume entitled *The Chemistry of the Radio-Elements*.[34] It was more an encyclopedia than a conventional textbook, for it listed all the radioelements and, in sequence, gave their known physical and chemical properties. The value of this exercise lay in the completeness of the information, in Soddy's analysis, and in the direction in which it pointed.

Even more than his predecessors, Soddy was struck by the analogies between the three decay series. The thorium and actinium lines were almost exactly alike in their succession of products, except that there was no counterpart in the actinium series for the two mesothoriums between thorium and radiothorium. The uranium series likewise contained an "extra" product, uranium X, between uranium and ionium (Soddy's chart showed just one uranium, although two were long considered to exist; their sequence with uranium X was uncertain), and had three more radioelements than the other series at its very end. But in both chemical nature and type of emission the matched properties were striking. Reading across Soddy's chart, ionium, radioactinium, and radiothorium all emitted alpha particles and were chemically identical to thorium, or in the case of radioactinium believed to be so. In the next line radium, actinium X, and thorium X likewise were alpha emitters and chemically identical. Chemically identical also, and alpha emitters, were the three emanations. Short-lived alpha emitters had recently been found immediately following both thorium and actinium emanations, and these corresponded to the known radium A. The second, or B, product in each series after the emanation was rayless, or more properly an undetectable beta emitter, and the longest-lived component of each series' active deposit. Subsequent analogies, while not so exact, were nevertheless compelling. The C members gave alpha particles and were complex, probably due to branching. The D members, other than radium D, were short lived and emitted betas and gammas. After these came a relatively or completely stable region, around bismuth or inactive lead. With more information about the nature of these later products, they too might fall precisely into a parallelism of properties.[35]

To Soddy these patterns suggested the "successive transit of matter from group to group of the Periodic Table." The table, in fact, was merely the consequence of the production of elements through disintegration. And even the apparently stable elements were suspected of being continuously transformed.[36]

There were more immediately applicable insights as well. Every product whose chemistry was known well enough to classify had even valency. Ionium and radiothorium were tetravalent. Upon emission of an alpha particle they transformed into radium and thorium X, respectively, both divalent. Further alpha emission brought them to the emanations, members of a nonvalent family. The transitions did not have to be so straightfoward, however. Thorium (group IV) decayed into mesothorium (group II) (the known existence of two mesothoriums was omitted here), then into radiothorium (group IV), and next thorium X (group II). Odd valency was to be found nowhere. The alpha particle, while nonvalent as helium in its uncharged state, bore two positive charges in the former role and accounted for the valency changes described above. Alpha decay, therefore, signified a jump of two boxes in the periodic table.

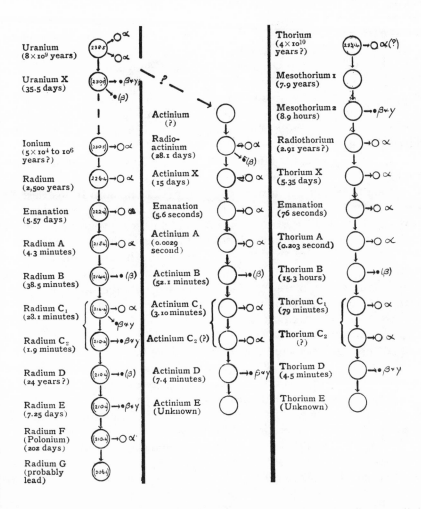

SOURCE: Frederick Soddy, *The Chemistry of the Radio-Elements* (London: Longmans, Green, 1911), p. 93.

Soddy was, however, pressed to explain how the move occurred from the emanations (group O), the beginning of one horizontal period in the table, to the end of the last period. The matter was complicated by the known fact that both the emanations and the A products were alpha emitters. He chose to ignore this last point and suggested that the path led through each series' chain of active deposit (generally the B, C, and D products), whose chemistry was sufficiently murky. In so doing he breathed some life back into the Lucas-Lerch rule, which characterized decay as a movement toward ever-increasing electrochemical nobility, or left-to-right in the periodic table. Attention was clearly focused now upon these short-lived constituents of the active deposits.[37]

Soddy never trained as many research students as did Rutherford, but developed high talent in two at least, A. S. Russell and Alexander Fleck. The latter began as his laboratory boy at Glasgow, was encouraged to take his degree, commenced research and planned to continue under Boltwood in 1914.[38] However, he went instead into industry, where he rose to head Britain's industrial giant, Imperial Chemical Industries, for which service he was raised to the peerage. The task Soddy placed before Fleck early in 1912 was to determine the chemical properties of those radioelements yet unidentified, especially those in the active deposits.

In an extensive investigation, Fleck deliberately mixed trace amounts of radioelements with larger quantities of elements with which they had shown similarities. Fractions were systematically taken by numerous methods of precipitation, crystallization, volatilization, and electrolysis, and their radioactivity tested. The unknowns were present in amounts so small as to be detected only by their radiation, yet electroscopes were thousands of times more sensitive to changes in concentration than any conventional chemical test. Because of the time required for chemical manipulation and subsequent examination in an electroscope, substances with half-lives less than three minutes could not be accommodated. Despite the laborious separations attempts made in the above processes, no alterations in the proportions were found; the trace radioelements remained uniformly distributed throughout their respective carriers. Fleck was able to present some early results at the British Association's Dundee meeting at the summer's end in 1912, and more detailed and complete conclusions in print the following year. He proved that uranium X and radioactinium were chemically identical to thorium; mesothorium II to actinium; thorium B to lead, and with slightly less assurance radium B and actinium B to lead; thorium C, radium C, and actinium C to bismuth; and that radium E was the only product between radiolead (radium D) and polonium (radium F) and was identical to bismuth.[39]

The effect of Fleck's work was not only to show more cases of chemical identity among the radioelements, but to extend the ties more strongly to the inactive elements. For a time, early in the history of radioactivity, there was a reluctance to believe that uranium and thorium were intrinsically unstable; their apparent activity must be due to a constituent. When this interpretation was abandoned, only lead remained as a link with the stable elements, through the chemical similarity—then identity—with radiolead. Now, Fleck was extending the connections, of all the B products with lead, and all the C products, plus radium E, with bismuth. Such efforts did much to overcome resistance to the placement of more than one element in a single box of the periodic table.

But this raised again the question whether radioelements had any "right" to appear at all in the periodic table. Rutherford and Soddy had coined the name "metabolon" in 1903 to indicate the impermanence of radioactive substances,[40] and though Marckwald used the term as late as 1910,[41] it had as little success as Rutherford's proposed "diplogen" for deuterium three decades later, and even its creators hastily abandoned it. Nevertheless, the name itself was an indication of early uncertainty of the "nature of the beast." Charles Baskerville, of the City College of New York, was typical of many chemists with an interest in radioactivity but little real comprehension, in his belief that only radium could be considered an element, thanks to its determined spectrum and atomic weight. As for the others, untested in these fashions and especially because many had such short half-lives, he felt "the greatest dignity we may assign them is meta-element."[42]

There was another side to the coin. Was the periodic table itself something of profound significance and therefore an object to venerate and trust? After all, it was not entirely periodic—the rare-earth elements did not fit and had to be placed in a separate grouping. Should the radioelements be similarly separated? Additionally, there were problems regarding the atomic weights of some adjacent elements, wherein the body filling the left-hand box of the two was heavier. And there were no guidelines setting limits on the number of possible elements lighter than hydrogen and heavier than lead.

On top of all this, A. S. Russell many years later revealed a pervasive attitude that further inhibited resolution of the problem. "Chemistry, we were always being told, was an experimental science. No good ever came from pontificating on the ways of Nature from the comfort of an armchair. The laboratory bench, not the sofa, . . . was where the truth would be found."[43] Nevertheless, the data upon which speculation would be based were accumulating and new techniques were aiding the process. The constituents of the active deposits after the emanations, for example, volatilized at different temperatures, indicating they possessed real physico-chemical properties, and by the method of radioactive recoil were separable one from another. Given the evidence, a small but courageous group of chemists was willing to break with tradition.

14 The Suicidal Success of Chemistry

The Ruling Passion

In 1909, shortly after Strömholm and Svedberg published their work on isomorphism, one of Ramsay's colleagues at University College, London, similarly tried to fit the radioelements into the periodic table. A. T. Cameron's designations, based primarily and conventionally upon atomic weights, contained serious errors—surprisingly, for example, he placed ionium in the box with actinium instead of thorium—but, in extenuation, there was still much uncertainty about chemical identity. The significant feature of this work lay in his explicit statement of the following rules:

> (1) It is possible that two elements of nearly the same atomic weight may occupy the same place in the table.
> (2) The emission of an α particle is accompanied by the production of an element which occupies the adjacent space of lower atomic weight.
> (3) The emission of a β particle, or a rayless change, may or may not be accompanied by a remove to a space of lower atomic weight.[1]

These rules were not very helpful, since they permitted movement in only one direction in the periodic table (which was not illogical, since daughter products might be presumed to have lower atomic weights) and, with hindsight, we know them to be inadequate. But they did combine a "traffic plan" for both alpha *and* beta decay with a clear expression that a box in the periodic table could hold more than one element.

Soddy's 1911 book came next, chronologically, and he more clearly stated alpha decay as a jump of two boxes, generally in the direction of lower atomic weight, but possibly also in the opposite direction. The beta-emitting radioelements, usually of shorter periods than the alpha-emitters, were of too uncertain chemistry to characterize their process then, but toward this goal he placed Fleck on the job of identifying them. For Soddy, chemical identification was the key to the puzzle. In August 1912, Russell, his own research with Rossi on the spectrum of ionium nearly completed, organized his thoughts about the disintegration series.[2] As an alumnus of the Glasgow laboratory he was well-informed of Fleck's progress; as a research student in Rutherford's laboratory he was intimately acquainted with Hevesy's electrochemical work, and with the branching experiments of Fajans and later of Marsden and C. G. Darwin.

Armed with this latest of information, Russell, like Cameron, placed radioelements in boxes of the periodic table. As might be expected, his arrangement was superior to that of 1909, but chemical identification still was not complete and he erred in assigning to several products their groups. Nevertheless, he felt able to offer a complement to Soddy's rule that in alpha decay there is a group change of two units. In beta decay (or rayless transformation), he said, there is a group change of one unit, and the change may progress in either direction in the periodic table. His system, based on radiochemical information, and specifically excluding valency data such as obtained by Hevesy, allowed him to predict the properties of some lesser-known radioelements and also the existence of another product between uranium X and ionium.

Russell showed his draft to Soddy and gave a copy to Rutherford, but neither expressed much interest and Russell himself was reticent to publish it.[3] Soddy presumably wished Fleck to complete his work before spinning grand schemes, or perhaps his mind was occupied with raising funds for the international radium standard and seeking a position at Oxford. Rutherford simply was not attuned to this sort of chemical speculation and in addition his taste ran to theories based on evidence he considered more air-tight.

During the last few months of 1912, Hevesy wrote up his work on valency, described earlier, and sent it to Russell, who improved its English for him. At the same time, Hevesy visited Fajans in Karlsruhe, where the latter had become a *Privatdozent,* and learned that he planned to publish a scheme for fitting the radioelements into the periodic table. Though Fajans' experimental investigation was not nearly as extensive as Fleck's, Hevesy had great respect for his insight and expected his intended publication to be significant. He therefore wrote to Rutherford, recommending that he urge Russell to publish his own ideas before they might perhaps be anticipated.[4] As 1912 closed, it is apparent that Soddy, Fajans, Hevesy, and Russell each was aware that at least one other was moving toward resolution of a major problem. Yet, despite the interconnections and our sometimes hazy understanding of the knowledge they had of the details of the others' work, the priority question seems straightforward.

Russell responded to Rutherford's directions and hurriedly revised his paper for publication.[5] He mailed it to Sir William Crookes' *Chemical News* on 15 January 1913, and it appeared in print the last day of that month. There were few changes since he formulated the original draft the previous August; recent work by Fleck confirmed some of his placements and most of the alterations made improved the fit. He now placed both uranium X_1 and his hypothetical new product, uranium X_2, between the two uraniums, and he moved all the D products into the fourth group of the periodic table. These appear to be changes based upon considered thought rather than new information. The justification for moving thorium D (now thorium C'') and actinium D (now actinium C'') into the same group with radium D (definitely known to be lead) probably was an esthetic desire for symmetry among the three decay series. No doubt it was also influenced by his decision simply to omit indications of branching at the C members, since their identities and hence placement were uncertain. Unfortunately, his original assignment of them to the thallium group, when he sought to incorporate branching into his scheme, was correct. Russell's arrangement of the radioelements showed decay descending from the higher groups of one period, through zero, and then ascending through the groups of the next period above. Most importantly, his rules for decay remained unchanged: alpha expulsion meant the daughter product was two groups higher or lower; beta expulsion, one group higher or lower.[6]

Group Displacement Laws and Concept of Isotopy

Neither Russell's rules nor his placement of all the radioelements in the periodic table were correct. But they were close enough to make it apparent that the "true" interpretation must be something similar. On 15 February 1913 two papers by Fajans appeared that fit everything magnificently in place. They were in the hands of the editor of the *Physikalische Zeitschrift* the last day of the preceding year, so could not have been influenced by Russell's published paper. They were, in fact, different in approach, being based primarily on electrochemical data, and only partly on radiochemical identities.[7]

In Fajans' first paper he organized all the evidence from electrolysis and from dipping metals in solutions that he and other investigators had generated. From these experimentally established relationships he saw that an alpha emission left the daughter more electropositive, while a beta emission yielded a daughter more electronegative or nobler. The Lucas-Lerch rule was thus true for beta decay alone. He then applied his rules and also analogies between the series to the radioelements whose chemical natures were yet untested, assigning them electrochemical relationships to products already identified. In his second paper Fajans used these relationships to place all the radioelements into the periodic table. An electropositive transformation meant movement from right to left in a horizontal period; Fajans accepted Soddy's conclusion that the jump was of two boxes. Conversely, an electronegative transition indicated left-to-right movement, and here Fajans decided it was of just one box. The few rayless transformations were treated as beta emissions. Both his rules and his arrangement of the radioelements were better than Russell's—more was explained, in a more cohesive manner.

Russell had painted himself into a corner by assuming the decay series moved through the periodic table from high numbered groups down to zero, and then rose through the next horizontal row above. Soddy earlier had claimed the active deposits as not-clearly-defined intermediaries between the emanations, at the far left of one row, and the final products of each series, placed toward the right side of the next higher row. Fajans now indicated that the procession went from higher to lower groups in *both* periods, and the transition between periods occurred from emanation (group 0) to an A product (group VI). The sequence could best be visualized if the two-dimensional table were rolled into a helix.

He was faced with a problem, however. His scheme was based mainly upon electrochemical considerations which he attempted to apply to all transformations, including those involving the emanations. Since noble gases were not known to form cations it appeared appropriate to consider them as highly electronegative. Yet, their parents (radium, thorium X, actinium X) were all considered to be very electropositive, alpha-emitting alkaline earth elements whose daughters should be even more electropositive. Fajans resolved the paradox by assuming the existence of a beta-emitting alkali metal in each series, situated between the emanation and its parent and less noble than both.

However, this led in his second paper to the contradiction that the alpha transformation of, for example, radium to the hypothetical radium X would cause a displacement of only *one* box to the left. Fajans left this problem unresolved at the time; later he came to feel that consideration of the noble gases as highly electronegative was unjustified because they were not known to form anions either.[8] The hypothetical elements were not found and, indeed, need for their existence soon was considered not compelling; the group displacement laws achieved primacy over electrochemical logic.

Fajans' proposal of a uranium X_2, advanced independently by Russell, met with more success, as will be mentioned below. His recognition that group six follows group zero placed the A products in the sixth group, important because their short periods made chemical testing of the thorium and actinium members virtually impossible. His further understanding that thorium D (now Th C″) is in the third (thallium) group was another significant advance, and crucial also to the question of priority.

Of the thirty-odd radioelements the chemical identity of but little more than half had been experimentally determined. Fajans' rules therefore assigned to a considerable number of these bodies their chemical natures. He also assigned to each an atomic weight, based upon the measured values for uranium, thorium, and radium, and calculating the differences from these as four units for an alpha (helium) decay and zero units for a beta decay. This reaffirmed earlier belief that one box of the periodic table often had to hold more than one body, each having a different atomic weight. To the plurality of bodies Fajans gave the name "pleiad" (after the star group) in March, more than half a year before Soddy's friend, Dr. Margaret Todd, coined the better-known term "isotope."[9] Fajans added his voice to those who had suggested the ordinary elements also were mixtures of nonseparable elements of different atomic weight. He further pointed out that the calculated atomic weights of bismuth and thallium, and especially lead, differed from those experimentally determined. The final product of the uranium series should have a weight of 206.5 and, if the thorium series ended in lead, that product should weigh 208.4. The measured value of ordinary lead was 207.1 and was possibly an average. Clearly atomic weight determinations of lead from the different radioactive series could serve to confirm the existence of pleiads/isotopes.

Less than two weeks later, on 28 February 1913 a paper by Frederick Soddy appeared containing essentially the same information. Like Fajans, he placed all the known radioelements into the periodic table; unlike Fajans, his approach was through chemical identities, not electrochemistry, so he saw no need to suggest the existence of radium X, thorium X_2, and actinium X_2. The reality of uranium X_2, however, was seen to be necessary, and Soddy correctly indicated it would fall below tantalum in group five. He also reemphasized that radioelements located in the same box would be chemically nonseparable even though they belonged to different decay series, and he too found the atomic weight of lead a promising key to confirmation of these identical-but-different substances.[10]

The puzzling thing about this paper is why it ever was published. It can hardly be called a case of simultaneous and independent discovery, for Soddy actually acknowledged in the text of his article that he had read Fajans' paper and accepted its ideas. The stated reason for publishing was that his paper was already drafted when Fajans' came to hand; presumably, in addition, Soddy felt that Fleck, who was furnishing so much of the experimental evidence that chemically identified the short-lived radioelements, was working in his laboratory at his own suggestion, and Soddy himself had been intimately connected with the subject. His paper, further, approached the question from a different point of view than had Fajans', namely chemical identity.

Had Soddy offered it as confirming Fajans' ideas with different evidence, and even extending the consequences to include the probable lack of spectroscopic distinctions between substances in the same box, this would have been acceptable. Had he criticized the errors of Russell and Fajans, which instead he tacitly rejected, this too would have been acceptable. But presenting it as an independent discovery was a surprising violation of scientific etiquette. Possibly it was even more. On 13 February 1913, two days before Fajans' papers were published, Fleck sent a letter to the *Chemical News* commenting upon Russell's earlier paper. Russell had indicated that thorium D (now Th C″) should be analogous either to lead or mercury. Fleck now described experiments proving it could not be the former. "It thus appears," he concluded, "to be much more nearly allied to the zinc, cadmium, mercury sub-group than to the germanium, tin, lead sub-group. So far, however, it has not been possible to prove that it is completely analogous to any known element."[11] Yet, five days later, on 18 February, Soddy penned his paper, described above, and correctly identified thorium D with thallium. Fleck *may* have immediately repudiated his published conclusion, but we have no indication of this; he returned to this substance only after the Fajans and Soddy papers were published. Soddy *may* have had an inspiration to place thorium D in group three (perhaps recalling such an identification in Russell's draft of the previous summer), but his earlier behavior indicated reluctance to speculate beyond Fleck's results. In fact, in his Nobel Prize address, Soddy remarked that "till then I had scrupulously avoided even thinking about these results [Fleck's] until they were completed, for fear of giving an unconscious bias to the work in progress in the laboratory."[12] It is thus likely that Fajans' paper did indeed cause everything to fall into place in Soddy's mind, and the latter was so close to the material that he could not later sort out what was borrowed from another.[13] Which reaffirms the conventional wisdom of respecting the priority of others and refraining from publishing what has already appeared in print (unless offering confirmatory evidence).

Beyond Radioactivity

The group displacement laws served as a fine organizing principle for the radioelements. Not entirely by coincidence, for there were many topical and personal connections, the year 1913 also saw enormous advances in knowledge about the inactive elements.[14]

Since J. J. Thomson's discovery of the electron in 1897, several models to incorporate this subatomic particle into the heretofore homogeneous ("billiard ball") atom had been conceived. Philipp Lenard's "dynamids" and Hantaro Nagaoka's "saturnian atom," of 1903 and 1904, respectively, were notable examples of this interest. More significant, because it was further developed, was what came to be called Thomson's "plum pudding" model. First proposed by Lord Kelvin, then refined by Thomson and expressed in detail in 1904, it viewed the electrically neutral atom as consisting of a number of negative corpuscular plums (electrons) moving rapidly about in a spherical pudding of positive electrification. The corpuscles were believed to orient themselves in concentric shells, and their number varied from element to element. By 1906, x-ray scattering and absorption experiments and dispersion of light calculations led Thomson to equate the number of electrons with the atomic weight of the element, which meant that this weight was due almost entirely to the positive electrification, and which implied that the positive electrification, if quantifiable, might also be granular. The atom could consist of neutralized positive and negative charges.

Alpha particle scattering experiments carried out by Rutherford, Geiger, and Marsden, which showed that deflection of the relatively heavy alpha could not be explained by a succession of encounters with electrons, led in 1911 to Rutherford's conception of the nuclear atom. The alpha projectile was turned from its path by the electrostatic forces concentrated on a tiny nucleus. J. W. Nicholson, in England, pursuing a modified Thomson atomic model that same year, felt the positive charge to be about half the atomic weight. Experiments by Geiger and Marsden in 1912 confirmed Rutherford's picture and offered evidence on the size of the nuclear charge in keeping with Nicholson's calculations.

The Dutch amateur physicist A. van den Broek, whose hobby for several years had been the periodic table, in 1913 saw that a better fit of the data resulted when not half the atomic weight but the sequential number of the element in the table was taken as the nuclear charge. While extremely promising, however, charges were known only with poor accuracy, and the numerical place of elements high in the table was also uncertain.

Niels Bohr, a young Dane who had absorbed the enthusiasm of Rutherford's laboratory in the infancy of the nuclear atom, was next to relate the atoms of the physicist and the chemist. He carefully explained how atomic structure was responsible for observed effects: Radioactive phenomena, the alpha, beta, and gamma emissions, originated in the nucleus. Spectral lines were due to external electrons jumping from one orbit to another; chemical behavior was due to the number and arrangement of these same electrons, whose total negative charge balanced the positive nuclear charge. Characteristic x-rays resulted from transitions between the inner orbits of the external electrons. Like van den Broek, he felt that nuclear charge corresponded to the element's place in the periodic table.[15]

X-ray research, which had contributed earlier to atomic model testing, now furnished additional help. Max von Laue and associates, in Germany, had shown that x-rays could be diffracted by crystals and would yield information about these targets. A former student of Rutherford's, H. G. J. Moseley, adapted the technique to his purposes, measuring the frequency of the characteristic x-ray spectral lines of a number of elements. This classic work, deeply influenced by Bohr, proved that nuclear charge increased regularly through the periodic table. The place of each element, its atomic number, was determined by its charge; the atomic weight, though it had served long and well, was seen to be only the basis of an *ad hoc* organizational scheme.[16]

It is obvious from the foregoing that by 1913 the atom's nuclear charge, not its weight, came to be seen as fundamental. Charge determined chemical identity. The group displacement laws, therefore, were more rules for moving down two elements (two charges) or up one element in the periodic table, than schemes for reconciling measured atomic weights. This is not to say that weight or mass determinations were no longer of interest. Quite the contrary was true. J. J. Thomson had made historic progress studying the electrons produced in his cathode ray tubes. If, however, holes were drilled in the metal cathode, positive bodies of atomic dimension moved through them at the same time cathode rays traveled in the opposite direction. These ions, charged atoms or molecules of the residual gas in the tube, could be studied in much the same fashion as Thomson had examined electrons. Tracks of positive rays, deflected by electric and magnetic fields, yielded parabolic-shaped curves when photographed. This was to be expected, but when Thomson tested neon, of known atomic weight 20.2, he found two curves. The second, much fainter than the first, corresponded to a mass of 22. Thomson himself was very reluctant to acknowledge that he had found neon isotopes. He admitted that neon must be a mixture of two gases, but felt the heavier one might be an unknown compound of the single neon species. His assistant, F. W. Aston, who later perfected the mass spectrograph which measured masses with great precision, reported at the 1913 British Association meeting a small but significant separation of neon by gaseous diffusion into portions of different weight. Thomson's reticence to proclaim the existence of isotopes among the non-radioactive elements was not shared by others.[17] The nineteenth-century idea of Prout about single unit building blocks of matter seemed nearly verified.

Confirmation of the New Ideas

A scientific theory that explains existing knowledge is worthwhile; one that is pregnant with the unknown is exciting as well. Predictions, moreover, furnish the means by which the theory may be tested. In the case of the group displacement laws verification came by several means.

Important, though not striking, were Fleck's additional results with the active deposits. In mid-May 1913 he reported to the Chemical Society the identities of the three remaining substances of average life greater than three minutes whose chemistry was heretofore experimentally unknown. Radium A was identical with radium F (polonium), he said, while actinium D and thorium D were nonseparable from thallium. These determinations corresponded to the placements made the previous February by Fajans and Soddy and reinforced belief in the accuracy of the group displacement laws.[18]

A more exciting and even earlier confirmation of the laws' power was the experimental discovery of uranium X_2 by Fajans and his student in Karlsruhe, Oswald Göhring. The existence of this radioelement had been predicted by both Russell and Fajans, who placed it in group five, below tantalum. It was especially interesting because no other "eka-tantalum" was known—at a time when pleiads/isotopes were recognized for virtually all the heaviest elements, here was a body filling its box (element 91) in the periodic table alone. Fajans and Göhring announced their discovery in early April, pointing out that the hard beta rays found in uranium X came from the X_2 product. Further work showed that the soft beta rays came only from uranium X_1 (a thorium isotope). This meant, according to an extension of the Geiger-Nuttall relation to beta transformations, that U X_2 could not be the parent of actinium, since that unknown body required a long half-life, while hard beta rays were emitted by short-lived elements. Uranium X_2 was electrochemically nobler than U X_1, as expected by its formation through beta decay, and could be isolated by placing a solution of U X in a lead dish. The lead, being more noble than thorium and less noble than eka-tantalum, became radioactive in their very first experiment. Greater separations could be achieved through standard analytical procedures for separating tantalum from thorium, such as precipitation with tantalic acid. Because of its brief half-life (1.1 minutes), they named this new body "brevium." When actinium's lineage was better resolved and its immediate parent found to be a long-lived isotope of brevium, Fajans agreed that the former's name, "protactinium," should thenceforth designate this pleiad.[19]

Such confirmations as described above related mostly to the group displacement laws. Welcome as they were, they did not remove all doubts that isotopes existed only in the fertile imaginations of a few radiochemists and positive ray specialists. The iron-clad proof required, as clearly seen by Fajans and Soddy, was a series of comparative atomic weight determinations of isotopic elements. Unfortunately, scarcely any radioelement was obtainable in both isotopic purity and quantity sufficient for its atomic weight to be measured. This was part of the problem surrounding whether they deserved to be called elements at all. There was, however, the inactive lead end product of the uranium decay series, available in satisfactory amounts, and the possibility that the thorium chain also ended in lead. Ordinary lead had an established atomic weight of 207.15. Taking that of radium as 226.0 and deducting five alpha particle expulsions of four units each, the uranium series lead should weigh 206.0 (or, 206.5 if calculated as eight alphas down from uranium at 238.5). Similarly, six alpha transitions from thorium at 232.4 indicated that thorium series lead should have an atomic weight of 208.4.

Prior to this time atomic weight experts had had little to do with radioactivity. Indeed, these painstaking, precision chemists may be viewed as logical opponents to ideas of isotopy, for had they not devoted their whole careers toward ascertaining the very best *single* value for each element? Still, vested interest or not, the experimental test was one within their capabilities and little further stimulus was needed. Fajans and Soddy each tried to make the atomic weight determinations themselves. Fajans soon realized that the task required such precision that only the results of recognized experts would be widely accepted. Soddy persisted in his efforts, only to have his results viewed skeptically.[20]

The recognized expert Fajans chose for this work was the world's leading atomic weight specialist, Theodore W. Richards (1868–1928) of Harvard University. Two years older than Boltwood, he too had pursued postgraduate study in Germany under Krüss and Ostwald, as well as others, but his main inspiration came from his senior Harvard colleague, Josiah P. Cooke. With this mentor, and then independently, Richards published an impressive series of measurements of hydrogen, copper, the alkaline earth metals, the alkali metals, chlorine, and other elements. Infinite care was the hallmark of this trade, for impurities would render experimental results valueless. The action of acids upon

THERDORE W. RICHARDS (1868–1928)

glass, the occlusion of gases by other substances, the need to prevent dried substances from absorbing moisture before being weighed—all had to be meticulously controlled. By the century's second decade, Richards was at the height of his powers and fame, such that he was possibly the only American scientist of his generation to be offered a professorship in Germany. He was elected president of the American Chemical Society and of the American Association for the Advancement of Science, and received the Nobel Prize in chemistry for 1914. Indeed, Rutherford was surprised that this award had not come at least three years earlier.[21]

Fajans arranged for Max Lembert, a Karlsruhe student just beginning doctoral research, to work in Richards' laboratory. When Lembert arrived in Cambridge, Massachusetts, in the autumn of 1913 and proposed to determine the atomic weight of lead from radioactive sources, Richards was skeptical though enthusiastic, for Sir William Ramsay, at Soddy's request, had recently urged the problem on him. Ramsay, though widely considered a scientific adventurer for his tendency to enter a field and claim priority for questionable discoveries based upon inadequate experiments and comprehension of the subject, was nevertheless respected for his real accomplishments and chemical wisdom.

Associates of Richards had perfected the technique for the determination of ordinary lead's atomic weight just half a dozen years earlier. He and Lembert followed this procedure, which consisted of preparing a chloride of great purity by extensive preliminary treatment to remove extraneous substances, followed by recrystallization in quartz and platinum vessels. Recrystallization was a recognized means of purifying lead. The lead chloride was then placed in a desiccator to dry and fused in a stream of hydrochloric acid gas and nitrogen. Finally, this compound of extreme purity was dissolved in water and the chlorine precipitated by silver nitrate, silver serving as the comparison standard. Precision weighing of both the amount of silver required and the precipitate allowed calculation of the atomic weight of lead.[22]

Control experiments on ordinary lead were carried out and yielded the accepted value (atomic weight 207.15). The weights of the radioactive samples, however, were strikingly lower. Fajans had provided lead from Colorado carnotite, which he obtained from Giesel in Braunschweig (206.59); another sample he sent was extracted from Joachimsthal pitchblende (206.57). Ramsay furnished residues from Cornwall pitchblende (206.86). Boltwood provided lead from Ceylonese thorianite, as did H. S. Miner, chief chemist of the Welsbach Light Company, in New Jersey (206.82). This mineral contained some three times as much thorium as uranium, but since the latter decays about three times as fast as the former, the amounts of lead from each series were expected to be roughly equal. Ellen Gleditsch, then working with Boltwood, gave Richards a particularly pure and hence valuable lead from North Carolina uraninite (206.40).[23]

Experimental comparisons of other elements, from widely scattered sources on the globe and even from meteorites, had never before yielded discordant results. In this case, Richards, who presumably composed the paper he published with Lembert, studded it with the words "amazing," "revolutionary," "extraordinary," and "striking." There was no doubt that radioactivity-derived lead was lighter than ordinary lead. Spectroscopic examination, moreover, revealed no differences whatever. Yet Richards was too much a chemist of the old school to drop his reservations and proclaim aloud the truth of isotopy. Instead, he spoke of an "admixture" having similar if not identical reactions to lead, though of a lower atomic weight. The admixture might have the same spectrum as lead, or have no spectrum in the ultraviolet region examined, or have its spectrum masked by the greater quantities of ordinary lead. Another possibility, he conceded, with reference to the work of Fajans and Soddy, was that ordinary lead was a "medley" of substances of different origin but similar properties, in varying proportions.

These results were reported in late May and June 1914. Slightly earlier Soddy announced the results of his own work with Ceylonese thorite. In this mineral there was 55 percent thorium and only 1 or 2 percent uranium. But, again because of their different periods, the amount of thorium-lead was about ten times that of uranium-lead. Purification by precipitation as sulphate and sulphide, and then crystallization as iodide, preceded calculation of the lead's atomic weight. This was done volumetrically by titration with a silver nitrate solution, against a sample of ordinary lead. The procedure was less accurate than Richards', yielding more a relative estimation than a measurement. It was, however, a qualitative success, for the value of 208.4 was significantly higher than ordinary lead's 207.15, and only a bit higher than the theoretical weight deduced for this mixture of leads from thorium and uranium. To the extent that Soddy's atomic weight work was regarded as accurate, it was further verification of the concepts of Fajans and Soddy.[24]

Maurice Curie, nephew of Marie, was next to join the ranks of the "true believers." Though his manipulations differed from Soddy's, his too was a relative estimation. In early June the Paris newspaper *Le Matin* headlined a page-one story "*Il y a plusieurs espèces de plomb*," describing Curie's report to the Académie des Sciences. He had examined the lead from three uranium minerals, one thorium mineral, and common galena, finding lower and higher atomic weights as expected.[25]

The most convincing confirmation of the Harvard work came, however, from another atomic weight expert. Just a few years earlier Otto Hönigschmid had learned the techniques in Richards' laboratory.[26] Now he was one of the bright young men in the field, making a name for himself in Prague and Vienna with measurements of radium and uranium. With his student, Stephanie Horovitz, Hönigschmid determined the atomic weight of lead from Joachimsthal pitchblende, obtaining a value of 206.736. This was published a week after Maurice Curie's paper, although the results had been presented on 23 May at a congress of the Bunsen Gesellschaft in Leipzig, at which meeting Fajans also reported upon the work of Richards and Lembert.[27] Thus, within about a single month the existence of isotopes was confirmed independently in four laboratories in four countries.

Some, such as Samuel Glasstone, recall the astonishment—"almost consternation"—of the period. But the evidence was so strong that prejudices were swept aside.[28] Others, such as James B. Conant, were earlier convinced by the radioactivity arguments and awaited the Richards and Lembert results with equanimity.[29] Regardless of preconceived notions, it was a time of intellectual excitement, not entirely overcast by the outbreak of World War I.

Consequences

Hönigschmid and Horovitz, and particularly Richards, continued investigating the atomic weight of lead from several sources. To avoid variations due to the mixture of different leads, the ideal materials were thorium-free uranium minerals and uranium-free thorium minerals. The former were easily obtainable, hence uranium-lead was more extensively examined and its weight more confidently determined. The Austrian team obtained lead from pure crystalline pitchblende mined in Morogoro, East Africa and from bröggerite found in Moss, Norway, with atomic weights of 206.046 and 206.063, respectively.[30] At Harvard, Richards and his student C. Wadsworth determined atomic weights of 206.08 and 206.12 from crystalline Norwegian cleveite and bröggerite, respectively.[31] Except for the last of these four values, the agreement with the predicted atomic weight (206.0) of radium G, the uranium series' end product, was so good that there could be no doubt of the theory's correctness. Even the further lowering of this predicted weight for lead, calculated from the mass equivalent of the energy lost in alpha particle expulsion, seems not to have shaken anyone's confidence: it amounted only to 0.03 units.[32]

That the thorium series also ended in lead was generally accepted in the latter half of the century's second decade, although the evidence was not as tidy as that for the uranium series. Part of the problem was that lead-thorium ratios were inconsistent; there was not the clarity such as Boltwood had found for lead and uranium. This was due to the unavailability of uranium-free thorium minerals. Another factor was the uncertainty whether series which branched later rejoined, or whether each branch might have its own end product. Nor was it certain that the lead was stable; it might decay with a very long period.

The last-named concern was disposed of by attrition: no real evidence in its favor was produced. Branches in series were seen to rejoin, once the type particle each product emitted was ascertained clearly and the group displacement laws were applied. In every case, if one branch had a sequence of alpha-beta decays, the other branch had the reverse. Both branches, therefore, returned to the same box in the periodic table. And if the mineralogical problems did not permit unequivocally consistent atomic weight determinations for thorium-lead, the values obtained, including the highest of 207.90, by Hönigschmid on a sample furnished by Fajans, departed so obviously from ordinary lead at 207.15 that its reality could not be denied.[33]

The matter of actinium remained uncertain far longer than that of thorium. For years it had been regarded as a uranium product. Boltwood, for example, made a typical comment in 1920 when he conceded the descent from uranium but confessed ignorance of the point it departed from that series.[34] Its end product, actinium D, was assumed to be a lead isotope, but since actinium was derived from uranium and always found with it, actinium-lead was always associated with uranium-lead. Moreover, since actinium's short half-life prevented the accumulation of large amounts, such that its atomic weight could be measured, and there was no datum from which to calculate it by the emission of alpha particles, the atomic weight of actinium-lead was equally unknown. This introduced a measure of uncertainty into the calculated weight of any sample of uranium-lead, albeit a small once since the proportion of actinium-lead was only about 3 percent.[35]

More interesting and significant was the resolution of actinium's origin. After their discovery of element 91, Fajans had Göhring look for another member of the brevium pleiad which might be the parent of actinium, for if the latter were formed by an alpha particle decay its parent would be element 91. Brevium itself, as mentioned earlier, could not be the parent since its half-life was too short and it emitted beta particles as well. Though he searched in the correct places—pitchblende and uranium Y (he also looked at ionium)—Göhring failed to find the product.[36] Just the year before Soddy had tested the alternative. If actinium were formed through beta decay its parent would be an isotope of radium. However, examination of one of Giesel's radium preparations that had been chemically untreated for ten years showed the presence of no actinium whatever.[37] The only possibility was that the actinium parent had escaped Göhring's investigation.

Delayed by the war—Hahn, for example, squeezed in experiments during his infrequent leaves from active duty—two teams independently reported success in 1918. With Lise Meitner, his physicist colleague in Berlin, Hahn examined the insoluble residues from the extraction of radium from uranium, following standard procedures for the extraction of tantalum. Tantalum, in fact, was added as a carrier to help separate eka-tantalum (brevium).[38] Soddy and John Cranston, in Aberdeen, also found what Hahn and Meitner called "protactinium." They had learned that brevium ($U X_2$) could be volatilized in a stream of carbon tetrachloride and air at incipient red-heat, and thereby separated from uranium X_1. By treating pitchblende in the same fashion, they obtained a sublimate which eventually generated detectable quantities of actinium.[39] The low range of protactinium's alpha particle indicated a half-life in the tens of thousands of years, which meant that enough existed in equilibrium to eventually make an atomic weight measurement and spectrum examination likely.

The remaining question was the sequence of radioelements leading to actinium. It was now accepted that Antonoff's uranium Y decayed into protactinium, which in turn yielded actinium. But was uranium Y a product of uranium I or uranium II? Or, was it, as A. Piccard suggested in 1917, the daughter of a third and entirely independent isotope of uranium?[40] This matter was finally settled in Piccard's favor in 1929, when Aston's work with his mass spectrograph showed the existence of lead-207, which was accepted as the end product of the actinium series. This discovery finally permitted atomic weights to be assigned to the members of this series, whose fountainhead was seen to be a new isotope of uranium, U-235, named "actinouranium." Boltwood and Rutherford thus were correct that uranium and actinium were genetically related, but had understandably erred, in the pre-isotope days, in believing actinium derived from a branch of the main-line uranium-238 series.[41]

A few other consequences of the concept of isotopy are worthy of mention. Not that there was reason to doubt the accuracy of work on lead, but ionium served to confirm the reality of multiple atomic weights for a given element. The only other radioelement with a period sufficiently long for suitable equilibrium amounts to accumulate was uranium II, but it could not be separated chemically from uranium I, of course, and no one took the lengthy trouble to grow weighable quantities. Ionium, however, was intrinsically more interesting and its identity recognized earlier, so that separations efforts were undertaken. It could also be prepared pure (from a thorium-free mineral) by allowing the only other thorium isotope, uranium X_1, to decay away with its half-life of less than a month. The great Austrian chemist Auer von Welsbach had worked with Joachimsthal pitchblende, which contained some thorium, and separated but a tiny amount from thirty tons of the mineral. The magnitude of the operation was reminiscent of the isolation of radium, although about forty times as much ionium was to be found per ton. Its atomic weight could be calculated by subtracting eight units (two alpha particles) from the recently remeasured value of uranium: 238.18 − 8 = 230.18, or by adding four units (one alpha) to its daughter, radium: 226.0 + 4 = 230.0. The calculated average, 230.09, differed significantly from thorium's accepted atomic weight of 232.4. Hönigschmid and Horovitz in 1916 redetermined the value for pure thorium, obtaining 232.12, and then tested Welsbach's ionium-thorium mixture. This showed a weight of 231.51, meaning the preparation contained about 30 percent of ionium. Spectroscopic examination revealed no impurities and no differences between the lines of ionium and thorium, the latter a confirmation of the work of Russell and Rossi and others. Hönigschmid and Horovitz, therefore, furnished the second case of chemical and spectroscopic identity but different atomic weights.[42] Besides uranium, thorium, ionium, and radium, the only radioelement extracted in large enough amounts and in satisfactory purity for atomic weight measurements was emanation, but determinations of the weights of its different isotopes was not attempted in this period.[43]

Other work that followed as a consequence of the group displacement laws and concept of isotopy included extensive studies of spectra, thermoelectric effects, refractive indices, solubilities, melting points, atomic volumes and atomic densities. As Soddy remarked about isotopes in his Nobel lecture, "put colloquially, their atoms have identical outsides but different insides."[44] These studies, therefore, were motivated by the understanding that certain atomic properties depended upon the "outside" electron configuration, and others were due to the "inside" mass of the nucleus. Small differences in line spectra and larger variations in band spectra of isotopes eventually were discovered. Similarly, slight differences in reaction rates affected by mass also were found. Besides that coming from the Richards circle, the only original work of this sort done in the United States was by Percy Bridgman (1882–1961) and William Duane at Harvard and by William Draper Harkins (1873–1951) and his students at the University of Chicago, from 1915 onward. With apparently little or no contact with McCoy or Millikan, the Harkins group examined energy relations in the formation of atoms, mass changes, methods of isotopic separation, atomic structure, and a host of related problems. But just as radioactivity was something of a hybrid between physics and chemistry, so too was this new field. It was sometimes called "modern chemistry," though more often "atomic physics." Regardless of the name, however, because it was no longer radioactivity its description is beyond the limits of this volume. The radiochemistry in which we are primarily interested essentially was no longer pursued.[45]

This evolution (or revolution, according to tastes) of a science into another, as old problems were solved and new ones arose, was made easier by the touchstone of traditional chemistry. What were radical ideas about the nature of elements but a few years before could be accepted, provisionally at least, because of the classical, wet-chemistry atomic weight evidence in favor of the hypotheses of Fajans and Soddy. Until 1919 it was the only technique available for such determinations. Then, Aston developed his mass spectrograph, which also provided precision measurements of atomic weights, through the deflection of ions in a magnetic field. The method was satisfactory to physicists, but chemists felt it yielded the weights by a process of calculation, and thus was little better in principle than subtracting four units for each alpha particle emission. Chemists did not desire such arithmetic, they much prefered laboratory measurement.[46] The initiation and acceptance of the many investigations of isotopic properties had already relied substantially upon the imprimatur given by the traditional work of such chemists as Richards and Hönigschmid. Increasingly they would come to respect the value of the mass spectrograph, but during the century's second decade classical chemistry was a more-than-sufficient bedrock for the new ideas.

Conclusion

For several years following the introduction of the idea of pleiads/ isotopes there raged a subdued controversy over their interpretation. The work of Aston made it clear that the multiplicity of forms for a body with a given atomic number was a general phenomenon, not restricted to radioelements. Fajans, seeking to disturb the age-old definition of "element" as little as possible, looked upon each new isotope as a different element. Uranium-238, U-235, and U-234, therefore, were individual elements because of distinctions in such properties as mass and half-life. Most other chemists and physicists, however, regarded chemical combination as the predominating yardstick in the definition of an element, and were content to accept isotopes as varieties of the same element. This new interpretation, which removed atomic weight as an indicator of the identity of an element, was smoothed by Moseley's work showing the significance of atomic number. By 1922 Fajans came around to the popular view.[47]

Another problem of interpretation that arose concerned precisely what occurred in the atom during radioactive decay. As early as February, 1913, Fajans declared it to be a surface phenomenon, because changes in the chemical nature of the atom due to alpha and beta particle expulsion were the same type as resulted from electrochemical changes of valency.[48] It was a view difficult to reconcile with Rutherford's model of a nuclear atom, however, for the only "occupants" of the surface region, where chemical reactions occurred, were orbital electrons. These might furnish the beta particles, yet where else but the nucleus could the alphas arise? There was, in fact, strong circumstantial evidence that radioactivity was solely a nuclear phenomenon. Not only was it logical to expect that atomic-size alphas would be expelled from the material nucleus, but this core was seen as composed of differing numbers of positive and negative particles, with a net positive sum equal to the element's place in the periodic table. The ideas of van den Broek, supported by the importance given the sequential arrangement of elements by the displacement laws, by Niels Bohr's relegation of radioactivity to the nucleus, and by Moseley's illustration of the significance of atomic number, all made it appear logical that beta particles were nuclear electrons, those negative bodies that allowed the hydrogen nuclei to comprise the atom's measured atomic weight, but reduced their positive charge to the appropriate atomic number. By the end of 1913, Soddy offered further argument against viewing radioactivity as a surface phenomenon. Uranium, in its uranous form, is tetravalent, he pointed out, but in its more common uranyl form is hexavalent, a loss of two orbital electrons. In a similar loss of two negative charges uranium X, upon the emission of two beta particles, is transformed into uranium II. Should electron and beta loss be identical processes, Soddy argued, uranous salts and compounds of uranium X would be chemically nonseparable. But uranium X is an isotope of thorium, and clearly separable from uranium.[49]

Beyond such matters of theoretical significance there was an abundant range of experimental investigations into the effects of radioactivity. Just as in the case of the consequences of isotopy, much of this work occurred in the United States. Yet, this too was a departure from the main line of radioactivity studies. While both physicists and chemists participated, a large portion of it was conventional chemistry, merely using radioactive materials. "The influence of radium on the decomposition of hydriodic acid,"[50] and "The speed of oxidation, by air, of uranous solutions," with a note on the volumetric determination of uranium,"[51] are somewhat early examples of this genre. With increased sophistication and recognition of the role played by different radiations, alpha particle studies predominated. William Duane, earlier mentioned as the first American to work in Marie Curie's laboratory, and his Harvard student Gerald L. Wendt produced a chemically active modification of hydrogen by alpha particle bombardment,[52] while their Ivy League colleague at Princeton, Hugh S. Taylor, examined the interaction of hydrogen and chlorine.[53] By far the greatest contributions in this area were made by Samuel C. Lind (1879–1965), another alumnus of the Curie laboratory, as well as of Stefan Meyer's Radium Institute in Vienna. After a number of years on the University of Michigan faculty he entered government service, primarily with the Bureau of Mines. His work on the commercial production of radium, under Richard B. Moore, was mentioned briefly in chapter ten; his fame, however, rests upon his alpha particle investigations. These included the coloration of glass and porcelain, the formation of ozone from oxygen, the luminescence produced by alpha bombardment, and, most especially, attempts to understand the mechanics of such reactions through his ionization theory. Whereas other types of reactions depended upon some sort of resonance effect between the energy source and the target molecule (e.g., wave length of light), alpha particles possessed so much kinetic energy that they always would produce ionization and frequently chemical change.[54]

As already indicated, these were investigations of the effects of radiation. While intrinsically interesting, they led to no major insights or revelations. It was a respectable, if not exciting, scientific specialty.

There were, however, some exciting lines of research leading off from radiochemistry, but they were European developments and centered upon applications of the phenomenon. Fajans, Hahn, and Paneth illuminated the nature of precipitation and adsorption by using such small quantities of matter that only radioactive means could serve for their detection. But the leading pioneer in tracer techniques was Hevesy, who extended the application of such indicators to biological (and medical) systems.

These uses of radioactivity were the wave of the future; the past had yielded the basic science of radiochemistry upon which such applications rested. By the early 1920s, then, almost all the radioelements were known, their sequences in the decay series determined, fundamental understandings such as the disintegration theory of radioactivity, the group displacement laws, and isotopy enunciated and verified, and innumerable quantitative relationships established. The direction of physical research will be discussed in the next chapters. Chemical research, in the commonly defined field of radiochemistry,[55] took no direction—it effectively ceased to exist! The questions this science asked had been satisfactorily answered. The opportunities for basic research in radiochemistry were now limited. Radiochemistry was suicidally successful. Not until the mid-1930s, following the discoveries of artificial radioactivity and the neutron, was the subject resurrected, then to be called nuclear chemistry. The prophesies of the completeness of science, discussed in chapter one, were not entirely wrong.

15 Physical Investigations

Alpha Particles in the Second Dozen
Years of Radioactivity

Physics in America

For perhaps eight years after Becquerel's discovery of radioactivity in 1896, physical investigations of the emissions predominated. Then, as already described, chemical studies of the emitters came to the fore. Physical research continued, of course, with such notable advances as Bragg's measurement of alpha particle ranges, Rutherford's increasingly precise values for the alpha's charge-to-mass ratio, proof by Rutherford and Royds that the alpha was a charged helium atom, and the Geiger-Marsden work on alpha scattering which led to Rutherford's concept of a nuclear atom.

Discussing these advances in greater detail while reviewing Rutherford's 1913 book, *Radioactive Substances and Their Radiations*, Robert A. Millikan noted that there were now included several areas of physical research that had not appeared in the 1905 second edition of Rutherford's classic text, *Radio-Activity*. [1] The Chicago physicist, already famous for his work on the electron and a figure of growing stature in the American scientific community, listed the following:

1. Range of the alpha particle, and the laws of its scattering and absorption when passing through matter.

2. The phenomenon of radioactive recoil.

3. Methods of directly counting alphas, especially by the scintillations they cause when striking a fluorescent screen.

4. The laws of scattering and absorption of beta particles when passing through matter.

    5. The relationship between beta particles and gamma rays.
    6. Study of the thorium and actinium series.
    7. Evidence for and against the activity of ordinary matter.
    8. The extended age of the earth in light of radioactivity data.

Were Millikan writing at the end of the century's second decade, he might have added questions of atomic and nuclear structure, although these subjects, while deeply influenced by the investigations in radioactivity, were different from it. Of the eight listed above, Boltwood's major contributions to geochronology have already been discussed, as have the advances in understanding the thorium and actinium series. This latter work, which increased the number of known radioelements from about twenty to over thirty in the years between Rutherford's two books, has been treated in the chapters on radiochemistry in this volume, but its discussion by Millikan—and Rutherford—illustrates the fuzziness of the line between physical and chemical research in radioactivity.

The other subjects all were examined by American scientists. Their work was overwhelmingly experimental and quantitative—the scientific style exemplified by F. K. Richtmyer's remark, "Look after the next decimal place and physical theories will take care of themselves."[2] This attitude has already been discussed in chapter one, though there more in its European context. That it also existed in the United States is unquestioned; indeed, the tradition had solid grounding in the Yankee-ingenuity, mechanical-technology, applied-science orientation of the New World and was not counterbalanced by any "schools" of theoretical science.[3] Henry A. Rowland, whose precision ruling of diffraction gratings played no small part in the American—and worldwide—advances in spectroscopy, was emblematic of this ideal. "It is in the realm of the definite rather than in the semi-speculative region beyond the boundary of the tangible," a colleague noted, "that Rowland's contributions to science lie. He selected by preference problems the solution of which must consist of precise quantitative results obtained by means of measurements with perfectly designed and accurately constructed apparatus, and in such work he had no superior."[4]

Rowland was the first professor of physics at the Johns Hopkins University, opened in 1876 as an institution designed to bring German-style graduate education (meaning research) to the United States. He and his university succeeded outstandingly, for by 1910, out of 2,513 doctorates awarded in all sciences since Yale granted the first to Arthur W. Wright in 1861, Hopkins had conferred 434. Its closest competitors, grouped in the mid-200 range, were Chicago, Yale, Columbia, Harvard, and Cornell.[5] In this sense, Rowland, perhaps the most eminent experimental physicist of his generation in the United States, clearly set his mark upon the tastes of his profession. His counterpart as the most distinguished mathematical physicist, and a man who, like Rowland, died shortly after the turn of the century, was Yale's Josiah Willard Gibbs, who had but few students. There was, not surprisingly, no continuing tradition of theoretical physics.[6] On occasion someone as brilliant and able as William Draper Harkins of Chicago (a chemist, and also an experimentalist) might appear, but most other theorists seem to have been of the armchair-dabbler genre, almost throwbacks to the nineteenth century and earlier type of gentlemen amateurs of science.

Nor should this necessarily be considered unusual, for the editor of *American Men of Science* listed eighteen men in 1910 who pursued science without occupying formal positions.[7] The physics profession, especially, was then quite young and not overly robust. The *Physical Review* was founded only in 1893, as a private venture supported by Cornell University. Not until 1913 did the American Physical Society, itself created in 1899, assume financial responsibility for the journal. Another measure of the adolescent state of domestic physics is the circumstance that over half the advertisements for apparatus in the 1915 *Physical Review* were placed by manufacturers in Europe, even during wartime.

Yet despite this clear depiction of something less than physical hegemony, the total of all work accomplished in radioactivity was a respectable contribution to the subject. In that it extended to all of Millikan's listed topics of interest, it was an accurate reflection of the worldwide advances; in that the work was published in a number of European journals, as well as domestically, but that few if any notable discoveries were made, it properly mirrored the position of the United States in the international physics ranking.

We are not interested in making a quantitative study of changing tastes in physics, but it does seem clear from an examination of the contents of the *Physical Review* over the century's first two decades that American physicists increasingly directed their interests to the "modern" areas of their profession. Studies in electricity and magnetism, spectroscopy, and optics were most popular around 1900. Typical of major and minor concerns were such papers in the 1903 *Physical Review* as "An explanation of the false spectra from diffraction gratings," "Derivation of equation of decaying sound in a room and definition of open window equivalent of absorbing power," and "A study of the effects of temperature upon a tuning fork."[8] Indeed, diffraction gratings and acoustics were topics to which American physicists made particularly valuable contributions, but they were soon destined to be somewhat disparagingly denoted as "classical" studies, for fascination with various forms of radiation and particles grew steadily. By 1919 the *Physical Review* still had its share of papers dealing with diffraction figures, vibrations, and tuning forks, but one senses a significant movement closer to the cutting edge of physics. The operative word, however, is "closer," for most papers on modern science were spin-offs or reviews of the major advances abroad. This period, then, is important not so much for its original contributions as for nurturing the concepts and techniques of modern physics to a level of quite acceptable professionalization, for serving as the nursery for those who, in the 1920s and 1930s, would raise the United States to the first rank in physics.

Alpha Particle Studies of Ionization, Absorption, Range

Even before the alpha particle's identity was ascertained, ionization, absorption, and range measurements had been widely conducted; they continued long after Rutherford and Royds proved in 1908 that alphas were charged helium atoms. The reasons for this concern are not difficult to apprehend: ionization studies formed a field of interest apart from the manner of production of the phenomenon, as the early work in electronics shows, while data on the range, and hence energy, of emitted particles, in air and in other absorbers, were seen as a key to knowledge of the structure first of the atom, and then of the nucleus.

As mentioned in chapter three, Bragg and Kleeman in Australia had shown in 1904–1905 that alpha particles from different radioelements possess specific ranges, a discovery confirmed the next year by Princeton's E. P. Adams, who used a different detection system and measured the "stopping power" of additional gases and vapors. This work illustrated the complex behavior of alpha radiation, and overturned the original belief by Rutherford and McClung that it followed an exponential absorption law. Since McCoy's 1905 study of the radiation from thin layers and films of uranium compounds, discussed in chapter six, lent support to the idea of exponential absorption, it too bowed to the new interpretation. Similarly, Boltwood's comparison in 1906 of his relative activity measurements of different products with the Bragg-Kleeman data, also in chapter six, and his conclusion from their close correspondence that the ionization produced by each alpha is proportional to its range, could not stand before the increasingly precise data being collected. The picture was not at all as simple as originally conceived.

Typical of those taking a more careful look at the details of alpha particle behavior was Samuel J. Allen, the Nova Scotian who had earned his master's degree at McGill and served as Rutherford's demonstrator from 1900 to 1903. He had since received his doctorate at Johns Hopkins and in 1906 joined the faculty of the University of Cincinnati, where he remained the rest of his career. Allen was long concerned with ionization measurements and, since he did not have access to highly radioactive sources—indeed, he frequently used the active deposit obtained from air—he logically sought to use his resources to best advantage. This involved measuring not the maximum range of an alpha particle but the distance from the source of maximum ionization current. The two, he learned in 1908, were not identical, the maximum current position being only about half the range when the radiation from a flat layer of material was allowed to travel in all directions. When, however, the rays were collimated such that they traveled more and more nearly perpendicular to the surface of the source, the position of maximum ionization approached that of maximum range. Numerous tests showed that, for a given geometry, a definite ratio existed between the maximum ionization and range distances, such that the range of an unknown source could be determined by measuring the other characteristic and applying the conversion factor. Since this method was especially suited for radioactive substances of low activity, it was useful for a variety of radioelements, as well as an assortment of absorbing gases and foils whose stopping power could be ascertained.[9]

Allen was already a published investigator when he began his Ph.D. studies and apparently neither needed nor received any meaningful supervision in radioactivity while at Hopkins. Indeed, had he chosen to pursue his career in the British Empire, such a degree would have been superfluous; Cambridge, for example, did not award the doctorate in science until the 1920s, the designation "B.A., Cantab." heretofore carrying sufficient distinction. In America, however, this imprimatur of accomplishment became increasingly necessary for an academic position.

For a graduate student who wished to investigate radioactivity, there were few physics departments in which the subject was as energetically pursued as at Yale. Paradoxically, this seems to have been minimally due to Boltwood's initiative, for though he belonged to the physics, not chemistry, department during most of the period under consideration, as has already been mentioned he had but few students, and these were chemists. Rather, the teaching interest in radioactivity resided in Henry Bumstead, whose own research in the subject was not overly extensive, but who recognized its importance and rejoiced in the proximity of his good friend and colleague.

Such students of Bumstead as Haroutune Dadourian and Perry B. Perkins have appeared earlier in this story. Both began promising research careers, Perkins even working in Rutherford's Manchester laboratory during the 1913–1914 academic year, but eventually joined faculties of small colleges where teaching responsibilities prevailed. At the same time in 1908 that Perkins completed an investigation of the diffusion of radium emanation, another student, Thomas S. Taylor (1883–

1970), published the first of a series of studies on alpha particle absorption in foils and gases.[10] Taylor provided yet another link with Rutherford, sharing the John Harling Fellowship in his laboratory with Henry G. J. Moseley in 1912–1913. Prior to this Taylor was an instructor at the University of Illinois, and following his bright year in dark Manchester, he joined the Yale faculty. But before long he left for industry, becoming a research physicist first for Westinghouse and then for Bakelite and other corporations.

However, in that half decade or so prior to Sarajevo, Taylor was among the most productive of American physicists studying radioactivity. For his doctoral dissertation research Bumstead directed him to one of the outstanding problems concerning alpha rays. When Bragg and Kleeman in 1905 plotted their ionization curves for the alphas from different radioactive sources, they noticed that particles of different range (or energy) were retarded to different degrees when passing through a given metal sheet. This led to a number of absorption studies in Europe and Canada. The retardation was measured by the decrease in range, compared with the distance the alpha would normally travel in air, and was called the "stopping power" (a ratio) or the "air equivalent" (a distance) for that particular absorbing material. Some observers found the diminution in range due to a foil dependent on the distance of that foil from the source. They also reported that absorption through two dissimilar metals was the same irrespective of which foil the alpha struck first.[11] These results, and those of Bragg and Kleeman were seemingly contradicted by experiments performed by Rutherford and others who reported a uniform loss of energy by alpha particles as they passed through layers of aluminum: each sheet of foil diminished the range by an equal amount. Further, the order in which alphas penetrated dissimilar metals did produce different reductions in range, an effect not satisfactorily explained by assuming a difference in scattering of the projectiles.[12]

Taylor recognized that this confusion about the quantitative manner in which alpha particles lost their energy might be a result of the way measurements had been made. Air equivalents had been determined by the difference in maximum range, with and without the metal screen in place. But since the air equivalents varied so slightly while the positions of maximum range were much larger numbers, difficult to determine precisely, the method did not lend itself to accuracy. Like Allen, who measured the position of maximum ionization rather than range, Taylor sought to examine a characteristic giving the largest effect.

With a sheet of gold foil over a polonium source, he took a series of careful ionization readings (deflections of the electrometer needle) for quite small increments of source-to-ionization chamber distance. Then he placed the source at a fixed distance from the chamber and varied the source-to-foil distance, a less refined measurement because of the unevenness of the sheets. By comparing plots of these two types of data he was able to relate the somewhat gross movable foil values to small variations in the movable source figures, thereby obtaining air equivalents. When he did this with several different metals and with different thicknesses of the same metal, he was able to show conclusively that air equivalents did indeed decrease as the foil was moved away from the source. Rutherford and the others had erred in believing each layer of aluminum had an identical air equivalent because they happened inadvertently to experiment with a metal whose values changed but slightly and concentrated upon a rather flat portion of this flat curve. Moreover, their range measurements were not sufficiently refined for the true picture to emerge. The matter of absorption differences depending upon the sequence of passage through two different metals was a real effect and could be explained by the connection of retardation with the energy of the alpha as it entered each foil.[13]

Taylor subsequently went to great lengths to assure that his results were not due to the scattering suffered by alpha particles in their collision with atoms of gases and solids. He did indeed confirm Geiger's discovery of scattering by metal foils, but by showing that the ionization curves plotted with and without such screens were parallel, he proved that the phenomenon had negligible effect upon his work. He also extended his investigation to radium C, whose alpha particle has a greater range than that of the polonium used earlier, and examined absorbers of hydrogen, paper, and celluloid. For the new substances of atomic weight near that of air (really the average of the atomic weights of the constituents in paper and celluloid) he found the air equivalents remained constant with distance from the source; for hydrogen with a lower atomic weight the air equivalents actually increased. The alpha absorption characteristics of different materials thus showed a spectrum of behavior.[14]

Complementary work was done by Hans Geiger, who measured the number of ions produced at different distances from a radioactive source, a quantity that increased with decreasing velocity and then dropped rapidly to zero. From his studies it appeared that the ionization produced at any point by an alpha particle and the energy it lost while passing through matter were proportional, a result that meshed satisfactorily with the interpretation of Taylor's data.[15]

Taylor in 1911 confirmed this. He plotted the Bragg ionization curves in several gases, finding close agreement between theoretical and experimental values. Since the theoretical formula included a constant for each gas that was dependent upon the total ionization produced (or, alternatively, upon the energy needed to produce an ion in each gas), the matching of its curve with the experimental data showed the continuous proportionality of ionization and energy loss at any point. Geiger had shown the ionization at any point to be inversely proportional to the cube root of the alpha's remaining range. Taylor integrated this to find the total ionization to be proportional to the two-thirds power of the average range. The different gases showed a considerable variation in the energy required to produce an ion, with far more required to ionize simple molecules than heavy or complex ones. In a brief, but, for an American, rare willingness to venture an interpretation, Taylor suggested that the electron structure in the latter type of molecule is less stable and thus more easily ionized.[16] The next year he quantified this work further. Division of the measured total charge carried by the ions in a given period of time by the standard value of the ionic charge gave the total number of ions. Division next by the number of alpha particles emitted by a polonium source in the same time period, as measured by the counting of scintillations, gave 164,000 as the number of ions produced by a single alpha particle, a value in good agreement with that found shortly before by Geiger. It was but a short step more to calculate the alpha's kinetic energy by accepted values of initial velocity, charge, and charge-to-mass ratio, and divide it by 164,000 to obtain the energy necessary to produce an ion in air of 5.3×10^{-11} erg.[17]

It was also but an extension of his previous approach to encase his apparatus in an electric furnace such that the ionization curve for mercury vapor might be ascertained. In keeping with his previous results, Taylor found that the energy to produce an ion in this mist of heavy atoms was but 0.72 of that required in air, lower than that for any lighter substance yet tested.[18]

Part of the attraction Rutherford's laboratory may have held for Taylor was the presence of Geiger, who had worked there for years on related alpha particle studies. But two weeks before Taylor's arrival in Great Britain, Geiger departed for his native Germany, bearing a gold watch and Rutherford's good wishes for his future academic career.[19] Taylor nonetheless completed two good investigations during his visit, one in collaboration with Ernest Marsden and the other alone. In the first he learned the technique of deflecting charged particles in a magnetic field in order to obtain measures of their velocities. Such data were of considerable and immediate value, for Rutherford's picture of the atom, having a tiny nucleus surrounded by electrons, which was advanced in 1911, had in the next two years been further detailed by the analyses of his colleagues C. G. Darwin and Niels Bohr. These two had attempted to learn more about the number and distribution of the external electrons by studying the velocity losses by alpha particles in passing through matter.[20] However, the only data they had to work with consisted of velocity curves in aluminum and mica, determined by Rutherford and Geiger, respectively, and these were in some cases flawed by the earlier assumption that air equivalents remained constant with distance from the radioactive source.

Marsden and Taylor therefore measured the velocity of alphas before and after passing through sheets of gold, copper, aluminum, mica, and air, and also tested varying thicknesses of the absorbers. Their work provided a body of useful information, particularly an empirical relationship between velocity and remaining range, and shed further light upon the curious minimum velocity at which the alphas could be detected. In these experiments measurements could be made no lower than 0.41 of the initial velocity of alphas from radium C; Rutherford earlier had obtained similar results, while lower values by Geiger were properly regarded as spurious. Alpha particles traveling below this cutoff velocity produced neither ionization nor scintillations. Photographs in C. T. R. Wilson's new cloud chamber of the many large angle deflections near the end of alpha tracks suggested some other process than single scattering was occurring. Alternatively, the alpha particle might acquire an electron as it slowed down, thus changing its characteristics—a supposition shown to be correct (though actually more complicated) several years later.[21]

Taylor's other investigation in Manchester was far more an extension of his previous work. Though the behavior of alpha particles in elementary gases was of particular theoretical interest, no one until this time had examined the range of radium C alphas in helium or plotted an ionization curve. Taylor did both, comparing the ranges of alphas from radium C and polonium, in hydrogen, oxygen, and air, as well as helium, showing a constant ratio in all the gases, and determining from the ionization plots that an alpha particle produces about 5 percent more ionization in helium than in hydrogen or air. As in all his earlier studies, the ionization curve for helium fit well with Geiger's formula relating ionization to the inverse cube root of the remaining range. The scintillation-counting technique used for these range measurements was well suited for a close look at the manner in which the number of alphas diminished at the end of the range in different gases, something Geiger had done only for air. The curves did not drop precipitously, but rather fell off smoothly, those for air and oxygen suffering a more gradual decline than the hydrogen and helium plots. This Taylor interpreted as conforming to Rutherford's theory of alpha scattering, the heavier the gas through which the particle passes, the greater the deflection from its original path.[22]

Taylor's 1911 results showing that the total alpha particle ionization was proportional to the two-thirds power of the average range were reinforced two years later by McCoy and Viol. The Chicago chemists, although believing the empirical relationship to be only a good approximation, nevertheless found their measured values of the relative activities of thorium products to be in accordance with the two-thirds power of their respective ranges.[23] With another student, Edwin D. Leman, McCoy later reached the same conclusions for the members of the actinium and radium series.[24] Besides providing something more than a rule-of-thumb connection between range and activity, such close agreement gave reassurance that these series were well understood, reassurance that perhaps was superfluous in view of the recent publication of the group displacement laws. A somewhat similar sort of reassurance was provided about the same time by Alois Kovarik in the physical laboratory of the University of Minnesota. It had long been assumed that the range of an alpha particle depended on the number and type of atoms in its path. Range might therefore be expected to be inversely proportional to pressure and directly proportional to temperature. Kovarik tested the range of the alphas from a polonium source at temperatures from 90° to 362° absolute and as anticipated reported a constant ratio.[25]

Delta Rays

Taylor's results in another area intersected additional investigations, this time at his alma mater, Yale. As described above, he investigated the amount alpha particles were stopped by different materials placed at varying distances from the radioactive source. For metals having atomic weights greater than that of the average of air, the alpha's range was reduced less as the foil was moved away from the source. In other words, the air equivalent decreased with distance. Taylor showed that the phenomenon was proportional to atomic weight, and for a substance lighter than air, such as hydrogen, the effect was reversed and the air equivalent increased with distance. In hydrogen, therefore, slow moving alphas were more retarded than fast moving ones. A graph of this effectiveness of retardation with distance bore some similarity to a Bragg curve in hydrogen of ionization with distance, and it was natural to assume that the energy lost by an alpha particle must be consumed in the production of ions.

Since the retardation effects appeared in a continuum across the range of atomic weights, it occurred to Bumstead that energy losses in solids might take place by a process analogous to gaseous ionization. He therefore bombarded metal foils with alpha particles of different velocities, and showed that the emission of slow-moving secondary electrons, or delta rays as they were named by J. J. Thomson in 1904, varied in precisely the same manner as the production of ions in a gas. The number released from the metals increased as the alpha particle neared the end of its path, and then rapidly decreased. In conformity with Taylor's work, fewer delta rays were yielded from metals than were ions created in a gas, but then Bumstead's analogy broke down. Gold and aluminum foils yielded results that were extremely close, whereas their great atomic weight difference led to the expectation that far fewer deltas would be released from the gold. By contrast, marked differences in ionization were known among gases and vapors: the lower the atomic weight, the more pronounced the maximum of the curve. If metals do not suffer ionization, he concluded, perhaps the effects observed were due to a film of adsorbed gas on the metals.[26]

To explore this uncertainty, Bumstead asked Alexander Graham McGougan (1881–1924) to join him. McGougan had received his first two degrees from post-Rutherfordian McGill, while for his part in this collaboration he earned his Yale Ph.D. in 1912. He subsequently served as an instructor in New Haven, and ultimately became a professor at the University of Saskatchewan. Using copper, platinum, and lead, in addition to the aluminum and gold of the earlier experiment, and taking more readings to obtain plots of greater precision, they found that the production of delta rays decreased as a vacuum was produced. This was in keeping with the supposition that the deltas were due to a film of gas on the surface of the metals and that this adsorbed gas was gradually being removed. But the circumstance that the shape of the "ionization curve" did not change during evacuation meant that the layer was not being removed entirely. Support for this view was seen in the virtual congruence of the curves for different metals; such a result could not be due to the metals themselves. However, this interpretation was rendered questionable when they raised a strip of platinum to a red heat in the vacuum, and the assumed film was not removed as expected.[27]

The matter was left unresolved for the moment, as they chanced upon a new phenomenon in the course of their experiments and wisely chose to examine it. As the number of absorbing foils (to slow the alphas) was decreased, the delta ray current also decreased, in a regular Bragg-like curve. Yet, when there remained no obstacle between the polonium source and the target, the current more than doubled. This they found was due to a very absorbable radiation having at least two components: one part consisted of electrons possessing velocities considerably greater than those associated with delta rays, while the other was comprised of recoil atoms from the radioactive source, an explanation offered by Ludwik Wertenstein in the Curie laboratory, who made the discovery simultaneously.[28] The swift-moving electrons were secondaries from the source, and both they and the recoil atoms were able to produce deltas when they struck the metal target; alpha bombardment was not the only mechanism. Indeed, Bumstead and McGougan wondered aloud how much of the delta production was due directly to the alphas and how much to swift secondary electrons released by the alphas from the metal. If this were the process of delta emission, it might hold true also for ordinary ionization of gases. Perhaps, they speculated, the column of ionization surrounding an alpha track in C. T. R. Wilson's cloud chamber was due largely to the action of secondary electrons created by the alpha's passage.[29]

Following McGougan's graduation, Bumstead tried to bring some order to the subject of delta rays. He chose to include in this designation all electrons produced by the direct action of alpha particles, whether these electrons had virtually no velocity or approached within about 10 percent of that of light. The newly discovered swift electrons, therefore, were embraced by the revised definition, while the slow electrons produced by any deltas, and in much greater number than the deltas, were called tertiary electrons. The tertiaries were recognized as a source of confusion, since they originated in the same places as deltas and their presence in a beam of deltas made it impossible to know the number of true deltas of very slow velocity.[30]

Before he left Yale in 1917, McGougan returned to the problem of the production of deltas by metals. By scraping the surface of various metals while in a high vacuum, and by breaking the surface film by overflowing a container of mercury, he obtained clean metal targets for the alpha particles. In the former case the emission of deltas was indeed reduced, though the surface recovered its normal condition shortly. No change in delta emission was observed for mercury, whose ionization curve had the same shape as those from other metals. By interpreting the results as due to deposition of residual gas in the chamber upon the scraped surfaces, and to molecules of gas within the liquid mercury, McGougan considered the concept of deltas arising from a film of gas on the surface as well supported.[31]

Columnar Ionization

A topic related to delta rays which evoked considerable interest in the century's second decade and to which Yale scientists made notable contributions was that of columnar ionization. It had long been known that the maximum, or saturation, current through a gas ionized by alpha particles required a fairly large potential gradient between the electrodes of the ionization chamber. This voltage was greater than would be needed if the positive and negative ions were uniformly distributed in the volume, with some recombining due to their attractive forces instead of being separated by the pull of the electrodes. In 1906 Bragg and Kleeman suggested that the ions were not evenly situated in the gas; rather, there was a likelihood that many recombined shortly after formation, when they were yet quite close to one another.[32] Kleeman, however, subsequently showed such variability in the phenomenon under certain conditions that "initial recombination" was viewed skeptically.[33] Marcel Moulin, at the École de Physique et de Chimie de Paris—the institution where the Curies did their early work on radioactivity—then followed a suggestion by Paul Langevin and investigated the possibility that the ionization was concentrated, at least initially, about the paths of the alpha particles in many "columns." Recombination of ions within these ionized tubes would, of course, be greater than elsewhere in the chamber. If the electric field was applied perpendicular to the alpha trajectories, the columns would be broken up faster and a saturation current achieved at a lower voltage than in the case where the field was parallel to the alpha tracks, expectations that Moulin confirmed.[34]

F. E. Wheelock, a graduate student working under Bumstead, in 1910 examined columnar ionization in considerably greater detail than had Moulin, finding general agreement with him. In particular, he showed that the ratio of current produced by strong and weak sources did not depend on the strength of the parallel electric field, a circumstance easily explained if an alpha particle from a given source always created the same number of ions in a column. Increasing the voltage would not attract any more ions. But his experimental results differed somewhat from calculated values and he was inclined to attribute the discrepancy both to variation of ionization within the columns (since the alphas ionize best when moving slowly) and to some real initial recombination of the sort Bragg and Kleeman had proposed.[35]

Yet another of Bumstead's students looked at columnar ionization when the columns were distributed randomly through a volume. To do this, William Barss used as his alpha source emanation mixed with the gas to be ionized. He argued that previous experiments showed that with such geometry the voltages normally used to measure ionization current could not have produced saturation. But the point, while valid in theory, and really significant since most radioactivity readings were thereby suspect, in practice was academic since Barss' own data showed that the ratio of one weak source to another remained constant over a wide range of voltages. In other words, accurate comparative measurements were possible under conditions below saturation. The ratio would be more likely to vary when strong sources were used, since more ions could recombine with ions in other columns, but Barss, while recognizing the likelihood, did not perform the experiment. His work, therefore, while advancing an understanding of the phenomenon, in no way changed procedure.[36]

The Yale commitment to delta rays and columnar ionization may have been influenced by the circumstance that for several years Boltwood gave the physicists his entire growth of polonium, a pure alpha emitter. It was also encouraged by the fact that Bumstead had succeeded in establishing a "school" or "tradition" of radioactivity, probably the only one in the country aside from that at Minnesota and McCoy's. Indeed, though McCoy's own work was of greater significance, it was Bumstead's group that far more attracted scholars from other institutions and sent them on for further work to foreign laboratories. In this case, at least, the nineteenth-century tradition of scholarly mobility, stronger at that time in chemistry, was reversed in the twentieth century in favor of physics.

The Yale tradition was next carried on by Edward M. Wellisch (b. 1882) and Jay W. Woodrow (b. 1884). Wellisch, an Australian who had graduated from the University of Sydney and then done advanced work in Cambridge, was an assistant professor in New Haven; later, he returned to the Antipodes. His student, Woodrow, had been a Rhodes Scholar at Oxford, a physics instructor at the University of Illinois, and then a graduate student at Yale; he later held faculty positions at Colorado and Iowa. While at Illinois, Woodrow and T. S. Taylor were instructors in the same department. Conceivably, Taylor, already a Yale Ph.D., portrayed New Haven's virtues to his colleague.

Wellisch and Woodrow not only carried on Wheelock's line of research, they used his abandoned apparatus to refute one of his conclusions, by showing that the initial recombination between an electron and its parent atom, of the sort proposed by Bragg, was of minimal importance. Recombination between ions formed from different atoms was of greater significance. They further confirmed Langevin and Moulin's view that a slight slope of the alpha ionization curve, under the influence of a field parallel to the tracks, was due to recombination within the ionized columns and not to ionization by collision, as Wellisch himself had earlier suggested in work performed with Howard Bronson.[37]

Research on columnar ionization continued through the century's second decade, and investigations of delta rays even longer, but the work served to increase precision, not refine interpretation.[38] The capstone to innovative examination was placed by Bumstead in an elegant experiment in 1916. As we have seen, Moulin executed experiments which convinced his colleagues that alpha particles leave columns of ionization behind them, a fact made strikingly visible in C. T. R. Wilson's famous cloud chamber. Bumstead and McGougan, in their pursuit of delta rays, were led to suggest that the ionized columns were due to the tracks of swift secondary electrons torn loose from gas atoms by the alphas' passage, and not due to the alphas directly. If this were true, the cross-sectional radius of a column might be of the order of 10^{-3} centimeter, the range of these electrons. The alternative, i.e., ionization due to the alphas alone, was a radius of about 10^{-5} centimeter, the mean free path of an ion in air at atmospheric pressure. Wellisch and Woodrow, in the work mentioned above, calculated a value close to 10^{-3} centimeter, although they attributed much of that to rapid diffusion. To avoid ambiguity, Bumstead decided to photograph the alpha particle tracks in a Wilson cloud chamber, which he procured from the Cambridge (England) Scientific Instrument Company. At atmospheric pressure the diameter of the tracks was so small and the ion density so great that no structure could be seen. But at reduced pressure, and after overcoming many technical difficulties that tried his experimental skill, Bumstead obtained photographs showing projections from the alpha tracks that in all respects resembled beta particle trails. In keeping with his previous observation of a wide range of velocities of the delta rays, there were tracks of definable length and others that appeared as just knobs or projections from the main trail. Bumstead thereby had proven that a substantial part of the ionization in an alpha particle column resulted from the action of an intermediary radiation: delta rays.[39]

16 Physical Investigations

Beta and Gamma Radiations in the Second Dozen Years of Radioactivity

Beta particles were recognized as J. J. Thomson's corpuscles, or electrons, as early as 1900, while alpha particles were accepted as massive bodies by 1903, although their identity with charged helium atoms was not proven until 1908. Gamma rays, though long regarded as a form of electromagnetic radiation, did not succumb to experimental proof until 1914, when Rutherford and Andrade diffracted them using a crystal as a grating. Knowledge of what the alpha particle was did not, of course, end research about its behavior, as shown in the last chapter. Similarly, the effects and implications of beta and gamma radiations were investigated long after they were removed from the list of unidentifiables. A major motivation for such studies lay in the understanding, slowly dawning after 1911, that radioactivity was a nuclear phenomenon. How, then did these radiations coexist in the nucleus, and what could be learned about the nucleus from the transformations (via the group displacement laws) and the emissions observed? Prior to this relatively sophisticated approach the impetus lay in the desire for better comprehension of the interaction of radiation and charged particles with matter, and what this could tell about the structure of the atom (J. J. Thomson's model, for example), as well as a primitive belief that atoms might deliberately be disintegrated through bombardment by the radioactive radiations.

Beta-Induced Radiation: Secondary or Scattered?

It was mentioned in chapter three that Samuel Allen turned from the study of atmospheric radioactivity to beta rays when he unexpectedly found it difficult to measure their magnetic and, especially, electrostatic deflections. Photographic proof, as obtained by Becquerel, gave certain values, but when the first-year graduate student at Johns Hopkins sought to construct a lecture demonstration using an electrometer or electroscope to measure the deviation, he was at a loss to explain his poor results. This was all-the-more embarrassing, since the experiment was for use by his professor, Joseph S. Ames, who knew little about radioactivity. An urgent appeal was mailed in early 1904 to his former mentor, and while Rutherford's reply is not preserved, it is likely that he urged Allen on to what became his Ph.D. research topic.[1]

The difficulty, in a nutshell, was due to the electrostatic disturbances set up by the action of secondary and tertiary radiations, of velocities nearly those of the primary beta beam, and with the same charge-to-mass ratios for corresponding velocities. It was to be expected that a true secondary (or lesser) radiation would have precise characteristics that were different from the incident beam, since it would involve a specific atomic rearrangement, energy level change, or work function.[2] That such radiations were produced by bombardment of various targets by beta rays was just at this time becoming apparent; indeed, A. S. Eve's early investigation of the phenomenon at McGill may have been stimulated by Allen's problem, although his published explanation was the analogy to secondaries produced by x-rays.[3] This was followed shortly by a report from H. F. Dawes, a student of John C. McLennan, of Toronto, who did little more than confirm Eve's finding that gamma rays also produce secondary radiation,[4] and by a series of papers by Professor J. A. McClelland, of University College, Dublin. McClelland, whose research student days in the Cavendish Laboratory had overlapped those of Rutherford, found the greatest amount of secondary radiation was emitted normal to the surface, as would be expected if the rays coming from some depth in the plate were to have the shortest distance to traverse. Superposed on this secondary radiation was another radiation of negatively charged particles, which he concluded were reflected primaries, since the maximum intensity was measured when the angles of incidence and reflection were equal. McClelland also showed that the intensity of the secondary radiation (commonly taken to be the combined "true" secondaries and the reflected primaries) increased with increasing atomic weight of the emitting material. This led him to suggest that the production of secondaries by the impact of beta rays involved the disintegration of atoms which are normally stable, a process somewhat analogous to the spontaneous radioactivity of the heaviest elements.[5]

Further progress was slow, both because it was difficult to obtain unambiguous results with the beta particle, and because the alpha particle offered a more interesting object. Rutherford's 1905 Silliman Lectures at Yale, for example, contained but a few pages on the beta when published the following year.[6] When a new ripple of interest occurred as the decade ended, the experimenters seemed still plagued by irreducible data. Clinton Davisson (1881–1958), a graduate student and physics instructor at Princeton, showed some of the technical finesse that brought him the Nobel Prize for electron diffraction nearly thirty years later, yet failed to determine clearly whether the impact of beta particles upon solids produced a nondeviable radiation such as x-rays.[7] Likewise, William B. Huff (b. 1866) was conscious of the problems of unwanted alpha and gamma primary radiations, nonparallel beta paths, and air absorption of the betas, and endeavored to eliminate or minimize the undesired phenomena, but his results could by no means be called conclusive. At Bryn Mawr, a women's college modeled after Johns Hopkins, which offered the doctoral degree in each of its departments, he did confirm that the intensity of the secondaries produced by beta rays of uranium X increased with the atomic weight of the elements from which they emerged, but felt frustrated by the completely different ionization readings obtained for varying arrangements of the apparatus. If his labors were not entirely satisfactory, much blame may be placed upon the refractory beta particles, for the experiments were ably conceived and executed, a measure of the growing competence in physics in America. Huff, with his graduate education at Chicago and Hopkins, and a year abroad in Cambridge, may be considered somewhat typical of the class of American physicists who were experimentally adept and increasingly sophisticated in the ramifications of their investigations. His paper's closing remarks, for example, noted that no selective absorption had been detected, which "may be held to indicate that the secondary is largely returned or scattered primary, rather than independent, though characteristic, radiation emitted merely as a result of incident primary."[8]

Before this modification of McClelland's composite radiation had time to become widely appreciated, there appeared several other papers on secondary betas by Samuel Allen. After graduating from Hopkins he took a post at the University of Cincinnati, an institution which pioneered the concept of "co-operative" education, wherein students gained industrial experience before completing their academic requirements for the bachelors degree.[9] Allen must have felt comfortable in this straightforward and practical atmosphere, for it was in harmony with his own research tastes. "It is not the intention of the author in this present paper," he wrote in 1909, "to advance any new theory of secondary radiation, or to reconcile the facts therein described with the present theories. . . . it is too early yet to put forward any hypothesis to account for the author's results, even if a logical one were at hand." Instead, he

SAMUEL J. ALLEN (1877–1966)

advocated more work, more data, and then perhaps the solution might unfold. Caution was preferable, for "we are perhaps too liable at the present time to advance incomplete and illogical theories, and it is an open question whether our true scientific knowledge is increased by such a procedure, if indeed we do not in many cases delay the truth."[10]

If Allen was not a dreamer, he was a thorough, hard-working, and intelligent experimentalist. If he was reluctant to spin theories, he was not just a magpie collector of facts. His research had a stated goal, the same as McClelland's: to decide if the atom "can be disintegrated by outside agency."[11] In this period any alteration of the atom might be called a disintegration; the clearest heads recognized that a true disintegration was more than just the removal of an electron, producing ionization. It was associated with chemical change, as in the Rutherford-Soddy theory of radioactivity. Neither extremes of temperature and pressure, nor different chemical combinations affected the rate of radioactivity of a given substance. The only apparent means of disrupting atoms seemed to be to "fire" at them energetic radiations, be they corpuscular or electromagnetic.

Allen consequently fired beta particles at his targets. He too found the intensity of secondaries in the same order as the target's atomic weight and was cognizant of reflected primaries in the secondary beam. In 1909 he decided to examine the radiation stimulated from salts, salt solutions, and pure liquids, almost all earlier investigations having been limited to metals. The mass of data produced was, however, unmanageable, no doubt a major reason for Allen's unwillingness to theorize. His assumption, strongly criticized by W. H. Bragg, of Leeds,[12] that the secondary radiation from a salt should be a function of its molecular weight left too many anomalies among the substances examined. Bragg argued for an interpretation wherein the encounter was between the primary beta beam and individual atoms, regardless of their chemical composition, but Allen felt that his results, particularly with liquids, showed that no simple atomic theory was satisfactory, and that an explanation somehow involving molecular geometry or forces was required.[13]

Half a year later Allen reported upon experiments involving variation in the angle of incidence of the primary betas. Here again, he extended the work of others to solutions and liquids. While some substances still yielded anomalous results, great consistency was seen in increased radiation from lighter atoms as the angle of incidence was increased. This suggestion of strong reflection, buttressed by the fact that at large angles of incidence the penetrating power of the secondaries was independent of the atomic weight of their source, induced Allen to join those who suspected that secondary radiation was nothing but scattered primary, and not the key to atomic disintegration.[14]

By early 1911, this once promising corner of radiation physics was recognized as mythological. Criticism by Bragg and others and some self-doubts led Allen to reexamine his methods and materials. The presence of vapor above some highly volatile liquids had given incorrect data; more damaging was the mislabeling of liquids given him by chemical colleagues. With these defects corrected, his anomalous results disappeared. The intensity of the secondary radiation depended not on some molecular geometry, but on the atomic weight and number present of the target atoms, while the change in secondary radiation with angle of incidence was the same for compounds as for pure elements. There was no need to interpret secondary beta radiation as anything but primary betas deflected from their paths by collision with atoms of the "radiator."[15]

Beta Reflection, Absorption, and Scattering

Recognition that secondary rays were merely scattered primaries might have been expected to aid in understanding other beta phenomena. The transferable benefits were slight, however, for the subject remained confused and filled with pitfalls. Whereas an alpha particle produced an unambiguous flash on a scintillation screen, even in the presence of strong beta and gamma radiation, beta counting was more open to question. Not only did the gamma rays impair electrical registration of betas, it was uncertain whether only one particle was emitted in each beta decay. Moseley, in 1912, showed that this was probably so in two cases, and the group displacement laws of the following year mandated such a decay pattern, yet considerable puzzlement remained because a single type of radioactive atom seemed able to produce betas of widely varying velocities. By contrast, alpha particles were known to have identical velocities of emission from a given source and to suffer precise reductions in this velocity after passage through an absorber. If, like the alphas, beta particles were considered a primary effect of transmutation, why were they not also homogeneous? Experiments conceived to yield the most valuable information about betas required their magnetic deflection, in order to direct a beam with a single velocity into the detector. But the small energy of the average beta was thereby effectively attenuated by the low density of the extracted beam, a condition exacerbated by the presence of strong gammas, and the resulting data were not readily interpreted.[16]

The only American beside Allen to study beta particles extensively was Alois F. Kovarik (1880–1965). He seems to have been drawn into the investigation of radioactivity in much the same way as Rutherford—his initial research on mobility of ions under varying pressure and temperatures[17] introduced him to techniques useful also in radioactivity. Of additional importance, his professor at the University of Minnesota was John Zeleny (1872–1951), whose research student years in the Cavendish Laboratory (1897–1899) partially coincided with those of Rutherford, and whose own researches over many years were on ionization topics. Zeleny moved to Yale in 1915, and ultimately others, including Kovarik, followed him from the unexpectedly active physics department at Minnesota.

Given his interests and these connections, it is no surprise that Kovarik would wish to pursue advanced work either in Cambridge or Manchester. Thanks to what must have been a magnificent recommendation from Zeleny, he succeeded Boltwood as John Harling Fellow in Rutherford's laboratory during the 1910–1911 academic year. For his postdoctoral research, Rutherford suggested an examination of the interaction of beta particles and matter, then a topic of great controversy. Otto Hahn and Lise Meitner in Berlin had produced evidence in support of an exponential absorption law for betas. This was interpreted to mean that the radiation was homogeneous, a view quite in keeping with their early belief that radioactive atoms emit particles of a single veloc-

HANS GEIGER (1882–1945), ALOIS KOVARIK (1880–1965),
AND BERTRAM BOLTWOOD (1870–1927)

ity.[18] William Wilson in Manchester, on the other hand, determined not only that a given source emitted a spectrum of velocities but that the absorption was not exponential for a magnetically produced homogeneous beam. The connection between absorption and velocity of the beta rays, moreover, was anything but simple.[19] Work by others supported aspects of both these investigations,[20] such that the heterogeneous-homogeneous question clearly was cloudy.

Kovarik recognized that *total* beta absorption followed an exponential law to a good approximation, and based his study on this fact. He also believed that inconsistent results of others were due in part to insufficient regard to the amount of scattered radiation, and determined that his absorption coefficients would incorporate their effect. To this end he measured more carefully than before the radiation of several radioactive sources with and without a series of metallic plates reflecting the betas into the ionization vessel. His results showed that the percentage of reflected rays increased with the atomic weight of the reflector, a fact he connected to similar conclusions from secondary radiation investigations, and that swifter betas seemed, for the most part, to be more readily reflected.[21] Kovarik then collaborated with William Wilson in exploring the dependence of reflection upon velocity. In the experiments above, Kovarik noticed the effect using heterogeneous beta radiation of different energies from different sources. With Wilson he tested magnetically deflected homogeneous betas over a range of specific velocities, confirming his earlier finding that the amount of reflection, as measured by the ionization produced, rose with velocity, reached a maximum, and then fell for the most penetrating betas. But since they recognized the danger of assuming that betas would lose no velocity upon suffering reflection, or that ionization was not velocity dependent, they were reluctant to ascribe the measured variation in ionization to a simple increase and decrease with velocity in the number of particles reflected. At this point it seemed that the more one knew, the more one recognized how little was known.[22]

This set the stage for Kovarik's next investigation in Manchester, in collaboration with Hans Geiger. By assuming that there was little or no velocity-dependent ionization, they measured the ionization produced by a number of radioactive sources and derived from this that sixty-seven ions were produced per centimeter of path in air at atmospheric pressure by high speed beta particles. Certain anomalies in the results seemed inexplicable even if their initial assumption of no change in ionization with velocity was questioned, leading them to suggest yet unknown beta emitters present in their sources, or the simultaneous ejection of two beta particles in a single disintegration.[23] As mentioned earlier, the question was resolved within the next two years: only one particle is emitted in each beta decay. The data collected by Geiger and Kovarik, "contaminated" by extraneous activity, simply were not precise enough to support their assumptions and conclusions.

This work had concentrated largely upon the swifter beta particles. On his return to Minnesota after the summer of 1911, Kovarik examined the slower betas, seeking a relationship between the soft and hard radiations. He did not use magnetic deflection to sort out the different velocities, perhaps because his radioactive materials were weak and filtering out a limited band of velocities would give him an effectively weaker source still. He did compare the heterogeneous radiations from sources of different beta velocities, such as radium D and radium E, and actinium B and actinium D (now C''), and because this would have been difficult using a metal foil as the absorbing medium, he determined the absorption coefficients in a few gases, at pressures up to twenty atmospheres, the ionization being proportional to pressure. The gas technique was useful not only because soft betas might be stopped by sheets of metal, but also because the foils could not be assured to be uniform in thickness.

Kovarik thereby verified the close approximation to an absorption law into the realm of low velocity betas, and was able to determine a set of absorption coefficients. In a sense, he succeeded in his goal of relating the soft and hard radiations, but in a larger sense he failed to add significantly to an understanding of the phenomenon because his papers characteristically were long on data and short on interpretation.[24]

William Huff, who earlier had investigated secondary radiation, in 1912 continued his study, under their new name of reflected beta rays. As before, he was keenly aware of experimental shortcomings: magnetic sorting yields pencils of only approximately homogeneous betas, these particles are easily scattered, their range is long, they ionize relatively few particles per centimeter of path, and thin sheets of metal are never of uniform thickness. Yet, despite these constraints, his work provided evidence that beta particles lose speed in passing through matter. He achieved this result by reflecting the particles from thin and thick layers of metal, observing by absorbers placed in their paths that those from the thin sheets were more penetrating. William Wilson had earlier come to the same conclusion by measuring the radius of curvature of beta particles in magnetic fields.[25] In keeping with the by then common understanding that the interaction of such particles with matter was an atomic phenomenon, Huff also verified that, as measured by their ionization, greater numbers of betas were reflected from thicker plates.[26]

Kovarik, however, had been wise to maintain reservations about a direct relationship between ionization and number of beta particles. At Minnesota he seemingly became less interested in the phenomena of radioactivity than in the techniques and instrumentation used to examine them. The use of absorbing layers of gas, for example, was mentioned above. With his colleague Louis W. McKeehan (b. 1887), who received all his degrees at Minnesota, spent a year in the Cavendish Laboratory, and much later joined Kovarik at Yale, he became enthralled with the point counter. This was a device refined by Geiger from the earlier counter he and Rutherford conceived in which an insulated wire ran down the center of a tube. A potential was maintained between the point (or wire) and the tube just below that required for a continuous discharge. When an ion entered the tube it created numerous other ions by collision, such that a discharge occurred which was counted in some fashion. Since the system was triggered by the entrance of single particles, the device measured with reasonable accuracy their number rather than the amount of ionization produced by them.

Using such a point counter in 1914, Kovarik and McKeehan examined the number of betas reflected and transmitted by metal foils. In both cases they found fewer than had been suggested by earlier ionization measurements, meaning noticeable departures from exponential curves. The explanation, they felt, was obviously that slow beta particles, those that were reflected and transmitted, were more effective ionizers than the fast betas in the initial beam.[27]

The next year Kovarik and McKeehan expanded their work with the point counter, finding its results far easier to interpret than ionization data. In the Cavendish Laboratory, C. T. R. Wilson's cloud chamber photographs of cathode ray and beta particle tracks, and both in the Cavendish and at Minnesota, McKeehan's own experiments with cathode rays seemed to show that single scattering occurred.[28] Yet, whether such abrupt deflections through a large angle, as in Rutherford's picture of alpha particle encounters with atoms, or multiple, cumulative scattering through individually small changes in direction occurred for betas was still a matter of some uncertainty. The data collected by Kovarik and McKeehan, using two point counters to measure simultaneously the transmitted and reflected beams, were explained by them according to single scattering, but one senses that they were more concerned with proving the virtue of their apparatus than discussing how well the data fit the theory.[29]

Their joint work in 1916, using a point counter connected to a string electrometer, the recording device, was apparently the only examination in America of the magnetic spectra of beta emitters. The wide range of beta velocities from a single type of radioactive atom, mentioned earlier in this section, was in sharp contrast to the unique energy of a given type of alpha particle. By the end of the century's first decade distinct groups of beta velocities were found in some sources, chiefly by the Berlin team of Otto von Baeyer, Otto Hahn, and Lise Meitner, and as more precise focusing spectrographs were employed numerous lines were detected. Before World War I interrupted most European research the continuous nature of the beta ray spectrum was established, and the lined spectrum superimposed on it was explained as due to extranuclear electrons ejected from their orbits by gamma rays, in a process that came to be called internal conversion. The continuous spectrum was believed to result from the beta particles' loss of energy during collisions with outer electrons as they emerged from the atom, a view that received major modification in Enrico Fermi's theory of beta decay over a decade later.[30] Kovarik and McKeehan's experiment dealt with none of these concepts; they simply provided a mapping of a few spectra, using a technique other than the customary photographic plate. Indeed, their purpose was not elaboration of atomic processes, but something of a "warm-up" for a quantitative study of the ionization caused by a single beta particle and the dependence of this ionization upon velocity, a study very much in keeping with their former line of investigation.[31]

It may be mentioned at this point that the subject was not to be pursued exclusively by the younger generation, although it is true that new fields of research generally attract those early in their careers. One of the exceptions, who intermittently published theoretical articles on radioactivity and atomic structure, was Fernando Sanford (1854–1948) of Stanford University. Before Boltwood or McCoy left elementary school, Sanford had graduated from college; by the time they completed secondary school he had spent two years in Berlin. But his interest in "modern" physics seems to have developed only within the last decade of his career. He appears not to have made any memorable scientific contributions, and his worthwhile contention, that the beta ray absorption coefficient for different elements is dependent upon the electrical density of these absorbers rather than upon their mass density, appears to have excited no replies.[32] Similarly, his view that gamma ray energy is that of the beta particles, less a work function, was but a modification of an attempt by Rutherford to relate the two emissions, and Sanford's argument from such data that Planck's constant is *not* constant again seem to have caused no stir.[33] Yet, Sanford is interesting as probably the only one on the west coast of America concerned with atomic matters, and as one of the few persons throughout the entire country who sought to extract from experimental data different conclusions than the authors of those figures.

Another of these rare individuals, though of considerably greater theoretical ability, was H. A. Wilson (1874–1964). Unlike his countrymen James Jeans and O. W. Richardson, who came to the new world but briefly, Wilson spent thirty-five years, from 1912, at Rice Institute in Houston. He had earlier been a fellow of Trinity College, Cambridge, professor of physics at King's College, London, and professor of physics at McGill, just a few years after Rutherford's departure. His insights and skills had been honed under J. J. Thomson's tutelage and his contributions to science recognized by election to the Royal Society. By the time of his paper on beta scattering, in 1922, a period beyond the scope of this study, the subject was still unresolved and would remain so for at least another decade, but relativistic effects of the high-velocity particles were being considered, some combination of single and multiple scattering appeared necessary, and the papers were assuming a decidedly mathematical flavor.[34]

The beta ray investigations described in this chapter may be categorized as involving interactions between the radiation and matter, and energy. Secondaries were shown to be scattered primaries, an understanding which fit well with common conceptions about collisions between bodies, and made no special problems with the major atomic models. Had the other interpretation been sustained, then characteristic radiations, or resonances, among the secondaries would have indicated internal rearrangements within the target atoms. The other reflection, scattering, and absorption experiments pursued produced many data, some fairly specific ideas about beta particle behavior, and no profound insights. It was a blind alley for future development, however competent the work and however necessary it was to follow the path to see its end. The energy-related investigations, on the other hand, though barely pursued in America and indecipherable during this period, led a decade or so later to fundamental understanding of the weak interaction.

If American work in these areas cannot be neatly summarized and descriptively brought to a climax, this in itself illustrates that scientific efforts cannot always be tidily packaged. Significant advances, by the very definition of being significant, offer landmarks that give direction to the historian's narrative. But much activity in science consists of small steps which, unfortunately, lead to no great insights. These increments, moreover, while useful in filling in knowledge of a subject, often seem incommensurable with other work unless apparatus and procedure are duplicated carefully. This leads to the situation in which experimental and theoretical results are neither accepted nor rejected; they are simply bypassed. The questions posed by those experiments no longer seem vital, for the subject has turned in a slightly different direction. The authors of such scientific papers cannot be said to have been in the right place at the right time.

Gamma Ray Ionization and Absorption

As suggested in the previous section, beta particle studies were not conducted in isolation from gamma ray concerns. The fundamental questions of the latter's origin and relation to beta particles were barely considered in the United States, but the few gamma absorption, scattering, and ionization investigations mirrored examinations of the betas. And like the beta studies, gamma phenomena were sufficiently complex that full comprehension took several decades, extending well beyond the cutoff date, ca. 1920, of this volume.

The man from whom Kovarik probably learned his high pressure techniques, mentioned above, was Henry A. Erikson (b. 1869), his senior by a few years and chairman of the Minnesota physics department from Zeleny's departure in 1915 until 1938. In the tradition created by Zeleny, Erikson in 1908 published an extensive study of the variation in ionization due to gamma rays, in air and carbon dioxide, at pressures up to 400 atmospheres. Curves plotted from the data were similar for the two gases, suggesting that ionization was independent of intermolecular forces. These curves, however, were not simple functions of pressure, but rose rapidly to maxima and then slowly decreased, the maximum value depending on the voltage across the ionization chamber. Erikson argued that, in fact, the number of ions had not decreased, but that the higher pressures implied slower ionic velocities, giving them greater opportunity to recombine. Implicit was the common understanding that gamma rays could not be detected directly, but were manifest through the secondary electrons they produced. Erikson's paper was, therefore, really part of the corpus of alpha and beta particle literature on ionization, with special connection to the Bragg-Kleeman work on initial recombination.[35] Four years later Erikson found his experimental data fit a theoretically derived formula of J. J. Thomson for scattering of gamma rays and revised his conclusion. The diminution of intensity when gamma rays pass through matter, he said, was due mostly to scattering (initial recombination by this time being out of fashion).[36]

Work on a far more significant topic was done by Samuel Allen as the century passed its first decade. More attuned than many Americans to problems facing the international community of physicists, he was aware of discrepant results from investigations of gamma ray absorption. The story may be traced to A. S. Eve, who, along with his study of beta rays in 1904, found that secondary radiation was produced also by gammas. Of greatest interest, he placed absorbing materials in the primary beam and then in the secondary beam, finding the latter to be much less penetrating.[37] This, of course, suggested that the original gammas had been modified in some fashion, for example, by some unknown process, or, more realistically, by a change in wavelength if one followed the wave interpretation of gamma rays, or by a loss of energy if one preferred the corpuscular interpretation. The ensuing great controversy over the nature of gamma and x-radiation, in which C. G. Barkla and W. H. Bragg were the leading figures, will not be described here,[38] but some subsequent developments regarding gamma ray effects need to be mentioned.

In an effort to understand the reduction in hardness, or penetrability, of Eve's secondary gammas, D. C. H. Florance, in Rutherford's Manchester laboratory, measured the ionization of gamma rays from radium, both primaries and secondaries, as a function of the scattering angle. He was concerned also with secondary x-rays, on the assumption that, if the radiations both were electromagnetic and the effects similar, then the mechanism of change could be assumed to be the same. He found that the absorption coefficient increased, or the secondary rays were more readily absorbed, as the angle increased, and, since the results appeared not to depend on the nature of the scattering material, he concluded that the secondary was nothing but scattered primary radiation—analogous to the scattering of beta rays.[39]

Since absorption coefficients were constants, calculated on the assumption of an exponential decrease in intensity of the radiation with increasing absorber thickness, and this in turn was taken as the hallmark of homogeneous radiation, the fact of variable coefficients suggested a heterogeneous beam. The mental picture Florance invoked in 1910 to explain his results was one in which gamma rays were emitted from a source with a spectrum of energies. The apparent softening with angle that occurred as they passed through matter was due to the greater scattering of the softer component.

Allen's contribution was an extension of work by Soddy and Russell and involved measuring absorption coefficients. Despite variability in this supposed constant, the two Glasgow chemists maintained the homogeneity of gammas, arguing that true secondaries were responsible for the recorded departures.[40] Allen essentially confirmed their experimental results and reported in more detail that the " 'stopping power per unit mass' is a function of the atomic weight of the absorbing material," although it was by no means a straightforward relationship. Since liquids behaved in generally the same fashion, he felt the absorption of gammas was an atomic, not molecular, property. Allen stopped short of endorsing the interpretation of Soddy and Russell, noting that the issue was far from resolved. Toward this end he subsequently endeavored to link x-ray and gamma ray phenomena, requiring the production of x-rays that were as hard as the gammas normally obtained from radioactive sources.[41]

The historically important main line of work, however, was that of Florance: absorption as a function of scattering angle. J. A. Gray, another pupil of Rutherford who by 1913 was on the McGill faculty, believed that the primary gamma beam was not heterogeneous, but consisted of identical entities. These were scattered and transformed in quality, not as a function of the thickness of material they passed through (for by analogy homogeneous x-rays were reduced in number though not softened by thickness), but solely as a function of the scattering angle.[42] Gray's investigation employed absorption experiments to determine the character of gamma rays. Even after Rutherford and Andrade in 1914 measured the wavelengths of gamma rays with a crystal spectrometer,[43] this new technique posed such experimental difficulties and uncertainties that absorption studies continued to be the means of examining the primary and secondary gammas' quality.

According to conventional x-ray scattering theory, the primary beam excited electrons of the scattering medium atoms, causing them to oscillate. These electrons, accelerating, reradiated in all directions x-rays of the same frequency as the primaries. Presumably a similar process should have occurred for gamma rays. Gray's work effectively showed, however, that the frequency decreased upon scattering, which was a major step toward Arthur Compton's quantum interpretation of the phenomenon in 1922. The primary beam was softened, as Compton explained, because a portion of the photon's energy was transferred to an electron, setting it in motion.[44]

Investigation of the phenomenon of ionization by Robert Millikan and his colleagues may be mentioned briefly, since radiations from radioactive substances were far more tools for him than themselves objects of study. In 1911, he showed that gamma rays, beta rays, and x-rays all remove but one electron from a neutral air molecule, while nearly a decade later he determined that alpha particles behave similarly, detaching only one electron in encounters with various gas molecules. This work, employing his famous oil drop technique to measure charges, showed that the release of multivalent ions was extremely unlikely.[45]

The radioactivity-connected gamma ray work in America (which does not include Compton's superb contribution) went little beyond the few original investigations described in this section. Aside from a handful of unoriginal studies, and some attention to the presence of gammas in atmospheric radioactivity, and to detection instrumentation, both to be mentioned in the next chapter, this radiation apparently proved at the time to be too esoteric for widespread pursuit. The American gamma ray work was not intimately connected with the retrospectively important topics of electron models, atomic theories, radiation theory, or quantum theory, and indeed the alpha, beta, and gamma ray work taken together included no major accomplishments or areas characteristically "new world." Yet, the contributions were worthwhile and mature, subtle discoveries about the interaction of radiation and matter.

17 Physical Investigations
Completing the Picture of the Second Dozen Years of Radioactivity

Bodies in Motion

Several minor trends in American investigations, beyond those already described, involved the movement of radioactive particles and ions; for convenience they may be surveyed together. Most important, yet barely pursued, was the technique of radioactive recoil. Like the heating effect due to alpha particles and like delta rays, this is a secondary effect of the emission of alphas. When discovered and investigated in 1909, it was shown by Hahn and Meitner in Germany, and Russ and Makower in Britain, to be an excellent way of isolating certain radioactive products. According to Newton's third law of action-reaction, what remains of the parent atom (actually now the daughter) should have a momentum equal to that of the expelled alpha particle. Though the atom's mass is much larger than that of the alpha, its velocity is yet considerable, often of the order of 10^7 centimeters per second. After breaking away from the radioactive source, this velocity makes it behave like a projectile, capable of ionizing gases, gaining and losing charges, and having a range. Given such characteristics, recoil atoms can be collected by placing a negatively charged plate near a plate coated with a radioactive source. The method was especially applicable to the short-lived products of the active deposits, which could not readily be separated by wet chemical techniques, and in this fashion pure actinium C″ was collected from actinium C, thorium C″ from thorium C, radium B from radium A, and several others.[1]

E. M. Wellisch, just prior to his assistant professorship at Yale in

1911, indulged his mathematical bent by interpreting numerous experiments on recoil by others into a theory. Recoil atoms were, he claimed, uncharged except when associated with an ion of the gas in which they traveled. Since the active deposit was found almost exclusively on the cathode, the recoil atoms associated only with positive gas ions. Recoil atoms, in fact, behaved like positive ions in the same gas.[2]

When he was in Manchester, Kovarik learned the recoil technique and even used it to prepare samples for his joint work with Geiger, described in the last chapter. Additionally, he studied the process itself. After his return to Minnesota, he expected to continue this investigation, but was unable to do so. When interest in recoil kept rising, and following a gentle push from Rutherford, he decided in 1912 to publish his preliminary results.[3] He was interested principally in determining if recoil atoms could be collected on a negatively charged plate when a "barrier" of strongly ionized air lay between that electrode and the source. His reasoning, verified by experiment, was that the ionization would neutralize the charges on or associated with recoil atoms and fewer would reach the plate. In further studies, done without an external ionizing agent, Kovarik showed the expected increase in recoil atoms reaching the cathode as the voltage difference between the plates was raised, and an increase also when the plate spacing was decreased. The net result of this work was a confirmation of the anticipated behavior of recoil atoms based upon a substantial familiarity by then of the properties of ions.[4]

Kovarik's colleague at Minnesota, Louis McKeehan, a few years later performed another of these "tidying up" investigations. A number of people had looked at the diffusion of actinium emanation and the distribution of its active deposit, but the experimental conditions precluded a clear-cut explication of the process. McKeehan, by simplifying the geometry of the apparatus, was able to show that the laws of ordinary gaseous diffusion and of radioactive recoil sufficed to explain the effects observed.[5]

But prior to this work in 1917, there had been a fair amount of effort worldwide to study the diffusion and other motions of radioactive gases. Like recoil, these resulted in a redistribution of material; and again like recoil, the information gleaned generally fit well with commonplace predictions. At the Christmas, 1911, meeting of the American Philosophical Society, Wellisch and Howard Bronson, the latter by then at Dalhousie, reported upon the distribution of radium's active deposit in an electric field; three months later their joint paper appeared simultaneously in the *Philosophical Magazine* and the *American Journal of Science*. Rather like Kovarik's recoil experiment, they wished to learn if extraneous ionization, caused for example by an x-ray source, affected the distribution of the active deposit on the electrodes. They decided it did not, but in the process concluded—based on an extension of Wellisch's theory of uncharged recoil atoms—that there were no negative carriers. All of the small amount of radioactivity on the anode was due to the diffusion of uncharged particles of active deposit. Positive ions in the gas, formed by the x-rays or alpha particles, carried most of the uncharged "restatoms" of the active deposit in the opposite direction, to the cathode, though no explanation was given why these neutral particles were uniquely attracted to positive ions. Difficulty in drawing all the activity to the cathode—some 10 percent at least was deposited on the walls of the testing vessel—had in corresponding experiments by others been blamed upon initial recombination and columnar ionization, wherein too low a potential had been applied to keep ions formed apart. Wellisch and Bronson subscribed to this interpretation as far as it went, but added their own concept of an uncharged restatom of active deposit—called a "neutron" (unhappily, in retrospect)—to explain further the lack of saturation.[6]

Wellisch and his student, A. N. Lucian, continued this work for the next few years, mixing gases other than air with the emanation, changing the size of the testing vessel, examining carefully the effect of small potentials, and studying the behavior of actinium. His earlier views were reinforced, refined, and slightly modified in that he retained his belief in an essentially neutral deposit atom, but allowed that it could acquire and give up charges. He no longer wrote of an "association" with ions. Given a pressure high enough to prevent recoil on the vessel's walls, the percentage of active deposit attracted to the cathode was independent of that pressure, but did vary with the gas employed. This was, in effect, a measure of the fraction of active deposit particles with a positive charge, at the end of the recoil path and before columnar or volume recombination could take effect. This charge originated in the recoil motion, as the particles both produced ions by collision and usually became ionized themselves. An electrical field, stronger than for light ions, was necessary to draw these charged heavy particles of active deposit to the cathode, since they tended to recombine more readily with negative ions than did positive ions (e.g., hydrogen).[7]

The mobility of ions was yet another subject of study that added to an understanding of the redistribution of charged particles. As mentioned in the last chapter, this was the means of Kovarik's entrée to the specialty of radioactivity. As a student at Minnesota, he had investigated the effect of different temperatures and pressures on the movement of ions in several gases. His ions were produced by ultra-violet light, as in a corresponding 1898 experiment by Rutherford. Indeed, Kovarik used the equation developed by Rutherford and adapted his technique to the present purposes.[8] When Kovarik then went to work in Rutherford's laboratory, one of his pursuits towards the end of 1911 was to test mobility at high pressures, using the intense ionium source recently separated at Manchester by Boltwood to create the ions. Earlier work at pressures ranging downward from atomospheric had shown that the product of mobility (measured in centimeters per second per volt per centimeter) and pressure was a constant: at reduced pressure the ions could move faster. Kovarik, using gases under as much as seventy-five atmospheres, detailed the constancy of this product under the new conditions.[9] Also at Minnesota, Henry Erikson, a few years later kept the pressure constant, but tested the ions produced by a polonium source at dry ice and liquid air temperatures, where he found a reduction of the mobility.[10] This was probably due to the clustering of molecules on the ion, changing its characteristics, a subject pursued by Erikson throughout the 1920s.[11] Indeed, the behavior of gaseous ions was not felt to be understood with some confidence until the 1930s,[12] perhaps an inadvertent measure of the fertility physicists considered the field to possess.

Radioactivity Here, There, and Everywhere

In her very first paper on radioactivity, in 1898, Marie Curie called attention to a weak activity in potassium, cerium, niobium, and tantalum.[13] But to a large extent such "medium weight" elements were ignored as the more active radioelements at the high end of the periodic table were preferentially pursued during the next few decades. Yet, on occasion, other common and not so common elements were claimed to exhibit radioactivity. In an effort to survey a range of potential emitters, W. W. Strong (b. 1883), a recent Ph.D. from Hopkins, in 1909 tested eighteen different elements, mostly rare ones that the deceased Henry A. Rowland had acquired. Despite the speed and sensitivity of ionization measurements, Strong felt they required too much attention and chose instead the photographic technique pioneered by Becquerel—even though the exposures lasted six months! His admirable patience went unrewarded, or perhaps his procedure failed even to inspire confidence in himself, for the strongest prose he could muster was that potassium, rubidium, and erbium, "may" be radioactive.[14] Indeed, the first two do exhibit weak beta decay, but this was reasonably well established several years earlier in England, and since that time a number of other elements have been shown to have naturally occurring radioactive isotopes.[15]

   More interesting because it was more imaginative, and because it showed that the "dabbler" tradition was alive and well, sodium was declared to be radioactive. F. C. Brown had no direct proof of this, such as indications of ionization, but he argued that circumstantial geological evidence was strong. John Joly of Dublin had estimated the present salt content of the oceans and the rate at which it had accumulated, and from this deduced how long the process had gone on. This age of the earth, however, was much shorter than the age based on radioactivity data. Brown therefore reasoned that the "missing" sodium in the sea, which would extend the period calculated by Joly, had transformed into another element.[16]

Other efforts to discover radioactivity were not as unsophisticated (or unscientific), nor were they confined to searching through the periodic table. Its distribution geographically or environmentally was studied even more extensively. Looking in esoteric places, two students of McCoy's at Chicago, Terence Quirke and Leo Finkelstein, reported that metallic meteorites have virtually no radioactivity, but the average stony meteorite, while far less active than the average earthbound igneous rock, clearly contains radioactive matter.[17] Earthbound rock, of course, was known to have so much uranium, radium, thorium, and their products distributed in it that a radioactive crust only about forty-five miles thick was needed to produce the earth's temperature gradient, as R. J. Strutt had shown in 1906. Indeed, in that very paper Strutt also described his own examination of meteorites, reporting essentially the same results as Quirke and Finkelstein detailed with greater precision over a decade later.[18]

If objects from space that penetrated the earth's atmosphere were radioactive, there was no reason to believe that bodies *not* moving in our direction were any less so. Helium, spectroscopically located on the sun before it was ever found on this planet, suggested that our star contained radium once Ramsay and Soddy showed the noble gas' growth from the wonder element, and even more so after Rutherford and Royds proved that the alpha particle is a charged helium atom. Given the production of heat by radioactive decay, it was easy to speculate that this was the source of stellar energy.

The spectrum of radium was at first difficult to find in the heavens. But that of radium emanation, once Rutherford and Royds provided accurate measurements of its lines, seemed fairly well assured. At the Philadelphia Observatory in 1909, Monroe B. Snyder was an early exponent of this view, claiming to match the radioelement's spectrum with lines from the great nebula of Andromeda, other spiral nebulae, and several star clusters. He seems, however, to have been ignored, probably because his article in *Science* had the self-aggrandizing ring of a crackpot, and because he advanced a vague theory of "radioaction" which challenged Rutherford's statement that radioactivity was unaffected by any temperatures thus far examined. At the temperatures and pressures of a star, he maintained, even ordinary elements might transform. Had Snyder offered experimental evidence or a theory of a more precise nature, he might well have fared better, for others soon found the spectra of radioelements in celestial bodies, and Rutherford himself agreed that all elements might undergo transformation under stellar conditions.[19]

The fascinating possibility that intelligent beings on another planet were using the radiations from radioactive bodies to signal earth was proposed to Boltwood in 1920. He replied laconically that a simple electroscope of high sensitivity would do for the test, which might more easily be conducted on earth than above it.[20] Clearly, radioactivity was all around—even in men's minds—but its greatest pursuit was earthbound.

The subject of atmospheric radioactivity, here defined as radioactivity of the air, soil, and water, was surveyed briefly for the early period in chapter three. In later years such efforts continued to provide newspapers and journals with a wealth of articles—and a dearth of anything scientifically significant. Those who examined the air sought to discern the various radioactive constituents in a sample, or measure the strength of a particular radiation, or compare activities in different localities, or assay the contribution to activity of a pollutant such as smoke. Perhaps because the technique generally used to obtain their study samples was exposure of a negatively charged wire in the air, and this presupposed some experimental dexterity, the authors of such papers were often physicists and chemists with familiar names.[21]

As the chapters on medical and commercial uses of radioactive products showed, public interest was maintained at a high level during the second dozen years of radioactivity, and this encouraged the initiation of many soil and water investigations by amateurs. Reputable scientists also contributed, with the commercial potential of soils, coals, mineral drinking water, and hot springs spas usually in mind.[22] The most interesting and imaginative atmospheric studies unfortunately were will-o'-the-wisp attempts to establish a meteorological or astronomical connection to radioactivity. Like the other investigations, these were not initially or uniquely domestic, for among many others John Satterly, at the Cavendish Laboratory, and A. S. Eve, at McGill, had conducted major inquiries earlier.[23] Still, the attempts reported in 1915 by Leopold Lassalle to measure the variation in ionization due to gamma rays as a function of time of day, and by J. R. Wright and O. F. Smith to connect the quantity of radium emanation with weather conditions, if not innovative, at least show an ability to recognize potentially fruitful topics. Both studies were conducted at Manila, where Wright was chairman of the physics department and the others assistant professors on leave from Pennsylvania State College. Lassalle did find a diurnal variation, which he tentatively associated with the land- and sea-breeze patterns in the Philippine Islands. He did not find anything more fundamental to explain the well-known natural ionization that required investigators to deduct background readings from their data-producing tests. It would be a few years still before cosmic rays were added to radioactivity to account for this background.[24]

Instrumentation

The American penchant for mechanical invention extended to the apparatus employed in radioactivity studies. Although a large number of scientific instruments seems to have been imported even well into World War I, such items as electroscopes, scintillation screens, and electrical counters often were home-made, and the ionization chambers and other unique items almost always locally crafted, leaving the electrometer the only standard tool commonly obtained commercially. This allowed—indeed encouraged—individual initiative in adapting instruments to special purposes, and such examples as Boltwood's gas-tight electroscope and Kovarik's electrical counter have been noted.

Ionization chambers attached to electroscopes or electrometers were useful for measuring cumulative charges or rates of change, but inadequate for the detection of individual particles. They nevertheless provided the bulk of early quantitative information about radioactivity, photographic impressions being essentially qualitative.

The quadrant electrometers of the nineteenth century were regarded as quite inconvenient and difficult devices with which to take accurate data. With increased usage dating from the discovery of radioactivity, and especially with the design bearing the name of Dolezalek, it came to be a dependable instrument, both sensitive and well enough damped for quick readings to be taken. This latter quality was not only desirable but necessary when radioactive substances of short half-life were being measured. Still, the "rate of leak" type of observation suffered somewhat from the uneven movement of the indicator needle, caused by electrostatic induction effects that could not entirely be avoided by shielding. The solution to this problem was construction of a null-reading instrument wherein the *steady* deflection of the needle was proportional to the instantaneous level of activity. Howard Bronson in 1905 and S. J. Allen two years later succeeded in this by placing a source of known activity in parallel with the charging quadrants to "counterbalance" the unknown source's activity, Allen even providing a calibrated cover for his standard source. The electrometer in such an arrangement served merely to indicate when a balance was attained.[25]

For simplicity of construction, ease of use, and the greatest sensitivity (though restricted to a more limited range of activity), many preferred the gold leaf electroscope.[26] Not just the non-mechanically inclined chemists, but physicists as well found it highly convenient; Rutherford's "boys" learned to build their own electroscopes from biscuit tins, rubbing a pen on their sleeves or in their hair and then touching it to the post to charge the leaves.[27] Turning this "garden variety" tool into a precision instrument capable of yielding thoroughly reproducible results was a task in which C. T. R. Wilson (later of cloud chamber fame) figured prominently. Notable features he introduced in 1903 were a charged plate near the gold leaf and a tilt to the entire electroscope box, both variable to achieve optimum sensitivity.[28] Subsequent developments generally included one or both of these modifications.

John Zeleny, for example, in 1911 built a lecture electroscope in which the leaf was able to touch the charged plate nearby. The chief novelty of this arrangement was that the leaf became recharged cyclically after losing a definite portion of its charge, and its "period of pendular vibration" was to be measured instead of the time to pass a number of scale divisions in a reading microscope.[29] In the same year Henry Bumstead studied the production of electrons when alpha particles struck metal surfaces, using not one but two plates in a "double electroscope." Here the plates were on opposite sides of the leaf, and charged to equal but opposite potentials. The instrument's sensitivity was adjusted by raising or lowering the leaf, and compared with Wilson's tilted variety was claimed to be more stable.[30]

As the American uranium deposits rose in commercial importance during the century's second decade, the need for accurate determination of their radium content grew apace. S. C. Lind, of the Bureau of Mines Experiment Station in Denver, devised a rugged, aluminum-leaf electroscope that could be detached from one emanation chamber and placed upon another. Since the former units were far more expensive and delicate than the latter, the ability to move them about increased the number of assays a laboratory could conduct at a given cost.[31] This device was widely adopted, with the industrial need for accuracy apparently stimulating analyses of its capabilities and those of other designs.[32]

The recording of evidence, as distinct from the detection of it, was a problem many scientists faced, particularly in a statistical subject such as radioactivity where numerous events had to be counted or cumulated. William Duane, while working in Curie's laboratory, arranged a powerful lamp to cast the shadow of an electroscope's gold leaf upon a moving photographic film. Upon development, the edge of the leaf had produced a line whose amplitude corresponded to the leaf's deflection at any given time.[33]

American attention was, as we already have seen, not focused entirely upon radioactivity "in the large"; several scientists became interested in observing individual events, which incidentally allowed the use of weaker sources than did ionization measurements. Bumstead was probably the only one to attempt, for serious research purposes, to count the flashes of light produced when alpha particles struck a zinc sulphide screen, and in 1907 he gave up in despair. He found it "a very nasty eye-straining job and mixed up a good deal with the psychology of attention which unavoidably fluctuates."[34] It was a technique which he happily left to Rutherford and Geiger, who developed it in preference to their electrical counter. Various recording methods were used for the visual counts, a gentleman in Colorado contributing the idea of calling them out into a dictaphone (indicative of a low counting rate).[35]

It was apparent, however, that scintillation counting, good as it was in the hands of Rutherford and Geiger, was impractical for active sources or long counts, and accuracy could be plagued by eye strain. Despite numerous "bugs" in the system, ionization by collision at low pressure held much promise, and the central wire electrical counter these two invented, and its point counter successor, appeared to set the direction of future instrumentation. By 1910 Rutherford and Geiger had replaced visual counting of the sudden throws of their quadrant electrometer needle (the wire counter serving as an ionization chamber attached to the electrometer) with photographic registration of the deflections of the silvered quartz fiber of a string electrometer. This latter apparatus had been devised by Willem Einthoven more than a decade earlier, in his famous electrocardiogram studies; Rutherford characteristically saw how the tool could serve his own interests.[36]

While some chose visual counting,[37] most preferred photographic recording of the string electrometer's "kicks", and this technique was widely adopted without much need for improvement. Attention focused far more on the electrical counter and its infuriating variability. Geiger was converted to a fine point instead of the central wire about 1913, but he required the microscopic examination of his steel needles, cleansing of them in alcohol, and heating of them in a Bunsen flame to remove "natural disturbances."[38] Upon the suggestion of George Pegram, J. E. Shrader, of Williams College, in Massachusetts, investigated the conditions for consistent use of such a counter, including such variables as needle material, point sharpness, cylinder size, and quenching gas. Platinum points, he found, were without natural disturbances after heating, but they deteriorated in quality if exposed to moist air. Different gases worked well, if thoroughly dried. Most important, he showed that disturbances, when present, came not from the size of the ionization chamber but from the point, and that at constant pressure the potential at which the system discharged was a function of the point's sharpness.[39]

Kovarik and McKeehan also used platinum points, which they checked frequently in a standardized counter. By careful comparison with scintillation and ionization results, they concluded that the electrical counting method was satisfactorily accurate and reproducible, effectively ending a minor controversy in this matter. The point counter was also shown to be most useful in recording the simultaneity of events, when a pair of them was connected to two string electrometers—a far more comfortable technique than making such determinations visually on scintillations.[40] Kovarik continued to tinker with the system a number of years. In 1917 he reported that Zeleny's electroscope could admirably be substituted for the string electrometer, or the counter could be connected directly to a sensitive telephone whose output, when amplified, was clearly audible. By this time point counters were regarded as accurate for alpha particle detection, reliable for beta particles if sufficient care was given to the point's sensitivity, and uncertain for gamma rays.[41]

Two years later Kovarik directed the point counter's current through a triode amplifier and ultimately to the electromagnet controlling a chronograph. Both eye and ear strain were thus circumvented, making long counts more practicable, and the technique, using recently available vacuum tubes, was more convenient than self-recording on photographic film. With this pen-upon-paper trace Kovarik proved the long-assumed application of the law of probability to beta particle decay.[42]

The ability to test for simultaneity intrigued Kovarik, and the end of World War I allowed him time to examine, in addition to the above work, the emission of gamma rays. Point counters placed in varying positions around a radium sample recorded the same number of events per unit time, but the events did not occur simultaneously. On a corpuscular interpretation the explanation was straightforward: the gamma pulses carried a quantum of energy in a definite direction. The wave hypothesis did not fare so well: if the spreading wave front struck both counters at the same time but did not cause a discharge in both, one might argue that energy could be stored up in an atom, to be released in the ejection of a beta particle (the vehicle for ionization by collision) at a later time. A corollary to this concept was that atoms could have different amounts of energy initially. Kovarik's dilemma typifies the problems encountered before the Compton effect and wave-particle duality entered the lexicon of physicists.[43]

American scientists did not contribute any significantly new or profoundly novel pieces of apparatus for the study of radioactivity; indeed, there were few such creations anywhere. But they did both apply and modify the commonly used instruments of the field, and exhibited an appreciation for the quality of their tools. Minnesota's Anthony Zeleny, brother of John, articulated this feeling while calling for an encyclopedia of apparatus and research methods and publication of a journal of scientific instruments:

> The greater the advancement in any branch of science, the greater must be the development of the apparatus that is employed. The two are necessarily interdependent. The instrument is to a great extent an index of the state of the science.[44]

Standards

Two themes that stand out in the development of radioactivity are organization and quantification. Organization is basic to any science, for even the natural sciences which rely heavily upon description strive to classify their subject matter, thereby to uncover relationships. Indeed, without a belief in the fundamental order of Nature, much science would come to a halt. The search for organization in radioactivity led to its two major concepts: the transformation theory and the group displacement laws. With the former, the various radioelements were ordered into decay series sequences; the latter both explained these sequences and placed radioelements in the "real world" of the periodic table.

The thrust of quantification, while less prominent, nevertheless was vital. Constancy of the radium-uranium ratio indicated their genetic relationship. Relative activity measurements of the decay series members served as probes for undiscovered radioelements. Precise atomic weight measurements proved the reality of isotopes and lent credence to the group displacement laws. In applied science also, accuracy was not just desirable but necessary, as in establishing the proper intensity and duration of radium exposures in medicine.

But there is more to precision than merely measuring a quantity with greater care. Measurements must have a commonly accepted standard, or one inch may differ from another. Primary standards of length, weight, time, electricity, etc., have long been enshrined in the Bureau International des Poids et Mesures, at Sèvres near Paris, while secondary standards are held by most governments (the National Bureau of Standards in the United States). Against these are compared the derivative standards used in various laboratories, in manufacturing operations, and to test meat market scales.

The discovery of radioactivity presented something of a problem. Its sources, as was learned later, might not only contain inseparable components, but the amount of these constituents would vary over time. More basic, however, was the lack of any certain idea of *what* a measurement signified. It was perhaps like attempting to gauge current electricity shortly after Volta invented the battery. Until uranium radiation was classified as ray or particle, and the phenomenon of radioactivity explained, the measurements made described the effects but not the cause.

In Becquerel's early experiments he used photographic plates to record the effects of the radiation he discovered. Occasionally he attempted to quantify the intensity of the photographic images, but his efforts were far too subjective. Other pioneers in the field, such as Rutherford and the Curies, used electrometers or electroscopes whose rate of change could be timed, or whose mechanical effect could otherwise be determined. Such procedures were satisfactory when the same instruments were used to compare active samples, but a given source could provoke different readings in different pieces of apparatus. And even in a single laboratory, because atmospheric, electrical, and instrumental conditions could change from day to day, it was a normal concept to calibrate the electroscope or electrometer against a radioactive sample reserved for this use only. Each laboratory experimenting extensively in radioactivity consequently had its own working standard. These standards, often of a similar nature, naturally gave generally consistent results. But as the subject matured and accuracy became imperative, differences between standards of even a few percent became unacceptable. An international standard of undoubted purity, against which all measurements could be reduced, was urgently required.

The early standards were suitable for measuring the weak radioactivities most scientists had at their disposal. R. J. Strutt in 1903 took as his standard the activity of a four by twelve millimeter crystal of uranyl nitrate. A weighed amount of uranium oxide served Samuel Allen in his studies of atmospheric radioactivity, while Elster and Geitel used uranyl-potassium sulphate and G. A. Blanc preferred uranium nitrate. Strutt's standard could not be copied elsewhere, unless one had a crystal of similar dimensions. The others were reproducible, relying simply on weight, yet the degree of purity could differ, while physical characteristics such as particle size could vary, signifying unequal emanating ability. In short, such standards were too primitive for fine, quantitative work.[45] Much more satisfactory was an easily prepared standard recommended by McCoy, consisting of one square centimeter of a layer of uranium oxide thick enough to yield maximum alpha activity. The ratio of the activity of one gram of uranium element to such a layer was called the "McCoy number" and was used for many years. Boltwood, by contrast, preferred a film of the oxide so thin that no alphas would be absorbed in it.[46]

Since accuracy is enhanced by using a standard of approximately the same activity and the same substance as the preparation to be measured, and since radium was the virtually unchanging, highly active radioelement most useful in laboratory and medical work, it was natural that radium standards also were prepared. Otherwise, radium preparations were necessarily and unsatisfactorily labeled as so-many-million-times more active than uranium.[47] Boltwood's 1904 suggestion that, since the radium-uranium ratio was constant, the amount of radium emanation from one gram of uranium, from an old mineral, be taken as a standard, had much merit. Despite the chore of dissolving the sample, boiling off the emanation, and correcting for the amount lost naturally by the mineral, the method was widely adopted, especially for measuring the radioactivity of air, water, and minerals.[48] Thus, the first radium "standards" employed were emanation sources, and the gas was used in comparisons; later the technique of gamma ray comparison with the radium itself, sealed in a glass vial, was adopted.

The difficulties encountered by Rutherford and Boltwood, in preparing a standard consisting of a known weight of pure radium bromide dissolved in water, have been described in chapters five and eleven. They were plagued not only by incorrect textbook values of mineral composition, but adhesion of radium to the glass container, a problem solved by the addition of some acid. Despite careful preparation and repeated revision, such calculated values as the amount of radium in equilibrium with one gram of uranium in a mineral—based on this standard solution and, at 3.4×10^{-7} gram, called the "Rutherford and Boltwood standard"—nevertheless underwent a final revision when the international radium standard was adopted just prior to World War I and showed there was less radium in this solution than believed. Other values as well, such as the half-life of radium and the amount of helium produced each year from radium, would remain uncertain until a worldwide standard of confirmed purity was adopted.[49]

Not only the Rutherford-Boltwood solution, but others suffered from questionable purity. Stefan Meyer's standard in Vienna was about 20 percent weaker than the one in Boltwood's possession, and Marie Curie's approximately 10 percent weaker. The international community of scientists recognized the need for a single authoritative reference point, especially since Madame Curie was unreceptive to the idea of comparing all the laboratory standards accurately with one another. The growing commerce in radium for therapeutic use added further pressure from the medical profession, industry, and government.[50]

The International Congress on Radiology and Electricity, held in Brussels in September 1910, offered the opportunity for Boltwood, Curie, Debierne, Eve, Geitel, Hahn, Meyer, Rutherford, von Schweidler, and Soddy to meet as the International Radium Standards Committee.

With apparently little discussion, they decided to name the unit they would establish the "curie," in honor of the deceased Pierre Curie and, presumably, of his widow also. With considerably more debate, they adopted the curie as the quantity of radium emanation in equilibrium with 10^{-8} gram of radium. The rationale in choosing this size was a desire to create a practical standard, one that could be used in everyday laboratory work without prefacing the term with "micro" or "milli." Marie Curie participated in this session which lasted until near dawn, and agreed with the committee's decision. But "at an unearthly early hour the next morning, she arrived at the hotel where Rutherford and [Boltwood] were stopping and informed [them] that after thinking the matter over she felt that the use of the name 'curie' for so infinitesimally small a quantity of anything was altogether inappropriate." She demanded, and got, the curie defined as "la quantité d'émanation en équilibre avec un gramme de radium." Her colleagues, anxious to present their report to the next session of the congress, were no match for this woman who relied equally well on her will of iron and her feminine frailties. Boltwood, in particular, was as bitter at her gamesmanship as he was over a unit 10^8 times larger than desired. Yet, the result, while impractical, was not as horribly unwieldy as anticipated, for the quantities of radium in use increased dramatically in the century's second decade.[51]

Definition of a unit of radioactivity, however, was but part of this committee's task, in fact its minor function. Its major responsibility was to arrange for the preparation of a radium standard. This Marie Curie agreed to do. By the next summer she had sealed 21.99 milligrams of pure radium chloride in a glass tube for this purpose. At the same time, Otto Hönigschmid, through his atomic weight work an expert in the preparation of pure substances, had filled three tubes with different amounts of radium chloride in Vienna. In March 1912, the committee met in Paris and compared these new standards, finding agreement within the limits of measuring error. The Curie standard, financed through the generosity of Soddy's in-laws, Sir George and Lady Beilby, was deemed the International Radium Standard, and soon deposited in the International Bureau of Weights and Measures, while one of Hönigschmid's preparations was preserved in Vienna as a secondary standard. Duplicate standards for various governments were fashioned in Stefan Meyer's Radium Institute and compared by the increasingly used gamma ray method against both Viennese and Parisian standards.[52]

A new International Radium Standard had to be prepared in 1934 by Hönigschmid, because the old one had become dangerous—accumulated helium and chlorine in the original standard made the glass tube liable to explode, and radiation damage to the glass added to its brittleness. Nevertheless, establishment of the first International Standard in the period 1910–1912 was a major indicator of maturity.[53] Both in research and in commerce the science of radioactivity was well developed. Of course there would be further growth and new discoveries before the end of this decade, but the fundamental directions were established, the attack on radioactivity's remaining mysteries was conducted across a large front, and a fairly cohesive international band of scholars in the field had emerged. In standards as in other aspects of radioactivity, Americans made some enduring contributions, namely the Boltwood and McCoy measures, but the "center of gravity" of scientific accomplishment remained abroad. And by the time this focus transferred to the United States, radioactivity, as discussed earlier, was no longer pursued.

18 Conclusion

Radioactivity as a Science

The study of radioactivity in America was characteristic of the changing condition of science in this century's first two decades—and it was not. Such ambivalence need not be unexpected, however, for guiding principles, themes, or organizational schemes are rarely able to embrace their constituents totally. Rather than personifying perfect conformity, they merely indicate a direction.

What then was the direction of American science, its achievements and social structure? As earlier mentioned, worldwide physical studies of matter, light, heat, and electromagnetism, which had yielded such success that various spokesmen claimed the end of great discoveries, had dominated nineteenth-century tastes. When the concept of the completeness of science was then shown to be in error, the major orientation of "modern" physics was toward the study of numerous radiations and their interactions with matter, relativity, quantum theory, and the structure of the atom. Clearly, the American radioactivity investigations were part of this new physics, with their studies of alpha, beta, and gamma ray phenomena, and the subject was immediately accepted as appropriate in physical laboratories.

This, of course, does not attest to the vigor with which it was pursued. Rutherford's descriptions of his work at McGill, presented at annual meetings of the newly founded American Physical Society, were so well received as to make Columbia University's Michael Pupin regard them alone as satisfactory justification for the society's existence, as already noted in chapter three. But this exposure to radioactivity, while stimulating in smoke-filled rooms, largely decayed before the attending physicists returned to the fresher air of their laboratories. It was personal contact with Rutherford (and other foreigners), as in the case of Boltwood and Bumstead, or personal inclination that directed the small band of Americans into radioactivity research, not the widespread glamor of a "movement" drawing troops to its colors. This in itself may be paradoxical, since the findings of radioactivity were of great scientific and public interest. Yet, the translation of such concern into laboratory activity was not a choice made by many.

We must not, however, overestimate the contemporary importance of radioactivity. It is necessary to recognize that for many scientists radioactivity was not paramount, nor even highly placed, among the current topics in physics and chemistry. Harold W. Webb, who counted Pupin and Pegram among his professors, and who, indeed, as a student collaborated with the latter on radioactivity research already described, chose to investigate electrical waves after receiving his Ph.D. degree in 1909, hoping to explore the gap between these waves and visible light. "The transition to 'modern physics' had hardly started," he reminisced, "and I did not then find many attractive fields of research to choose from. 'Discharge through gases' had been pretty well exhausted by the Cambridge school and the current work on arcs, ion mobilities and ion recombinations did not attract me. Professor Pegram was investigating certain properties of radioactivity but again this field was attracting only moderate interest." For Webb, who remained at Columbia for the next forty-four years as a faculty member, American physics was then dominated by men with "classical" interests, a group led by Albert Michelson, A. G. Webster, Carl Barus, E. L. Nichols, Ernest Merritt, Dayton C. Miller, and others of that generation. It was Niels Bohr's atomic theory that provided the impetus for new research topics, delayed, however, until after World War I.[1]

The physical radioactivity work, like other sciences suffering from the inadequacies of manpower and funding in a developing nation, was further hampered by a lack of genius and of luck. The topics chosen were not the fruitful ones retrospectively, or, where the potential existed for significant work, as in Allen's investigation of beta particles, theory was so far removed from experiment that the latter suffered. The national aversion to theory took its toll. Yet, too much importance should not be placed upon national characteristics, for personal happenstance was more a determinant. England, before the return of Soddy and then Rutherford, had little to point to, despite the interests of Kelvin, Crookes, J. J. Thomson, and others. Had Rutherford, for example, accepted one of the several American positions offered him, then this country very likely would have become the world center of radioactivity studies. Nonetheless, national traits were not entirely without significance, for, to use the same example, Rutherford's rejection of these invitations was based upon his goal of leaving the scientific hinterland. The physical work done in the United States thus lacked the inspiration of extraordinarily able people; its threadbare respectability reflected the status of the domestic physics community.

The American radiochemical community also comprised but a handful of people, yet their contributions were more significant and enduring. However, the acceptance of radioactivity by the field of chemistry affords another paradox. Chemistry, like physics, biology, astronomy, and others, had known great triumphs in the nineteenth century, such as the detailed concept of atoms, laws regulating the ways elements combine, the periodic table of elements, the development of organic chemistry, and understanding of changes of state, to name but a few. Unlike physics, however, these advances seemed not to signify an end to creative work; the feeling of satisfaction paralleled that in physics, but the concept of completeness was less evident. And this was reflected in the continuing progress during the twentieth century of many chemical subjects developed in the nineteenth. Of course, this was true also of physics, but the difference is one of degree.

Chemistry naturally also developed new topics of interest in the twentieth century, but the science suffered something of an identity crisis. Nature, rebelling against the sometimes artificial, if convenient, divisions of science into academic departments, seemed intent on proving its continuity. This is manifested in the constant creation of "hybrid" sciences, formed at the intersection of the established sciences. Thus, electrochemistry, biochemistry, biophysics, astrophysics, molecular biology, and others come to the fore and then may recede. The problem experienced by chemistry, a science which matured a century after physics, yet was far larger due to its industrial applications, was that it seemed almost unwilling to recognize its relationship to certain areas of new science. Radioactivity, in particular, seems to have been enthusiastically accepted and supported in *no* chemical laboratories. Boltwood worked on it first in his private consulting practice, then in the Yale physics department. By the time he moved over to chemistry, his creative period had ended. McCoy conducted his research in Chicago's chemical laboratory, it is true, and did have a continuing, if small, line of students, but he lacked professorial colleagues and any apparent departmental commitment to his subject. In fact, though no work on radioactivity was conducted after 1904 under Millikan's aegis in the Chicago physics department, his group maintained far greater enthusiasm for the subject than did their chemical colleagues, because of its bearing on atomic physics.[2] Theodore Richards alone seems to have been a major force in his own chemistry department at Harvard, but his stature was due to his pursuit of classical atomic weight investigations, not its one application in radioactivity.

Abroad, Otto Hahn obtained his first position in Emil Fischer's institute at the University of Berlin thanks largely to Fischer's broadmindedness, for the numerous organic chemists there considered that radioactivity had "no direct connection . . . [to] normal chemistry." Indeed, one department head exclaimed in disgust, "It is incredible what gets to be a Privatdozent [i.e., university lecturer] these days!"[3] Similarly, the careers of Frederick Soddy, Kasimir Fajans, Georg von Hevesy, and even Marie Curie—originally a physicist—seem to suggest some isolation from the profession of chemistry and a struggle for support.

The chemists investigating radioactivity naturally took support where they found it, and they found it in the community of physicists, a group perhaps more open because they had but recently organized, and also because of greater revolutionary turmoil in the science.[4] Radiochemists more often than not published in physical journals, and physicists were genuinely interested in their work. Hahn has recorded the friendliness of physicists, resulting in his regular attendance at the physics institute's colloquium.[5] Aside from McCoy, who appears to have established few personal or professional relationships, the lines of communication for American chemists working on radioactivity also seem to have been with physicists.

The other side of the coin, a face no doubt perplexing to conventional chemists, was that their field was being invaded by physicists, and these upstarts could beat them at their own game.[6] Whereas chemists had been proud of their ability to analyze materials down to 10^{-5} gram with an ordinary chemical balance, and 10^{-9} gram with a quartz microbalance, spectrum analysis pushed the limit to 10^{-10} gram, an ordinary microscope to 10^{-13} gram, an ultra microscope to 10^{-18} gram, and radioactivity still further to 10^{-22} gram.[7] The apparatus of these latter techniques was relatively unfamiliar to chemists and many were disinclined to gain expertise with them. Additionally, physicists were discovering and investigating new elements (Rutherford with emanation, the Curies with polonium and radium), using x-rays to explain the basis of the periodic table (Moseley), employing electromagnetic bending of particles to determine the masses of isotopes (Aston with his mass spectrograph), and using both mathematical and physical techniques to elaborate the structure of the atom (J. J. Thomson, Rutherford, Lenard, Nagaoka, Langmuir)—all topics that chemists could logically claim as their own. And, to rub salt in the wound, for such work physicists were winning Nobel Prizes in *chemistry!* These natural associations and barriers, in America as well as abroad, determined that radioactivity be pursued as an adjunct of physics more than chemistry, and the subject's demise by 1920 gave chemistry departments insufficient time to reconsider. When the field was resurrected as nuclear chemistry, following the discoveries of the 1930s of the neutron and artificial radioactivity, chemistry departments had learned their lesson.

A few words about radioactivity's conformity to the manner in which science is believed to progress may be appropriate, since the subject of models has drawn the attention of historians, sociologists, and philosophers of science in recent years. Radioactivity seems to fit reasonably well with Thomas Kuhn's concept of growth based on a paradigm or set of ideas that the community accepts.[8] This unifying principle was the Rutherford-Soddy transformation theory of 1902–1903, which, except for a surprisingly small number of diehards, was readily accepted as the explanation of radioactivity.[9] Prior to this time the phenomenon was interesting, to be sure, but the fascination was based upon its novelty, astounding energy release, and promise to unlock some mysteries of Nature. With the Rutherford-Soddy conceptual framework, or, more precisely, with a mechanism to elucidate and a clearly defined path to explore, other scientists were provided with a structure for the development of the science. Because genetic connections between radioelements were a fundamental tenet of their theory, and the identity of these substances required determination, chemists were recruited into the field. Likewise, because the theory postulated the transmutation of one atom into another, with the emission of alpha, beta, or gamma rays, physicists interested in such topics as atomic structure and ionization also were recruited.

Derek de Solla Price has shown that quantitative indicators of science, such as personnel, publications, and funding, tend to follow what is called a sigmoid or S-shaped curve. After a scientific specialty is founded it grows slowly for a period of time. As Diana Crane has done, let us call this stage one, in which the paradigm appears, but there is yet little or no social organization of those pursuing it. A plot of the research papers by Americans on radioactivity (Fig. 1) shows that this "foot" of the S-curve extended to about 1903, which agrees well with the historical picture presented in this book.[10] The worldwide curve can have been but little different from the American in its critical points.

From 1903 to 1909 the field grew at a rapid rate, and between 1909 and 1915 continued at only a slightly slower pace. Grouped together these years may be called stage two, that of normal science. This is the time characterized by problem-solving within the framework of the paradigm. It is also a period in which groups of collaborators and groups in informal contact (invisible colleges) arise. Except that there was little collaboration beyond the teacher-student type, and the extent of com-

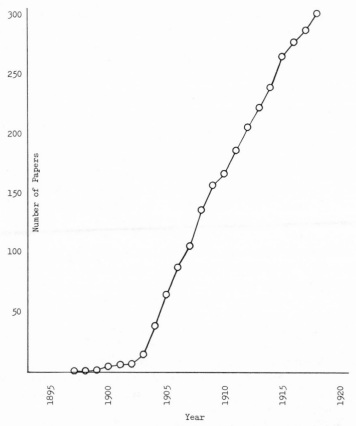

FIGURE 1—CUMULATIVE NUMBER OF RADIOACTIVITY RESEARCH PAPERS BY AMER-
ICANS

munication within the American cluster of scientists was nowhere as striking as that maintained by Rutherford, a virtual post office, this period may be regarded as a weak example of stage two. The second half of this stage, when enough papers were published to smooth out fluctuations on a semilogarithmic plot, shows that publication in radioactivity experienced a doubling in seven years, a rather rapid rate of exponential growth.

Stage three is one of gradually declining growth: the S-curve now bends over. It is seen as a period in which the subject's major problems are solved, leading participants to search for greener pastures or to pursue narrow specialities. Or it may be a time of increasing controversy and irritation with the inability of the field's concepts to incorporate experimental data. Radioactivity fits the former model, with physicists departing it (Allen investigating x-rays) or concentrating their interests (Kovarik on electrical counters), and chemists being left with little to investigate besides the effects of radiation on matter and using radioactivity as a tool (tracer studies). The new direction taken in 1915 appears but a slight departure on the graph, but it supports the argument in this book that the science experienced a profound change about that time, especially following the formulation of the group displacement laws and proof of the concept of isotopy.

If radioactivity had experienced anomalies in stage three, the fourth and final stage would have been one of crisis. The old ideas would have been recognized as inadequate for an understanding of all the data, and a new interpretation, a new paradigm, would have followed. With the solution of problems instead of the sprouting of anomalies, however, stage four is one of exhaustion and a decline in membership.[11] This conforms precisely to the theme of this book, though quantitative evidence is unavailable or unreliable. Because the subject changed significantly toward the end of the century's second decade, there is a problem in deciding what to include in a count. The papers found are too few to regard them as an accurate measure of the field, yet the difficulty in locating them lends conviction to the general conclusion. Moreover, the transition to narrow specialization, radiation effects, application, etc., would not be reflected in the count, while it is an indication of the field's stage of development. And finally, the effect of World War I, which must have been a decrease in publication, can not be quantified for radioactivity and incorporated because it can not be distinguished from coincident new trends. Nevertheless, on the principle that quantitative data are but one of several types of information used by the historian, and not a chain with which to lead him, it can comfortably be said that the growth rate of radioactivity in America, as presented, is necessary but not sufficient proof of the conclusions reached with other evidence. It corroborates, and does not contradict, but, given its limitations, certainly can not prove the validity of the growth and decay of this science as here described.

Radioactivity as an American Science

Besides this fitting of radioactivity into the physics and chemistry of its period, and fitting it into a model of scientific growth, generalizations which have validity worldwide, there is another dimension, one for which statements peculiar to the American experience can be made. This has to do with the place of the investigators of radioactivity within the scholarly community.

American science, judged by its quality, was not one of the world leaders during these years. Such a position was not attained until the 1930s, followed then by preeminence after World War II. In the late nineteenth century the United States was underdeveloped scientifically, in many ways still a colony of European science, while the early twentieth century saw a slow and sometimes unsteady rise in stature.[12] Indeed, science as a profession was a relatively new concept. Formerly, a natural philosopher was likely to be a financially comfortable physician, clergyman, or teacher, who pursued his studies more as avocation than vocation. Gradually, in the nineteenth century, the value of applied science was recognized and the merit of intellectual activity acclaimed, such that employment opportunities arose in government and industry, while colleges and universities took science teaching more seriously. By the time of Rowland, Gibbs, and Michelson, there had been a great expansion in science education, through creation of a number of land grant colleges, scientific schools at Harvard and Yale, graduate schools as at Hopkins, new private universities such as Chicago, and the multiplication of posts in other established institutions.

Scientists of this period were anxious to raise the standards of their disciplines and did so by encouraging students to obtain the Ph.D. degree, by publishing specialty journals, and by forming professional organizations. Thus, graduate training of Americans in Germany became quite common, the American Chemical Society was founded in 1876, and its *Journal* in 1879, while the *Physical Review* began publishing in 1893, six years before creation of the American Physical Society which eventually took it over.

The scientists were encouraged in their pursuits by increased interest among sections of the public. By the final quarter of the nineteenth century the hostility to all science that had been aroused by Darwin's *On the Origin of Species* (1859) had dissipated considerably. Attention was turned from science to social, economic, and political issues, particularly the closing of the frontier, the transformation from an agricultural to an industrial nation, the great fortunes being amassed, and the rising stature of a middle class which glorified technological advance. In reaction, the patrician, cultured elite, which heretofore had little doubt about its primary place in the scheme of things, now felt both a bit uncertain of its role and disdainful of the middle class' celebration of the country's size and wealth, the government's corruption, and the nation's crudity and practical-mindedness. Increasingly, they reiterated the theme that engineers and inventors only capitalize upon the disinterested contributions of basic science—and what great discoveries can America claim? Since the answer to their rhetorical question was few or none, they naturally urged expanded opportunity to pursue basic science.[13]

Such efforts undoubtedly influenced the creation of more science posts in academia and encouraged their occupants toward basic, not applied, research. They also assured the professors and their wives of a respected place in local society. Yet, scientists continued to enjoy too little prestige in government and industry, and were often unable, even on their campuses, to transmute the elite's good wishes into tangible research support. For all the fine words, research was not needed for promotion. It was a personal, self-satisfying endeavor, exemplified by Michigan's Henry Smith Carhart's statement: "The taste for scientific research is a passion which finds its gratification in the truth it seeks."[14]

If research in the nineteenth century was regarded as a calling, practiced by only a few of those termed scientists, in the twentieth it became the norm for the genus "scientist," and ultimately, for many, a nine-to-five job. As science, with its eruption of turn-of-the-century discoveries, became less intelligible to the cultured elite, its source of support changed. The message that technology depends on science was not only accepted but implemented: a scientifically trained engineer, Charles Proteus Steinmetz, impressed the electrical industry with his mathematical explication of alternating current phenomena; physicist Frank B. Jewett, aware of fruitful worldwide investigations on the conduction of electricity in gases, urged research on triodes in AT&T's Western Electric Engineering Department as a means of developing a telephone amplifier; and even such an atheist—or at least agnostic—towards science as Thomas Edison came around to acknowledging the value of higher learning.[15]

Reform, the banner of the new wave of Progressivism early in the century, saw science as one of its tools for desired change. Objectivity, truth-seeking, disinterestedness, a willingness to challenge old assumptions—these characteristics of science were to be applied to government and society. Experts, well-trained in different disciplines, were needed, a circumstance that gave not just moral value, as before, but economic incentive to the pursuit of science. Yet, it was not for bread alone that man became thus oriented; public service was also extolled. And the democratic values of the day implied that anyone could seek a college education. The public, forgetting its earlier distrust of science as an ally of aristocracy, now supported its state institutions, and the presidents of both public and private colleges and universities looked with greater favor upon the needs of science as one of their contributions to the welfare of society. This sympathetic attitude happily coincided with mushrooming opportunities to explore new avenues of science. Undergraduate college enrollment in all fields, per million population, rose from 736 in 1875, to about 1300 in 1900, and about 2200 in 1915. Graduate enrollment per million for the same years was 8, about 75, and about 145,

showing a correspondingly greater increase. The number of Ph.D. degrees awarded each year in physics hovered between one and four
through most of the nineteenth century, exceeding ten only in its last
few years. In the years prior to America's entry in World War I, this
figure fairly steadily climbed through the tens and twenties, into the
thirties. Chemistry, by contrast, started the century with well over twenty
Ph.D.'s per year and exceeded one hundred by the war.[16]

Ph.D. production, adapted from data presented by Daniel Kevles,
shows some interesting trends which stand out in the tabular form below:

	Sources of Ph.D. Degrees			
	U.S.	Germany	Elsewhere	No Ph.D.
Physicists over 40 in 1895	10	2	0	24
Physicists under 40 in 1895	19	20	0	13
Entrants to physics profession, 1895–1906	53	6	1	4
Entrants to physics profession, 1906–1916	43	1	5	4

For the generation that began their careers before the century's final
quarter, the research degree was irrelevant. For their students, however,
the degree meant increased professional and social mobility, and the best
training was to be had in Germany. By the twentieth century, the Ph.D.
meant economic security as well, and the changing population of
aspirants—from sons of lawyers, doctors, clergymen, and prosperous
businessmen, to children of small farmers, middle-class merchants, and
even the poor—patronized state institutions far more than the elite private schools in the East or the further remote (culturally as well as geographically) universities of Europe. Science attracted many of these
people as a profession in which ability, not lineage, mattered; physics
may have been more alluring still, as a subject in turmoil, and one with a
minimal academic tradition of deference to elders.[17]

The new entrants to the profession of science brought a spirit of
ambition and competition with them; they were in a race for jobs and for
priority in discovery. Publish-or-perish was not the name of the game for
a few decades more, since science teachers could yet carve out an
academic career. But professorial chairs increasingly were awarded to
proven scholars, those who had published significant work, and their
institutions accommodatingly were willing to place greatest reliance
upon this yardstick of creativity, since it could far more objectively be
measured than could teaching ability. In emulation of the German uni-

versity system, professors became known for their contributions to knowledge.[18]

The change was many-faceted, but inexorably pointed toward a stronger effort in science. To raise the stature of the domestic doctorate in foreign eyes, high and uniform standards were established for the degree in 1900, by the newly formed Association of American Universities. Eight years earlier the University of Chicago was founded under the leadership of William Rainey Harper, who had the almost blasphemous idea that research was more important than teaching. When Göttingen University in 1900 took the unique step of inviting an American, Theodore W. Richards, to a professorship, Harvard's president, Charles William Eliot, reconsidered his previous lukewarm support of research and urged Richards to remain. In recognition of its value to industry, and following similar moves abroad, the federal government in 1901 created the National Bureau of Standards; from the beginning its leaders regarded it as a research institution and not just a center to compare weights and measures. Other laboratories were constructed, both industrial and academic, and some even enjoyed modest endowments for research. Royal Society President Sir William Huggins enviously cited the United States as a country blessed with large "private benefaction[s] for scientific purposes and scientific institutions."[19] Membership in the American Physical Society rose from its 69 founding fellows in 1899, to about 100 in 1900, to nearly 1300 in 1920; the American Chemical Society nevertheless maintained its ten-fold lead and was even increasing the gap toward the end of the second decade. Membership in the country's umbrella organization of scientists, the American Association for the Advancement of Science, hovered around 2000 during the last two decades of the nineteenth century and then began a steep rise. This increase, impressive as it was, fell behind that of the ACS, which, by 1920, had *more* members than the AAAS (15,582 to 11,547, but the ACS counted executives, technicians, assistants, etc.).[20]

How well does the above picture characterize the disciplinary dynamics of radioactivity in America? First, a disclaimer must be made, to the effect that the population of this community was small and statistical conclusions are therefore unreliable. At any given time there were probably only a dozen or so scientists in this field who were doing significant work, and several times this number of dabblers, mostly on atmospheric radioactivity.[21] Still, we may generalize to the extent of saying that the older members of the group, such as Bertram Boltwood, Robert Millikan, George Pegram, William Duane, Henry Bumstead, Theodore Richards, Herman Schlundt, and Richard Moore, studied abroad, and most came from families that were financially and socially well-off, and able to send their sons to the better American colleges. Herbert McCoy seems to be the exception, being both poor and untraveled, though he undoubtedly benefitted by attending the University of Chicago in its exciting period.

The younger generation came increasingly from public institutions, especially Minnesota, but Yale continued to train a large portion of this group. Family social status and wealth are known for very few of them, but the trend seems to conform to American education as a whole. The place where the younger radioactivity scientists seem to have broken the mold, however, was in their advanced training. Wherever Ernest Rutherford held forth, *there* was the center of radioactivity research. Americans recognized this and eagerly sought the opportunity to spend a year or more in his laboratory. But because few, if any, institutions in the British Empire offered the degree of doctor of philosophy until after World War I, those who came to bask in his reflected glory more often than not had already earned their Ph.D.s back home. This category of scientist runs from Howard Bronson and Boltwood in the earlier years, through Alois Kovarik, Leonard Loeb, Perry B. Perkins, and Thomas S. Taylor. Not all Americans chose to do postdoctoral work under Rutherford, but English institutions still were more attractive than their counterparts on the Continent, Henry Bumstead, William Huff, and Louis McKeehan opting for J. J. Thomson's Cavendish Laboratory and its tradition of ionization studies. Of those who crossed the Channel, Duane, Samuel C. Lind, and Gerald Wendt studied with Marie Curie, Wendt before receiving his Ph.D. in chemistry from Harvard in 1916, and his two seniors some years after earning their German doctorates. Lind, it may be noted, was perhaps the only American to work in the Viennese Radium Institute, where he went in 1911, after Paris. Of course, a number of investigators of radioactivity never went abroad for advanced work. This list includes Haroutune Dadourian, William Barss, Jay Woodrow, and Alexander McGougan, all of Yale; William Ross and Leopold Lassalle, of Chicago; Henry Erikson, of Minnesota; and Princeton's Fay C. Brown and Clinton Davisson. As would be expected, some of these dropped the study of radioactivity after graduation, while the contributions of others were not overly impressive, circumstances which would prejudice their opportunities for research travel.

The reverse flow of scholars is another measure of the stature of a scientific community. Here we may point to Ellen Gleditsch and Max Lembert, who came to work briefly with Boltwood and Richards, respectively; Canadians, such as Samuel Allen and H. L. Cooke, who settled permanently in the United States; and Englishmen, such as O. W. Richardson and H. A. Wilson, who came for shorter or longer periods.

Frederick Soddy also might have been transplanted, to Rice Institute, but the post went to H. A. Wilson.[22]

In summary, radioactivity differed from the pattern of science as a whole in sending men to England rather than Germany, and continuing this circulation during a period when foreign experience became less common. Radioactivity violated the norm also in remaining stronger in American private institutions, such as Yale, Princeton, Harvard, Hopkins, and Chicago, than in the public universities of Michigan, Wisconsin, and California. Only in Minnesota did the subject receive enthusiastic attention. Socioeconomically, the community seems representative of the changes elsewhere in American science, and scientifically it was characteristic also, i.e., though there were a few contributions of great international distinction, such as Boltwood's, McCoy's and Richards', most of the results published were only of moderate interest, and the elders of the profession continued to decry America's poor showing generally in science.[23]

Financially, the science also seems unexceptionable. Despite Sir William Huggins' envy of American philanthropy and apart from the very real benefits such largess produced in the form of laboratory stone and cement, such benefactions were not extensive in the actual support of research. S. J. Allen received a grant from the Bache Fund of the National Academy of Sciences for the purchase of apparatus and supplies while investigating beta ray-induced secondary radiation, and the Carnegie Institution of Washington gave financial support to T. W. Richards and Max Lembert for their work on lead atomic weights, but the vast majority of papers on radioactivity bore no such acknowledgments.[24] Boltwood, for example, benefitted repeatedly from the enlightened policy of the Welsbach Gas Mantle Company chemists, who furnished him with thorium samples at no charge, but he was frustrated in raising cash to purchase radium and complained that the time was passing for workers using only "homeopathic doses."[25] When one heard of significant amounts of money to purchase apparatus, such as John Zeleny's $10,000 in 1902, it was almost certainly entirely for teaching purposes.[26] Andrew Carnegie in 1907 responded to the worldwide interest in radioactivity by giving Marie Curie $50,000 for fellowships and research, and the famed industrialist also funded positions in the United States, but the only radioactivity worker to be a "Carnegie Research Assistant" was Fritz Zerban, while with Baskerville at City College of New York.[27] Radioactivity research in general shared the same genteel poverty as other branches of science.

Sociologically, the radioactivity scientists had rather weak domestic institutions of an informal type. Herman Schlundt's suggestion to

Boltwood in 1907 that they form an organization of radioactivity workers apparently fell on deaf ears.[28] The lack of cohesiveness, of reinforcement by a professional community, struck Bumstead, too. A committed anglophile, who regarded England as "such a bully old place" that he would happily consider a job there, he was further impressed that "scientific men are so close together that they can really create an atmosphere."[29] But an atmosphere, if it developed at all in America, was rarified at best. The exchange of preprints, information, and ideas, and the scheduling of special sessions at society meetings occurred minimally. There were no charismatic leaders, and no laboratory seems to have created a vital environment sufficient to establish a tradition, with the weak exceptions of Yale and Minnesota.[30] Despite the success of *Le Radium* and the *Jahrbuch der Radioaktivität und Elektronik*, no English-language journal for this science was created, except perhaps the British publication *Ion*, which lasted but a few issues and whose birth defects doomed it. The consequences of such weak interaction among American radioactivity scientists went beyond the lack of encouragement and peer approval. It isolated intelligent and imaginative individuals, such as Fernando Sanford at Stanford, whose uncurbed flights of fancy led them to speculations too far removed from the evidence. Lacking the scientific judgment they might have obtained as students in an active research school, what they needed was effective communication with colleagues for the testing of ideas. First-rate contributions were, of course possible, as the work of Boltwood and McCoy shows. But significant advances are easiest when a new science is first mined, whereas continued leadership comes to depend increasingly on the social fabric of the nation's science.

Internationally, the Americans did better. If we use Gerald Holton's ingredients "for expressing and using the internationality of science," namely, travel abroad, attending and holding international meetings, providing hospitality to visitors, and publishing abroad, we find a more promising record.[31] Though they went without fellowships and grants, largely nonexistent then, Americans did spend valuable years in foreign laboratories. The 1904 St. Louis international meeting was not devoted to radioactivity, but the new science was on the program and a hit at the associated world's fair. When the International Congress on Radiology and Electricity was held in Brussels in 1910, Boltwood was the official United States delegate. Few Europeans had reason to come to America for stimulation or edification in radioactivity, but those that did come were well treated. Even Ellen Gleditsch, who had to endure jokes about Boltwood's determination to remain a bachelor, remembered her visit warmly.[32] And, finally, Americans certainly did publish in foreign journals, often simultaneously with their paper's appearance in this country.

The overall picture shows moderate conformity to Holton's model of internationalism.

The science itself, as has been discussed in an earlier chapter, was unusual in having no near-miss revelations prior to Becquerel's detection of the phenomenon in 1896. Only Silvanus P. Thompson, in England, could claim simultaneous and independent discovery, whereas precursors generally emerge for most natural phenomena.[33] The situation is somewhat different for theoretical milestones, such as quantum theory and relativity, for while the new interpretations may be radical, the data on which they are based are often quite familiar. Radioactivity differed from most other sciences as well in its ready application to therapeutic medicine and in its great popularity. Relativity also experienced widespread fame, but much of this was controversial, and it occurred after radioactivity's heyday. Radioactivity, further, was unusual in its scientific success, the chemical half dissolving itself by leaving no major questions unanswered, and the physical side evolving into atomic physics and, especially, nuclear physics. Yet, there is no evidence that this uniqueness and popularity were significant factors in drawing people to study radioactivity (excluding the atmospheric radioactivity dabblers). The subject never came to dominate American physics or chemistry, in breadth of effort or number of followers. Rather, the handful drawn to it came more in the long-honored tradition of experimentally-oriented data gatherers. Radioactivity was not a highly mathematical science, or still fewer Americans would have pursued it. Its practitioners, thus, relied much upon their physical intuition. Happily, Boltwood and McCoy possessed this intangible; no physicist involved with radioactivity had it in the quantity that allowed someone like Robert W. Wood (in optics) to rise to the first rank of the profession.

The leaders of American physics in the 1920s, men such as Robert Millikan, Harrison Randall, Arthur Compton, Theodore Lyman, Percy Bridgman, Richard Tolman, and the new faces of Rockefeller scholars returning from study abroad (Germany once again predominating), were not the radioactivity scholars of the century's first two decades. The ionization and radiation studies conducted in the earlier period had some connections with the later interests, but they were too few and too little mathematical. Just as radiochemistry disappeared and radiophysics transformed into other fields, its personnel also vanished. Not even their students remained to carry the banner before the next generation, for there was no radioactivity for them to pursue.

Was it a frustration or waste of effort? Clearly not. The chemical and physical investigations provided both major and minor insights into the natural world which were needed to assure full comprehension of the subject. There is no evidence of low morale within the radioactivity community, or disparaging retrospective remarks. Moreover, the effort provided far more than scientific understanding; it served to consolidate and expand further the concept and practice of research in American universities. Science was, after all, still in a condition of immaturity in this country and in these institutions. These decades were ones of continued construction and development of laboratories, professorships, organizations, and periodicals—the infrastructure of science. It would take these decades and more to erect the institutional framework of American science and make it function as a world leader. American contributions to radioactivity, both to the science and to the profession, were solid, if not especially enduring, achievements entirely in proportion to the technical and sociological stature of American science in general. They helped form the fountainhead for the pursuits of the next decades, even if the spring's source had dried by the time the next generation of students could partake of the waters.

Abbreviations

Am. Chem. J.   *American Chemical Journal*
Am. J. Phys.   *American Journal of Physics*
Am. J. Sci.   *American Journal of Science*
Ann. N.Y. Acad. Sci.   *Annals of the New York Academy of Sciences*
Ann. Physik   *Annalen der Physik (und Chemie)*
Ann. Repts. on Progr. Chem., Chem. Soc. London   *Annual Reports on the Progress of Chemistry* (Chemical Society of London)
Arch. Hist. Exact Sci.   *Archive for History of Exact Sciences*
Arch. Intern. Hist. Sci.   *Archives Internationales d'Histoire des Sciences*
BCY   Boltwood Collection, Yale University Library
Ber. Deut. Chem. Ges.   *Berichte der Deutschen Chemischen Gesellschaft*
Ber. Deut. Physik. Ges.   *Berichte der Deutschen Physikalischen Gesellschaft*
Biog. Mem. Fellows Roy. Soc. London   *Biographical Memoirs of Fellows of the Royal Society of London*
Biog. Mem. Nat. Acad. Sci. U.S.   *Biographical Memoirs of the National Academy of Sciences of the United States*
Bull. Am. Inst. Mining Engrs.   *Bulletin of the American Institute of Mining Engineers*
Chem. Abst.   *Chemical Abstracts*
Chem. Bull.   *Chemical Bulletin*
Chem. in Britain   *Chemistry in Britain*
Chem. & Met. Eng.   *Chemical and Metallurgical Engineering*
Chem. News   *Chemical News*
Compt. Rend.   *Comptes Rendus Hebdomadaires des Séances de l'Académie des Sciences* (Paris)
CPR   James Chadwick (ed.), *The Collected Papers of Lord Rutherford of Nelson* (New York: Interscience, and London: Allen and Unwin, 1962, 1963, and 1965)
DAB   *Dictionary of American Biography*
DSB   *Dictionary of Scientific Biography*
Elec. World   *Electrical World*

Geol. Mag. Geological Magazine
Graduate J. Graduate Journal
Ind. Eng. Chem. News Ed. Industrial and Engineering Chemistry, News Edition
Jahrb. Radioakt. u. Elektronik Jahrbuch der Radioaktivität und Elektronik
J. Am. Chem. Soc. Journal of the American Chemical Society
J. Chem. Educ. Journal of Chemical Education
J. Chem. Soc. Journal of the Chemical Society (London)
J. Franklin Inst. Journal of the Franklin Institute
J. Hist. Med. Journal of the History of Medicine and Allied Sciences
J. Ind. Eng. Chem. Journal of Industrial and Engineering Chemistry
J. Phys. Chem. Journal of Physical Chemistry
J. Phys. Radium Le Journal de Physique et Le Radium
Mem. Proc. Manchester Lit. & Phil. Soc. Memoirs and Proceedings of the Manchester Literary and Philosophical Society
Monatsh. Chem. Monatshefte für Chemie und Verwandte Teile Anderer Wissenschaften
Phil. Mag. Philosophical Magazine
Phil. Trans. Roy. Soc. London Philosophical Transactions of the Royal Society of London
Phys. Rev. Physical Review
Physik. Z. Physikalische Zeitschrift
Popular Sci. Monthly Popular Science Monthly
Proc. Am. Chem. Soc. Proceedings of the American Chemical Society
Proc. Am. Phil. Soc. Proceedings of the American Philosophical Society
Proc. Cambridge Phil. Soc. Proceedings of the Cambridge Philosophical Society
Proc. Nat. Acad. Sci. U.S. Proceedings of the National Academy of Sciences of the United States
Proc. Phys. Soc. London Proceedings of the Physical Society (London)
Proc. Roy. Soc. London Proceedings of the Royal Society of London
RCC Rutherford Collection, Cambridge University Library
Rev. Gén. Sci. Revue Générale des Sciences Pures et Appliquées
Rutherford and Boltwood Lawrence Badash (ed.), Rutherford and Boltwood, Letters on Radioactivity (New Haven: Yale University Press, 1969)
Sci. American Scientific American
Sci. American Suppl. Scientific American Supplement
Sitzber. Akad. Wiss. Wien, Math. Naturw. Kl. Sitzungsberichte der Akademie der Wissenschaften in Wien, Mathematisch-Naturwissenschaftliche Klasse
Trans. Am. Electrochem. Soc. Transactions of the American Electrochemical Society
Trans. Am. Inst. Elec. Engrs. Transactions of the American Institute of Electrical Engineers
Trans. Chem. Soc. London Transactions of the Chemical Society (London)
Verhandl. Deut. Physik. Ges. Verhandlungen der Deutschen Physikalischen Gesellschaft
Z. Anorg. Chem. Zeitschrift für Anorganische Chemie
Z. Elektrochem. Zeitschrift für Elektrochemie und Angewandte Physikalische Chemie

Notes

Preface

1. Lawrence Badash (ed.), *Rutherford and Boltwood, Letters on Radioactivity* (New Haven: Yale University Press, 1969) (hereafter cited as *Rutherford and Boltwood*).

Introduction

1. For discussion of such sociological aspects of scientific communities, see Joseph Ben-David, *The Scientist's Role in Society* (Englewood Cliffs, New Jersey: Prentice-Hall, 1971), especially chapter 1; and Edward Shils, *Center and Periphery* (Chicago: University of Chicago Press, 1975), especially chapter 1.

Chapter 1

1. Max Planck, "New paths of physical knowledge," *Phil. Mag.*, 28 (July 1914), 60–71, quotation on p. 61.
2. For discussion of this topic, see Martin Klein, "Albert Einstein," *DSB* (New York: Scribner's, 1971), vol. 4, pp. 313–15. Russell McCormmach, "H. A. Lorentz and the electromagnetic view of nature," *Isis*, 61 (Winter 1970), 459–97. Erwin Hiebert, "The energetics controversy and the new thermodynamics," in Duane Roller (ed.), *Perspectives in the History of Science and Technology* (Norman: University of Oklahoma Press, 1971), pp. 67–86.
3. The problem here, of course, is not that hindsight tells us these nineteenth-century scientists were wrong, but the extent to which they did, indeed, hold this *fin de siècle* feeling. Certainly, there were some who opposed the prophets of "doom" and saw a rosy future. Yet, I feel that there is a convincing amount of evidence, some from the participants but most from the next generation of scientists, that the negative attitude was widespread. See L. Badash, "The completeness of nineteenth century science," *Isis*, 63 (Mar. 1972), 48–58. For the other viewpoint, see Stephen G. Brush, "Thermodynamics and history," *Graduate J.*, 7 (Spring 1967), 522–23, and Alfred M. Bork, "The fourth dimension in nineteenth-century physics," *Isis*, 55 (Sept. 1964), 338. History may, in fact, be repeating itself, for a noted physicist has recently written that he considers it unlikely that our scientific under-

standing will become much deeper than it is today: Eugene Wigner, *Symmetries and Reflections* (Bloomington: Indiana University Press, 1967), p. 215.

4. Ernest W. Brown, "George William Hill," *Biog. Mem. Nat. Acad. Sci. U.S.*, 8 (1916), 275.

5. For a study of nineteenth-century astronomy, see Arthur Berry, *A Short History of Astronomy* (New York: Dover, 1961. Reprint of the 1898 edition), ch. 13.

6. Henry M. Leicester, *The Historical Background of Chemistry* (New York: Wiley, 1965), p. 220. Florian Cajori, "The age of the sun and the earth," *Sci. American Suppl.*, 66 (12 Sept. 1908), 174-75.

7. W. D. Niven (ed.), *The Scientific Papers of James Clerk Maxwell* (Cambridge: University Press, 1890), vol. 2, p. 244.

8. The name "classical" was applied, of course, only after the development of "modern" physics. To the scientists of the nineteenth century, their physics was quite modern.

9. Robert Kennedy Duncan, *The New Knowledge* (New York: Barnes, 1906), p. xvi. See also Robert A. Millikan, *Autobiography* (New York: Prentice Hall, 1950), pp. 23-24.

10. Thomas S. Kuhn, *The Structure of Scientific Revolutions* (Chicago: University Press, 1962, revised edition 1970).

11. F. K. Richtmyer, "The romance of the next decimal place," *Science*, 75 (1 Jan. 1932), 1-5.

12. *Ibid.*, p. 5.

13. Maurice de Broglie, *Les Premiers Congrès de Physique Solvay* (Paris: Éditions Albin Michel, 1951), pp. 10-12.

14. David L. Anderson, *The Discovery of the Electron* (Princeton, New Jersey: Van Nostrand, 1964). Not all scientists believed in this "billiard-ball" type atom, however, many feeling that there must be a degree of internal structure. Also, the atom of the chemist corresponded more to the physicist's molecule, while the latter's atom was a more fundamental particle. Yet, I feel that there is sufficient evidence to claim a widespread conception of the atom as hard, impenetrable, indivisible—the basic particle of matter. See John Heilbron, "A History of the Problem of Atomic Structure from the Discovery of the Electron to the Beginning of Quantum Mechanics" (unpublished doctoral dissertation, University of California, Berkeley, 1964), ch. 1.

15. Otto Glasser, *Wilhelm Conrad Röntgen and the Early History of the Roentgen Rays* (Springfield, Illinois: Thomas, 1934). Bern Dibner, *Wilhelm Conrad Röntgen and the Discovery of X Rays* (New York: Watts, 1968).

16. Henri Becquerel, "Sur les radiations émises par phosphorescence," *Compt. Rend.*, 122 (24 Feb. 1896), 420-21. Lawrence Badash, "'Chance favors the prepared mind': Henri Becquerel and the discovery of radioactivity," *Arch. Intern. Hist. Sci.*, 18 (1965), 55-66.

17. Alfred Romer (ed.), *The Discovery of Radioactivity and Transmutation* (New York: Dover, 1964); *Radiochemistry and the Discovery of Isotopes* (New York: Dover, 1970). A. Romer, *The Restless Atom* (Garden City, New York: Doubleday, 1960). T. W. Chalmers, *A Short History of Radio-Activity* (London: The Engineer, 1951).

18. H. Becquerel, "Sur les radiations invisibles émises par les corps phosphorescents," *Compt. Rend.*, 122 (2 Mar. 1896), 501-3. L. Badash, "Becquerel's 'unexposed' photographic plates," *Isis*, 57 (Summer 1966), 267-69.

19. H. Becquerel, "Sur les radiations invisibles émises par les sels d'uranium," *Compt. Rend.*, 122 (23 Mar. 1896), 689-94.

20. H. Becquerel, "Émission de radiations nouvelles par l'uranium métallique," *Compt. Rend.*, 122 (18 May 1896), 1086-88.

21. L. Badash, "Radioactivity before the Curies," *Am. J. Phys.*, 33 (Feb. 1965), 128-35.

22. G. C. Schmidt, "Ueber die vom Thorium und den Thorverbindungen ausgehende Strahlung," *Verhandl. Deut. Physik. Ges.*, 17 (4 Feb. 1898), 14-16. M. Curie, Rayons émis par les composés de l'uranium et du thorium," *Compt. Rend.*, 126 (12 Apr. 1898), 1101-3. See also L. Badash, "The discovery of thorium's radioactivity," *J. Chem. Educ.*, 43 (Apr. 1966), 219-20. Eve Curie, *Madame Curie* (Garden City, New York: Doubleday Doran, 1938). Robert Reid, *Marie Curie* (New York: Dutton, 1974).

23. P. Curie and M. Curie, "Sur une substance nouvelle radio-active, contenue dans la pechblende," *Compt. Rend.*, 127 (18 July 1898), 175-78. P. Curie, M. Curie, and G. Bémont,

"Sur une nouvelle substance fortement radio-active, contenue dans la pechblende," *Compt. Rend.*, 127 (26 Dec. 1898), 1215-17.

24. W. Crookes, "Inaugural address, by the president of the Association," *Nature,* 58 (8 Sept. 1898), 438-48.

25. *Ibid.,* p. 446.

26. J. Elster and H. Geitel, "Versuche an Becquerelstrahlen," *Ann. Physik,* 66 (11 Nov. 1898), 735-40.

27. J. Elster and H. Geitel, "Weitere Versuche an Becquerelstrahlen," *Ann. Physik,* 69 (14 Sept. 1899), 83-90.

28. J. J. Thomson and E. Rutherford, "On the passage of electricity through gases exposed to Röntgen rays," *Phil. Mag.,* 42 (Nov. 1896), 392-407; reprinted in James Chadwick (ed.), *The Collected Papers of Lord Rutherford of Nelson* (3 vols.; New York: Interscience, 1962, 1963, 1965), vol. 1, pp. 105-18 (hereafter cited as *CPR*).

29. E. Rutherford, "Uranium radiation and the electrical conduction produced by it," *Phil. Mag.,* 47 (Jan. 1899), 109-63; reprinted in *CPR,* vol. 1, pp. 169-215. Also see A. S. Eve, *Rutherford* (Cambridge: University Press, 1939).

30. P. Villard, "Sur la réflexion et la réfraction des rayons cathodiques et des rayons déviables du radium," *Compt. Rend.,* 130 (9 April 1900), 1010-12.

31. W. H. Bragg and R. Kleeman, "On the ionization curves of radium," *Phil. Mag.,* 8 (Dec. 1904), 726-38; "On the α particles of radium, and their loss of range in passing through various atoms and molecules," *Phil. Mag.,* 10 (Sept. 1905), 318-40.

32. E. Rutherford, "The magnetic and electric deviation of the easily absorbed rays from radium," *Phil. Mag.,* 5 (Feb. 1903), 177-87; reprinted in *CPR,* vol. 1, pp. 549-57.

33. F. Giesel, "Ueber die Ablenkbarkeit der Becquerelstrahlen im magnetischen Felde," *Ann. Physik,* 69 (15 Dec. 1899), 834-36.

34. H. Becquerel, "Déviation du rayonnement du radium dans un champ électrique," *Compt. Rend.,* 130 (26 Mar. 1900), 809-15.

35. E. Rutherford, "A radioactive substance emitted from thorium compounds," *Phil. Mag.,* 49 (Jan. 1900), 1-14; reprinted in *CPR,* vol. 1, pp. 220-31.

36. E. Rutherford and F. Soddy, "The radioactivity of thorium compounds, II. The cause and nature of radioactivity," *Trans. Chem. Soc. London,* 81 (July 1902), 837-60; "Radioactive change," *Phil. Mag.,* 5 (May 1903), 576-91; reprinted in *CPR,* vol. 1, pp. 435-56, 596-608, respectively.

37. L. Badash, "How the 'newer alchemy' was received," *Sci. American,* 215 (Aug. 1966), 88-95.

Chapter 2

1. J. J. Thomson, "The Röntgen rays," *Nature,* 53 (23 Apr. 1896), 581-83. This paper was abstracted in the American periodical *Science,* 3 (8 May 1896), 700-1.

2. I am indebted to Prof. A. T. Sprague, of Auburn University, for information about McKissick.

3. A. F. McKissick, "Becquerel rays," *Elec. World,* 28 (28 Nov. 1896), 652; *Sci. American Suppl.,* 43 (9 Jan. 1897), 17542; *Electrician,* 38 (1 Jan. 1897), 313.

4. O. M. Stewart, "A résumé of the experiments dealing with the properties of Becquerel rays," *Phys. Rev.,* 6 (Apr. 1898), 239-51.

5. M. Curie, "Les rayons de Becquerel et le polonium," *Rev. Gén. Sci.,* 10 (30 Jan. 1899), 41-50.

6. *New York Tribune,* 12 and 26 Dec. 1897. L. Badash, "Carnotite: What's in a name?," *Chem. In Britain,* 2 (June 1966), 240-41.

7. *New York Tribune,* 7 Jan. 1899.

8. P. Curie, M. Curie, and G. Bémont, "Radium: A new body, strongly radio-active, contained in pitchblende," *Sci. American,* 80 (28 Jan. 1899), 60.

9. *New York Tribune,* 17 Dec. 1899.

10. *New York Tribune,* 18 Dec. 1899.

11. Minutes of the meeting of 5 Jan. 1900, *Proc. Am. Chem. Soc.* (1900), 33, bound with the Society's *Journal,* 22 (1900).

12. H. C. Bolton, "An experimental study of radio-active substances," *J. Am. Chem. Soc.,* 22 (Sept. 1900), 596-604. This article also appeared in substantially similar form as "New sources of light," *Popular Sci. Monthly,* 57 (July 1900), 318-22.

13. *Ibid.,* p. 603.

14. E. G. Merritt, "Recent developments in the study of radioactive substances," *Science,* 18 (10 July 1903), 41-47.

15. W. J. Hammer, "Edison's tungstate of calcium lamp.—the Nernst lamp.—radium, polonium and actinium," *Trans. Am. Inst. Elec. Engrs.,* 19 (1903), 67-75. This paper also appears in *Sci. American Suppl.,* 55 (28 Feb. 1903), 22710-11.

16. *Ibid.,* p. 75.

17. W. J. Hammer, *Radium, and Other Radio-Active Substances; Polonium, Actinium, and Thorium, with a Consideration of Phosphorescent and Fluorescent Substances, the Properties and Application of Selenium and the Treatment of Disease by the Ultra-Violet Light* (New York: Van Nostrand, 1903). This lecture also appeared in a number of journals, including *Sci. American Suppl.,* 55 (23 May 1903), 22904-7.

18. *Ibid.,* p. 14.

19. See note 17. The book was simply his published lecture.

20. Information about Hammer was gathered from his papers, preserved in the Division of Electricity, Museum of History and Technology, Smithsonian Institution, Washington, D.C.

21. *Schenectady Star,* 14 Dec. 1903.

22. Robert Abbe, "Radium and radio-activity," *Yale Medical Journal,* 10 (June 1904), 433-47.

23. B. Boltwood, "Radioactivity—Lecture IV, radioactive change," undated (but *ca.* 1910), unpublished notes, Boltwood collection, Yale University Library (hereafter cited as BCY).

24. Numerous newspaper articles appeared on this subject, *e.g., Chicago Record,* 16 Sept. 1904. *New York Evening Journal,* 16 Sept. 1904. *New York World,* 17 Sept. 1904.

25. *New York American,* 20 Mar. 1904.

26. *New York World,* 17 Apr. 1904.

27. *New York Evening Journal,* 30 July 1904.

28. *Electrical World and Engineering,* 13 Feb. 1904.

29. *New York American,* 1 May 1904.

30. *Boston Herald,* 13 Dec. 1903.

31. Eve Curie, *Madame Curie* (Garden City, New York: Doubleday, Doran, 1938), p. 217.

32. John Hall Ingham, "Radium," *Lippincott's Magazine,* 73 (May 1904), 640.

33. Clipping with no date and no source indicated, Hammer collection, Smithsonian Institution, Washington, D. C.

34. *Potsdam* (New York) *Courier,* 9 Dec. 1903.

35. *New York Commercial Advertiser,* 16 Dec. 1903.

36. *New York Herald,* 19 June 1904. *American Inventor* (Washington, D.C.), 15 Apr. 1904. And numerous other newspaper articles.

37. *New York Daily Tribune,* 24 Nov. 1903.

38. *New York Daily Tribune,* 26 Nov. and 4 Dec. 1903.

39. *New York Times,* 20 Dec. 1903. *Troy* (New York) *Press,* 29 Mar. 1904. *Boston Journal,* 10 Apr. 1904. *Milwaukee Sentinel,* 13 Apr. 1904.

40. Clipping from unnamed New Orleans newspaper, 29 June 1904, Hammer collection, Smithsonian Institution, Washington, D.C.

41. *International University Lectures,* (12 vols.; New York: University Alliance, 1909). Howard J. Rogers (ed.), *Congress of Arts and Science,* (8 vols.; Boston and New York: Houghton, Mifflin, 1906).

42. "A list of the invited foreign delegates to the Congress of Arts and Science," *Science,* 18 (11 Dec. 1903), 764-66.

43. *Congress of Arts and Science* (note 41), vol. 4.

44. E. Rutherford, "Disintegration of the radioactive elements," *Harper's Monthly Magazine,* 108 (Jan. 1904), 279-84.

45. E. Rutherford, "The radiation and emanation of radium," *Technics* (July and Aug. 1904), 11-16, 171-75; reprinted in *CPR*, vol. 1, pp. 641-57.

46. E. Rutherford, "Radium—the cause of the earth's heat," *Harper's Monthly Magazine*, 110 (Feb. 1905), 390-96; reprinted in *CPR*, vol. 1, pp. 776-85. Also see L. Badash, "Rutherford, Boltwood, and the age of the earth: The origin of radioactive dating techniques," *Proc. Am. Phil. Soc.*, 112 (June 1968), 157-69.

47. This largely unrecorded aspect of Ramsay's career may be examined in *Rutherford and Boltwood*. For the other side of his life, by a devoted disciple and friend, see Morris W. Travers, *A Life of Sir William Ramsay* (London: Arnold, 1956).

48. *Philadelphia Public Ledger*, 1 May 1904.

49. *New York American*, 11 Sept. 1904.

50. *New York Herald*, 11 Sept. 1904.

51. *Washington Post*, 14 Sept. 1904.

52. W. Ramsay and F. Soddy, "Experiments in radioactivity, and the production of helium from radium," *Proc. Roy. Soc. London*, 72 (15 Aug. 1903), 204-7.

53. *Buffalo Courier*, 25 Sept. 1904. *Boston Herald*, 29 May 1904. *Boston Transcript*, 8 Sept. 1904. *Philadelphia Public Ledger*, 13 Sept. 1904.

54. R. A. Millikan, "Recent discoveries in radiation and their significance," *Popular Sci. Monthly*, 64 (Apr. 1904), 481-99.

55. E. G. Merritt, "New element radium," *Century Magazine*, 67 (Jan. 1904), 451-60.

56. C. R. Stevens, *Radio-Activity* (New York: Broadway Publishing Co., 1904).

57. *Ibid.*, p. 7.

58. W. J. Hammer (note 17).

59. C. Baskerville, *Radium and Radio-Active Substances* (Philadelphia: Brown and Earle, 1905).

60. B. Boltwood, *J. Am. Chem. Soc.*, 27 (Dec. 1905), 1569-70.

61. B. Boltwood letters to E. Rutherford, 22 Sept. 1905 and 4 Oct. 1905, Rutherford collection, Cambridge University Library (hereafter cited as RCC), and E. Rutherford letters to B. Boltwood, 28 Sept. 1905 and 10 Oct. 1905, BCY; printed in *Rutherford and Boltwood*, pp. 78-91.

62. R. K. Duncan, *The New Knowledge* (New York: Barnes, 1905).

63. R. K. Duncan, "Radio-Activity," *Harper's Monthly Magazine*, 105 (Aug. 1902), 356-66; *Current Literature*, 33 (Sept. 1902), 335-37; *Overland Monthly*, 44 (Oct.-Nov. 1904), 460-67, 548-57.

64. E. Rutherford letter to B. Boltwood, 28 Sept. 1905, BCY; printed in *Rutherford and Boltwood*, pp. 83-86.

65. E. Rutherford, *Radio-Activity* (Cambridge: University Press, 1904).

66. None of these volumes were by Americans. Both F. Soddy, *Radio-Activity: an Elementary Treatise, from the Standpoint of the Disintegration Theory* (London: Electrician, 1904), and R. J. Strutt, *The Becquerel Rays and the Properties of Radium* (London: Arnold, 1904) were popular, though technical, accounts. Rutherford's text was strictly a scientific product, as were the monographs of H. Becquerel, "Recherches sur une propriété nouvelle de la matière," *Mémoires de l'Académie des Sciences, Paris*, 46 (1903); Marie Curie, *Recherches sur les Substances Radioactives* (Paris: Gauthier-Villars, 1903); and a few others.

67. L. Badash, "Radium, radioactivity, and the popularity of scientific discovery," *Proc. Am. Phil. Soc.*, 122 (June 1978), 145-54.

68. This perception was a step toward the understanding today of four forces in nature: gravitational, electromagnetic, strong interactions, and weak interactions. See, e.g., Richard Feynman, "Structure of the proton," *Science*, 183 (15 Feb. 1974), 601-10.

Chapter 3

1. C. Baskerville, "On the existence of a new element associated with thorium," *J. Am. Chem. Soc.*, 23 (Oct. 1901), 761-74.

2. *Ibid.*, p. 762.

3. W. Crookes, "Radioactivity of uranium," *Proc. Roy. Soc. London,* 66 (10 May 1900), 409–23.

4. C. Baskerville (note 1), p. 773.

5. E. Rutherford and F. Soddy, "The radioactivity of thorium compounds. I. An investigation of the radioactive emanation," *Trans. Chem. Soc. London,* 81 (Apr. 1902), 321–50; reprinted in *CPR,* vol. 1, pp. 376–402.

6. E. Rutherford and F. Soddy, "The radioactivity of thorium compounds. II. The cause and nature of radioactivity," *Trans. Chem. Soc. London,* 81 (July 1902), 837–60; reprinted in *CPR,* vol. 1, pp. 435–56.

7. K. A. Hofmann and F. Zerban, "Ueber radioactives Thor," *Ber. Deut. Chem. Ges.,* 35 (Feb. 1902), 531–33.

8. G. F. Barker, "Radioactivity of thorium minerals," *Am. J. Sci.,* 16 (Aug. 1903), 161–68.

9. *Ibid.,* p. 167.

10. Henry D. Smyth, *Atomic Energy for Military Purposes* (Princeton: University Press, 1945), pp. 46–47.

11. G. B. Pegram, "Radio-active substances and their radiations," *Science,* 14 (12 July 1901), 53–59.

12. G. B. Pegram, "Radio-active minerals," *Science,* 13 (15 Feb. 1901), 274.

13. G. B. Pegram (note 11), p. 59.

14. G. B. Pegram, "Electrolysis of radioactive substances," *Science,* 16 (21 Nov. 1902), 825; "Secondary radioactivity in the electrolysis of thorium solutions," *Phys. Rev.,* 17 (Dec. 1903), 424–40.

15. An analogy may be made with the mid-20th century proliferation of fundamental particles, followed by the increasingly popular interpretation of these particles as energy states, which has introduced a measure of order.

16. G. B. Pegram (note 14), pp. 439–40.

17. C. Baskerville letter to G. B. Pegram, 9 Feb. 1904, Pegram collection, Physics Department, Columbia University.

18. C. Baskerville, "Thorium; carolinium, berzelium," *J. Am. Chem. Soc.,* 26 (Aug. 1904), 922–42, quotation on p. 923. See also "The complex nature of thorium," *Science,* 19 (10 June 1904), 892–93.

19. *New York Times,* 22 May 1904.

20. F. Zerban, "Notiz zur Mittheilung über radioactives Thor," *Ber. Deut. Chem. Ges.,* 36 (1903), 3911–12.

21. C. Baskerville and F. Zerban, "Inactive thorium," *J. Am. Chem. Soc.,* 26 (Dec. 1904), 1642–44.

22. *Ibid.,* p. 1644.

23. B. Boltwood, "On the ratio of radium to uranium in some minerals," *Am. J. Sci.,* 18 (Aug. 1904), 97–103.

24. H. Becquerel, "Déviation du rayonnement du radium dans un champ électrique," *Compt. Rend.,* 130 (26 Mar. 1900), 809–15.

25. P. Villard, "Sur la réflexion et la réfraction des rayons cathodiques et des rayons déviables du radium," *Compt. Rend.,* 130 (9 Apr. 1900), 1010–12.

26. E. Rutherford, "The magnetic and electric deviation of the easily absorbed rays from radium," *Phil. Mag.,* 5 (Feb. 1903), 177–87; reprinted in *CPR,* vol. 1, pp. 549–57.

27. P. Curie and A. Laborde, "Sur la chaleur dégagée spontanément par les sels de radium," *Compt. Rend.,* 136 (16 Mar. 1903), 673–75.

28. G. B. Pegram and H. W. Webb, "Energy liberated by thorium," *Ann. N. Y. Acad. Sci.,* 16 (1905), 328–29. A later report by Pegram and Webb appeared as "Heat developed in a mass of thorium oxide due to its radioactivity," *Phys. Rev.,* 27 (July 1908), 18–26. That a mass of thorium maintains itself slightly above ambient air temperature was actually determined a year before Pegram, in the Cavendish Laboratory, but the finding seems not to have been published. J. J. Thomson letter to E. Rutherford, 14 Apr. 1903, RCC.

29. G. B. Pegram, "The generation of electrical charges by radium," *Ann. N.Y. Acad. Sci.,* 16 (1905), 342.

30. W. Duane, "Emission of electricity from the radium products," *Science,* 24 (13 July 1906), 48–49.

31. Radium A (polonium-218) also decays by beta emission, but the branching ratio is very small and was unknown at the time.

32. Despite the similarity in small branching ratio with radium A, the dual means of decay for radium C (bismuth-214) were known because the smaller branch happened to be the alpha emission, which was more easily detected with the electrometers used.

33. H. A. Bumstead, "On the absence of excited radioactivity due to temporary exposure to γ rays," *Proc. Cambridge Phil. Soc.*, 13 (1905), 125–28.

34. S. J. Allen, "The velocity and ratio e/m for the primary and secondary rays of radium," *Phys. Rev.*, 23 (Aug. 1906), 65–94. S. J. Allen letter to E. Rutherford, 28 Feb. 1904, RCC.

35. W. Kaufmann, "Ueber die electromagnetische Masse des Elektrons," *Physik Z.*, 4 (10 Oct. 1902), 54–57.

36. W. H. Bragg and R. Kleeman, "On the α particles of radium, and their loss of range in passing through various atoms and molecules," *Phil. Mag.*, 10 (Sept. 1905), 318–40.

37. E. P. Adams, "The absorption of alpha rays in gases and vapors," *Phys. Rev.*, 22 (Feb. 1906), 111–12; also 24 (Jan. 1907), 108–14.

38. O. W. Richardson, "The α rays," *Nature*, 75 (3 Jan. 1907), 223–24.

39. *Ibid.*, p. 223.

40. Richardson's argument was not entirely accurate, for Ramsay's original speculation, that helium is a decay product of radium emanation, did not require the helium to be evolved as an alpha ray.

41. W. Rollins, "The cathode stream and x-light," *Am. J. Sci.*, 10 (Nov. 1900), 382–91.

42. Ronald L. Kathren, "William H. Rollins (1852–1929): X-ray protection pioneer," *J. Hist. Med.*, 19 (1964), 287–94.

43. W. Rollins (note 41), p. 387.

44. M. Metzenbaum, "Induced radio-activity and aluminum," *Sci. American*, 90 (14 May 1904), 383.

45. This means 300,000 times the activity of an equal weight of uranium.

46. *New York Herald*, 19 June 1904.

47. One wonders why the phosphorescence caused by radioactive bodies on zinc sulphide and other types of screens, of which there had been knowledge for some years, was insufficient to precipitate the investigation earlier.

48. G. F. Kunz and C. Baskerville, "Action of radium, Roentgen rays and ultra-violet light on minerals and gems," *Science*, 18 (18 Dec. 1903), 769–83.

49. *New York Daily Tribune*, 6 Oct. 1903.

50. C. Baskerville and G. F. Kunz, "Effects on rare earth oxides produced by radium-barium compounds and on the production of permanently luminous preparations by mixing the latter with powdered minerals," *Am. J. Sci.*, 17 (Jan. 1904), 79–80.

51. See, e.g., S. Avery, "Changes of color caused by the action of certain rays on glass," *J. Am. Chem. Soc.*, 27 (July 1905), 909–10. C. Baskerville, "Note on the coloration of didymium glass by radium chloride," *J. Am. Chem. Soc.*, 28 (Oct. 1906), 1511.

52. B. Davis and C. W. Edwards, "Chemical combination of knall-gas under the action of radium," *Ann. N.Y. Acad. Sci.*, 16 (1905), 356–57.

53. F. C. Brown and J. Stebbins, "The effect of radium on the resistance of the selenium cell," *Phys. Rev.*, 25 (Dec. 1907), 505–6.

54. *Yale Daily News*, 11 May 1903, p. 2.

55. H. A. Bumstead and L. P. Wheeler, "Note on a radio-active gas in surface water," *Am. J. Sci.*, 16 (Oct. 1903), 328.

56. Leigh Page, "Henry Andrews Bumstead," *Am. J. Sci.*, 1 (1921), 469–76; *Biog. Mem. Nat. Acad. Sci. U.S.*, 13 (1930), 105–24.

57. E. Rutherford letter to B. Boltwood, 18 May 1905, BCY; printed in *Rutherford and Boltwood*, pp. 73–74.

58. Norman Feather, *Lord Rutherford* (London and Glasgow: Blackie, 1940; reprinted London: Priory, 1973), p. 102.

59. H. A. Bumstead and L. P. Wheeler (note 55).

60. H. A. Bumstead and L. P. Wheeler, "On the properties of a radio-active gas found in the soil and water near New Haven," *Am. J. Sci.*, 17 (Feb. 1904), 97–111.

61. E. Rutherford and S. J. Allen, "Excited radioactivity and ionization of the atmosphere," *Phil. Mag.*, 4 (Dec. 1902), 704–23; reprinted in *CPR*, vol. 1, pp. 509–27.

62. S. J. Allen, "Radioactivity of the atmosphere," *Phil. Mag.*, 7 (Feb. 1904), 140–50.

63. H. A. Bumstead, "Atmospheric radio-activity," *Am. J. Sci.*, 18 (July 1904), 1–11.

64. *Ibid.*, pp. 3–4.

65. The excited activities, or active deposits, left by radium and thorium emanations were defined as the entire series of decay products, starting with radium A and thorium A and continuing through as many letters as products were known. The substance known today as thorium A (polonium-216) was yet to be discovered, because of its short half-life (0.16 second), so the thorium A of 1904 is our thorium B (lead-212), and their thorium B is our thorium C (bismuth-212). Since the half-life of lead-212 is the longest in this series (10.6 hours), this period predominated in measurements of thorium active deposit activity. Similarly, among the well-known products in radium active deposits, which were radium A, B, and C (polonium-218, lead-214, and bismuth-214, respectively), the 26.8 minute half-life of radium B tended to predominate.

66. H. M. Dadourian, "Radio-activity of underground air," *Am. J. Sci.*, 19 (Jan. 1905), 16–22.

67. H. M. Dadourian, "On the constituents of atmospheric radio-activity," *Am. J. Sci.*, 25 (Apr. 1908), 335–42.

68. B. Boltwood letter to E. Rutherford, 8 Aug. 1904, RCC; printed in *Rutherford and Boltwood*, pp. 33–37.

69. B. Boltwood, "On the radio-active properties of the waters of the springs on the Hot Springs Reservation, Hot Springs, Ark.," *Am. J. Sci.*, 20 (Aug. 1905), 128–32.

70. *Ibid.*, p. 132.

71. B. Boltwood, "On the radio-activity of natural waters," *Am. J. Sci.*, 18 (Nov. 1904), 378–87, quotation on p. 382.

72. H. Schlundt and R. B. Moore, "Radio-activity of some deep well and mineral waters," *J. Phys. Chem.*, 9 (Apr. 1905), 320–32.

73. *New York World*, 6 Sept. 1903. "Mr. Edison's ideas on radium," *Harper's Weekly*, 47 (29 Aug. 1903), 1421.

74. The word "transformation" is used here in a different sense than in the name of the Rutherford-Soddy theory, where it means a transmutation or disintegration. Here it suggests a change of energy from a nondetectable form to one that can be perceived.

75. H. Maxim, "Radio-activity—the secret of radium's light and heat," *Sci. American Suppl.* 56 (3 Oct. 1903), 23204–5.

76. J. C. Featherstone, "Radium a transformer," *Sci. American Suppl.*, 58 (23 July 1904), 23871.

77. M. Pupin, *From Immigrant to Inventor* (New York: Scribner's, 1925), p. 353.

78. H. L. Bronson, "Some reminiscences of Professor Ernest Rutherford during his time at McGill University, Montreal," in *CPR*, vol. 1, p. 163.

79. N. Feather (note 58), p. 95.

80. E. Rutherford, "Present problems in radioactivity," *Popular Sci. Monthly*, 67 (May 1905), 5–34.

81. G. F. Hull, "The new spirit in American physics," *Am. J. Phys.*, 11 (Feb. 1943), 23–30, quotation on p. 26.

82. Reported in Columbus, Ohio *Medical Journal* (May 1904), 231.

83. E. Rutherford letter to B. Boltwood, 9 May 1906, BCY; printed in *Rutherford and Boltwood*, pp. 135–36.

84. Reported in *Science*, 25 (1 Mar. 1907), 357.

85. E. Rutherford letter to W. H. Bragg, 24 Feb. 1907, Bragg collection, Royal Institution of Great Britain library.

86. E. Rutherford letter to B. Boltwood, 14 Oct. 1906, BCY; printed in *Rutherford and Boltwood*, pp. 139–40.

87. O. Lodge letter to E. Rutherford, 4 Jan. 1904, RCC.

88. N. Feather (note 58), pp. 109–10.

89. E. Rutherford letter to J. J. Thomson, 26 Dec. 1902, Thomson collection, Cambridge University Library.

90. B. Boltwood letter to E. Rutherford, 4 Oct. 1905, RCC; printed in *Rutherford and Boltwood*, pp. 87-89.
91. O. Hahn, *A Scientific Autobiography* (New York: Scribner's, 1966), p. 19.
92. O. M. Stewart, "A résumé of the experiments dealing with the properties of Becquerel rays," *Phys. Rev.*, 6 (Apr. 1898), 239-51; "Becquerel rays, a résumé," *Phys. Rev.*, 11 (Sept. 1900), 155-75.

Chapter 4

1. Bronson's thesis was entitled "The transverse vibrations of helical springs." See *Am. J. Sci.*, 18 (July 1904), 59-72.
2. This was done at the suggestion of his adviser, Bumstead. See A. S. Eve, *Rutherford* (Cambridge: University Press, 1939), p. 112.
3. H. L. Bronson, "Reminiscences," in *CPR*, vol. 1, p. 163.
4. *Ibid.*
5. A. T. Hadley letter to A. P. Stokes, 8 Feb. 1906, Hadley correspondence, book 12, Historical Manuscripts Room, Yale University Library.
6. For further biographical information about Boltwood, see his entry in the *DSB* (New York: Scribner's, 1970), vol. 2, pp. 257-60; the sketch by Alois Kovarik in *Biog. Mem. Nat. Acad. Sci. U.S.*, 14 (1930), 69-96; and *Rutherford and Boltwood*.
7. Presented at the 29 Nov. 1897 meeting of a group of Yale graduate students and young faculty who called themselves the "Inner Temple." I am indebted to Prof. Benjamin Nangle, of the Yale English Department, for permission to examine this group's Minute Book.
8. Alexander Classen, *Quantitative Chemical Analysis by Electrolysis* (New York: Wiley, 1898; revised 1903). Charles van Deventer, *Physical Chemistry for Beginners* (New York: Wiley, 1899; revised 1904).
9. B. Boltwood, "On a simple automatic Sprengel pump," *Am. Chem. J.*, 19 (Jan. 1897), 76-78.
10. H. B. Vickery, "The early years of the Kjeldahl method to determine nitrogen," *Yale Journal of Biology and Medicine*, 18 (July 1946), 511, 513.
11. Information contained in undated lecture notes (probably after 1910) entitled "Radioactivity and its bearing on chemical theories," BCY.
12. B. Boltwood letter to E. Rutherford, 23 Sept. 1907, RCC; printed in *Rutherford and Boltwood*, pp. 163-67.
13. Untitled thesis prepared by Clifford Langley, June 1899, BCY.
14. *Ibid.*, p. 7. Also see B. Boltwood, "On ionium, a new radio-active element," *Am. J. Sci.*, 25 (May 1908), 368-69.
15. The first entry date for radioactivity in his laboratory notebook number 2, p. 59, BCY, is 26 April 1904.
16. B. Boltwood letter to E. Rutherford, 11 April 1905, RCC; printed in *Rutherford and Boltwood*, pp. 57-61.
17. B. Boltwood letters to E. Rutherford, 11 May 1904, 8 Aug. 1904, RCC; printed in *Rutherford and Boltwood*, pp. 27-31, 33-37. That Boltwood began his radioactivity investigations before meeting Rutherford is inferred from remarks in Boltwood (note 14), p. 369.
18. B. Boltwood letter to E. Rutherford, 1 Feb. 1906, RCC; printed in *Rutherford and Boltwood*, pp. 117-21.
19. Biographical sketch of Boltwood in *A Record of the Class of Eighteen Ninety-Two, Sheffield Scientific School, Yale University, 1892-1917*, p. 78.
20. E. Rutherford, "Prof. Bertram B. Boltwood," *Nature*, 121 (14 Jan. 1928), 65.
21. A. Kovarik (note 6), p. 91.
22. A. T. Hadley letter to B. Boltwood, 6 Dec. 1909, Hadley correspondence, book 18, Historical Manuscripts Room, Yale University Library.
23. A. T. Hadley letter to B. Boltwood, 31 Dec. 1909, *Ibid.* See also George W.

Pierson, *Yale College; An Educational History, 1871-1921* (New Haven: Yale University Press, 1952), p. 655.

24. A. T. Hadley letter to B. Boltwood, 24 Jan. 1910 (note 22).

25. Edward Roberts, comptroller of Yale University, conversation with author, 12 Aug. 1962. Research funds, in any case, were not a budgeted item, and it is virtually impossible to obtain information of this sort from official documents. Boltwood seems to have been unable to tap the 1890 bequest of Thomas C. Sloane, who had left $75,000 as a research endowment to the laboratory bearing his and his brother's name. Likewise, there is no evidence that the $125,000 endowment (in addition to the building costs of $385,000) given by the remaining Sloane brothers in 1912 went for professors' research. The income from $75,000 of this amount was designated for maintenance of the laboratory; income from the remaining $50,000 was used to employ research assistants and for miscellaneous purposes. Carl R. Carlson, Funds Administration manager, Yale University, letter to the author, 10 May 1977. See Edward S. Dana, "Arthur William Wright, 1836-1915," *Biog. Mem. Nat. Acad. Sci. U.S.*, 15 (1932), 242-43.

26. Ellen Gleditsch, who previously had spent five years in Mme. Curie's laboratory.

27. Admittedly, World War I hindered European production, but the ores processed in Europe had largely been mined in America. The United States lost its hegemony when the Katanga fields were exploited in the early 1920s, for the ore was richer and the labor cheaper. See chapter 10.

28. B. Boltwood letter to E. Rutherford, 5 Dec. 1911, RCC; printed in *Rutherford and Boltwood*, pp. 259-61.

29. Leigh Page, "H. A. Bumstead," *Biog. Mem. Nat. Acad. Sci. U.S.*, 13 (1930), 105-24.

30. Rutherford, for example, noted that "on the outbreak of the war, the research laboratory was practically deserted." See his "Early days in radioactivity," *J. Franklin Inst.*, 198 (Sept. 1924), 288.

31. E. Rutherford, "Henry Gwyn Jeffreys Moseley," *Nature*, 96 (9 Sept. 1915), 33-34.

32. B. Boltwood, "The new Sterling Chemistry Laboratory of Yale University," *Ind. Eng. Chem.*, 15 (Mar. 1923), 315-19.

33. For further biographical information about McCoy, see G. Ross Robertson, *Herbert Newby McCoy* (Los Angeles: privately printed, 1964). Julius Stieglitz, "American contemporaries: Herbert Newby McCoy," *Ind. Eng. Chem. News Ed.*, 13 (10 July 1935), 280. H. McCoy, "Retrospect," *Chem. Bull.*, 24 (June 1937), 207-13.

34. H. McCoy, "On the ionization constants of phenolphthalein, and the use of this body as an indicator," *Am. Chem. J.*, 31 (May 1904), 503-21.

35. H. McCoy (note 33), p. 209.

36. H. McCoy and W. C. Moore, "Organic amalgams: Substances with metallic properties composed in part of non-metallic elements," *J. Am. Chem. Soc.*, 33 (Mar. 1911), 273-92.

37. Leonard B. Loeb letter to author, 30 Nov. 1962. See also R. A. Millikan and John Mills, *A Short University Course in Electricity, Sound, and Light* (Boston: Ginn, 1908, 2nd ed. 1935), pp. 340-53.

38. Anon., "Achievements in radioactivity: Dr. H. N. McCoy receives Willard Gibbs Medal of the American Chemical Society," *Chemical Age*, 37 (12 June 1937), 519.

39. V. T. Jackson letter to Ethel Terry McCoy, 21 May 1945, McCoy collection, Manuscripts Room, University of California at Los Angeles Library.

40. Charles D. Coryell letter to author, 5 Aug. 1963. Also, Glenn T. Seaborg, *The Trans-Uranium Elements* (New Haven: Yale University Press, 1958), organizational chart opp. p. 14.

Chapter 5

1. B. Boltwood, "Radium in uranium compounds," *Engineering and Mining Journal*, 77 (1904), 756; "Relation between uranium and radium in some minerals," *Nature*, 70 (26

May 1904), 80. *New Haven Register,* 12 May 1904. *Boston Transcript,* 18 June 1904. These are preliminary announcements of his results; the following is the full paper on the subject: B. Boltwood, "On the ratio of radium to uranium in some minerals," *Am. J. Sci.,* 18 (Aug. 1904), 97–103. In this paper, additional ores from Colorado and Saxony were reported upon.

2. R. J. Strutt, "A study of the radio-activity of certain minerals and mineral waters," *Proc. Roy. Soc. London,* 73 (7 Mar. 1904), 191–97.

3. B. Boltwood (note 1), *Nature.*

4. E. Rutherford, "Prof. Bertram B. Boltwood," *Nature,* 121 (14 Jan. 1928), 64.

5. B. Boltwood (note 1), *Am. J. Sci.,* p. 103.

6. B. Boltwood letter to E. Rutherford, 11 May 1904, RCC; printed in *Rutherford and Boltwood,* pp. 27–31.

7. E. Rutherford letter to B. Boltwood, 20 June 1904, BCY; printed in *Rutherford and Boltwood,* pp. 31–32.

8. H. McCoy, "Ueber das Entstehen des Radiums," *Ber. Deut. Chem. Ges.,* 37 (9 July 1904), 2641–56; also in *Chem. News,* 90 (14, 21 Oct. 1904), 187–89, 199–201.

9. *Ibid., Chem. News,* p. 200.

10. *Ibid.*

11. R. A. Millikan, "The relation between the radioactivity and the uranium content of certain minerals," in Howard J. Rogers (ed.), *Congress of Arts and Science, Universal Exposition, St. Louis, 1904* (Boston and New York: Houghton, Mifflin, 1906), vol. 4, p. 187.

12. *Ibid.*

13. R. A. Millikan letter to Ethel Terry McCoy, 28 Aug. 1945, McCoy collection, Manuscripts Room, University of California at Los Angeles Library.

14. See note 7.

15. B. Boltwood letter to E. Rutherford, 8 Aug. 1904, RCC; printed in *Rutherford and Boltwood,* pp. 33–37.

16. B. Boltwood letter to E. Rutherford, 23 Dec. 1904, RCC; printed in *Rutherford and Boltwood,* pp. 44–50.

17. *Ibid.*

18. *Ibid.*

19. *Ibid.*

20. B. Boltwood letter to E. Rutherford, 1 Jan. 1905, RCC; printed in *Rutherford and Boltwood,* p. 51.

21. B. Boltwood, "The origin of radium," *Phil. Mag.,* 9 (Apr. 1905), 599–613.

22. *Ibid.,* p. 599n.

23. In a rare example (for papers on radioactivity) of precision estimation, Boltwood claimed it was possible "to measure the leak of the electroscope with a fair degree of accuracy to the second place of decimals." (*Ibid.,* p. 610).

24. B. Boltwood (note 21), p. 608.

25. *Ibid.,* p. 613.

26. H. McCoy, "Radioactivity as an atomic property," *J. Am. Chem. Soc.,* 27 (Apr. 1905), 391–403.

27. *Ibid.,* p. 403.

28. E. Rutherford and B. Boltwood, "The relative proportion of radium and uranium in radio-active minerals," *Am. J. Sci.,* 20 (July 1905), 55–56; reprinted in *CPR,* vol. 1, pp. 801–2.

29. B. Boltwood letter to E. Rutherford, 18 June 1905, RCC; printed in *Rutherford and Boltwood,* pp. 75–76.

30. E. Rutherford and B. Boltwood (note 28), p. 56.

31. B. Boltwood letter to E. Rutherford, 10 Dec. 1905, RCC; printed in *Rutherford and Boltwood,* pp. 109–13.

32. *Ibid.*

33. E. Rutherford and B. Boltwood, "The relative proportion of radium and uranium in radio-active minerals," *Am. J. Sci.,* 22 (July 1906), 1–3; reprinted in *CPR,* vol. 1, pp. 856–58.

34. A. S. Eve, "The measurement of radium in minerals by the γ-radiation," *Am. J. Sci.*, 22 (July 1906), 4-7.

35. E. Rutherford and B. Boltwood (note 33), p. 3. The figure accepted today is 3.42 × 10⁻⁷ gram.

36. B. Boltwood, "The production of radium from uranium," *Am. J. Sci.*, 20 (Sept. 1905), 239-44.

37. See, e.g., the series of reviews and letters to the editor, by Soddy and Marckwald, in *Nature*, 69 (11 Feb., 17 Mar. 1904), 347, 461-62.

38. B. Boltwood (note 36), p. 244.

39. *Ibid.*, p. 239.

40. F. Soddy, "The life-history of radium," *Nature*, 70 (12 May 1904), 30; "The origin of radium," *Nature*, 71 (26 Jan. 1905), 294; "The production of radium from uranium," *Phil. Mag.*, 9 (June 1905), 768-79.

41. B. Boltwood (note 36), p. 241.

42. *Ibid.*, p. 241n.

43. *Ibid.*, p. 241.

44. *Ibid.*

45. *Ibid.*, p. 242.

46. E. Rutherford letter to B. Boltwood, 16 Sept. 1905, BCY; printed in *Rutherford and Boltwood*, pp. 76-77.

47. B. Boltwood (note 36), p. 243.

48. *Ibid.*

Chapter 6

1. E. Rutherford and F. Soddy, "Radioactive change," *Phil. Mag.*, 5 (May 1903), 576-91, quotation on p. 579; reprinted in *CPR*, vol. 1, pp. 596-608.

2. B. Boltwood letter to E. Rutherford, 18 Apr. 1905, RCC; printed in *Rutherford and Boltwood*, pp. 65-67.

3. B. Boltwood, "On the ultimate disintegration products of the radio-active elements," *Am. J. Sci.*, 20 (Oct. 1905), 253-67.

4. *Ibid.*, p. 263.

5. *Ibid.*, p. 256.

6. *Ibid.*, p. 257.

7. E. Rutherford and B. Boltwood, "The relative proportion of radium and uranium in radio-active minerals," *Am. J. Sci.*, 20 (July 1905), 55-56, quotation on p. 56; reprinted in *CPR*, vol. 1, pp. 801-2.

8. B. Boltwood (note 3), p. 267.

9. B. Boltwood, "On the ultimate disintegration products of the radio-active elements. Part II. The disintegration products of uranium," *Am. J. Sci.*, 23 (Feb. 1907), 77-88.

10. *Ibid.*, p. 77.

11. Arthur Holmes, "Radioactivity and geological time," in "Physics of the earth—IV. The age of the earth," *Bulletin of the National Research Council* (U.S.), number 80 (June 1931), 214.

12. B. Boltwood (note 9), p. 88.

13. *Ibid.*, p. 87. B. Boltwood letter to E. Rutherford, 18 Nov. 1905, RCC; printed in *Rutherford and Boltwood*, pp. 100-5.

14. B. Boltwood (note 9).

15. For this story in some detail, see L. Badash, "Rutherford, Boltwood, and the age of the earth: The origin of radioactive dating techniques," *Proc. Am. Phil. Soc.*, 112 (June 1968), 157-69. Also see Joe D. Burchfield, *Lord Kelvin and the Age of the Earth* (New York: Science History Publications, 1975).

16. H. McCoy, "Ueber das Entstehen des Radiums," *Ber. Deut. Chem. Ges.,* 37 (9 July 1904), 2641–56.

17. H. McCoy, "Radioactivity as an atomic property," *J. Am. Chem. Soc.,* 27 (Apr. 1905), 391–403.

18. H. McCoy, "The relation between the radioactivity and the composition of uranium compounds," *Phil. Mag.,* 11 (Jan. 1906), 176–86.

19. This was a rather rough calculation because it was assumed that the activity of any polonium, actinium, etc. in the ore was negligible; it neglected any activities between uranium and radium in assigning the value to radium and its products; the Rutherford and McClung measurement was upon radium and just its short-lived products; and the figure 1.35 × 10⁶ was taken from Rutherford and Boltwood's first joint paper, and was later revised to 2.78 × 10⁶.

20. B. Boltwood, "On the relative proportion of the total α-ray activity of radioactive minerals due to the separate radioactive constituents," *Phys. Rev.,* 22 (May 1906), 320; "The radioactivity of the salts of radium," *Am. J. Sci.,* 21 (June 1906), 409–14.

21. *Ibid., Am. J. Sci.,* p. 410.

22. Now called radium C' (polonium-214).

23. B. Boltwood (note 20), *Am. J. Sci.,* p. 414.

24. B. Boltwood letter to E. Rutherford, 17 May 1906, RCC; printed in *Rutherford and Boltwood,* pp. 137–38.

25. H. McCoy and W. Ross, "The specific radioactivity of uranium," *J. Am. Chem. Soc.,* 29 (Dec. 1907), 1698–1709.

26. *Ibid.,* p. 1705.

Chapter 7

1. B. Boltwood letter to E. Rutherford, 7 Nov. 1906, RCC; printed in *Rutherford and Boltwood,* pp. 141–43. Since the old laboratory was being redone at this time, it would appear that Mr. Sloane decided not to give funds for the construction of a new building when Rutherford turned down the professorship at Yale. A new laboratory finally was built in 1912.

2. B. Boltwood, "The production of radium from actinium," *Nature,* 75 (15 Nov. 1906), 54; "Note on the production of radium by actinium," *Am. J. Sci.,* 22 (Dec. 1906), 537–38, quotation on p. 538.

3. *Ibid., Am. J. Sci.,* p. 537.

4. B. Boltwood (note 2), *Nature.*

5. B. Boltwood letter to E. Rutherford, 9 Dec. 1906, RCC; printed in *Rutherford and Boltwood,* pp. 147–49.

6. E. Rutherford, "Production of radium from actinium," *Nature,* 75 (17 Jan. 1907), 270–71; reprinted in *CPR,* vol. 1, pp. 907–9.

7. B. Boltwood, "Radium and its disintegration products," *Nature,* 75 (3 Jan. 1907), 223.

8. E. Rutherford and B. Boltwood, "The relative proportion of radium and uranium in radio-active minerals," *Am. J. Sci.,* 20 (July 1905), 55–56, quotation on p. 56; reprinted in *CPR,* vol. 1, pp. 801–2.

9. H. S. Allen, "Radium, actinium, and helium," *Nature,* 75 (6 Dec. 1906), 126.

10. The original idea, however, seems to have come from Bumstead. See B. Boltwood letter to E. Rutherford, 4 Oct. 1905, RCC; printed in *Rutherford and Boltwood,* pp. 87–89.

11. Besides information from the *DAB,* I am indebted to Mrs. Charles D. McLendon, of Shreveport, La., for knowledge about her father.

12. *Who's Who in America,* vol. 20 (1938–39). Also, I wish to thank Miss Esther Schlundt, of Lafayette, Indiana, for much information about her father.

13. F. F. Stephens, *A History of the University of Missouri* (Columbia, Missouri: University Press, 1962), p. 358.

14. E. Schlundt letter to the author, 8 Jan. 1963.

15. For an account of these developments, see Daniel Lang, "A most valuable accident," *New Yorker*, 35 (2 May 1959), 49ff.

16. R. Moore and H. Schlundt, "Some new methods of separating uranium X from uranium," *Phil. Mag.*, 12 (Oct. 1906), 393–96.

17. As the frequent selection of uranium nitrate by numerous chemists suggests, it was the purest uranium compound commercially available in quantity.

18. E. Rutherford, "The origin of radium," *Nature*, 76 (6 June 1907), 126; reprinted in *CPR*, vol. 2, pp. 34–35.

19. *Ibid.*

20. F. Soddy, "The origin of radium," *Nature*, 76 (13 June 1907), 150.

21. *Ibid.*

22. E. Rutherford, *Radioactive Transformations* (New Haven: Yale University Press, 1911), p. 159.

23. F. Soddy (note 20).

24. *Ibid.*

25. B. Boltwood, "The origin of radium," *Nature*, 76 (25 July 1907), 293.

26. *Ibid.*

27. B. Boltwood, "The origin of radium," *Nature*, 76 (26 Sept. 1907), 544–45.

28. *Ibid.*, p. 545.

29. B. Boltwood, "Note on a new radio-active element," *Am. J. Sci.*, 24 (Oct. 1907), 370–72.

30. B. Boltwood, "The origin of radium," *Nature*, 76 (10 Oct. 1907), 589.

31. *Ibid.*

32. *Ibid.*

33. B. Boltwood letter to E. Rutherford, 23 Sept. 1907, RCC; printed in *Rutherford and Boltwood*, pp. 163–67.

34. E. Rutherford letter to B. Boltwood, 20 Oct. 1907, BCY; printed in *Rutherford and Boltwood*, pp. 170–71.

35. O. Hahn, "The origin of radium," *Nature*, 77 (14 Nov. 1907), 30–31; *Otto Hahn: A Scientific Autobiography* (New York: Scribner's, 1966), especially pp. 48–50.

36. O. Hahn, "Einige persönliche Erinnerungen aus der Geschichte der natürlichen Radioaktivität," *Die Naturwissenschaften*, 35 (1948), 67–74; *Otto Hahn: My Life* (London: Macdonald, 1970).

37. B. Boltwood letter to E. Rutherford, 21 Apr. 1906, RCC; printed in *Rutherford and Boltwood*, pp. 133–34.

38. O. Hahn letter to B. Boltwood, 14 May 1907, BCY.

39. O. Hahn (note 35), *Scientific Autobiography*, p. 49.

40. B. Boltwood letters to E. Rutherford, 18 Nov. 1905, 1 Feb. 1906, 17 May 1906, 7 Nov. 1906, RCC; printed in *Rutherford and Boltwood*, pp. 100–5, 117–21, 137–38, 141–43.

41. B. Boltwood letter to E. Rutherford, 30 Oct. 1907, RCC; printed in *Rutherford and Boltwood*, pp. 171–74.

42. *Ibid.*

43. See note 34.

44. N. R. Campbell, "The nomenclature of radio-activity," *Nature*, 76 (24 Oct. 1907), 638.

45. E. Rutherford, "Origin of radium," *Nature*, 76 (31 Oct. 1907), 661; reprinted in *CPR*, vol. 2, pp. 38–39.

46. B. Boltwood letter to E. Rutherford, 28 Nov. 1907, RCC; printed in *Rutherford and Boltwood*, pp. 177–79. The remark about getting Io tied up to U refers to the fact that he had not yet shown the amounts of Io to be proportional to the amounts of Ra or U in minerals.

Chapter 8

1. B. Boltwood letter to E. Rutherford, 8 Aug. 1904, RCC; printed in *Rutherford and Boltwood*, pp. 33–37.
2. E. Rutherford letter to B. Boltwood, 16 Aug. 1904, BCY; printed in *Rutherford and Boltwood*, pp. 40–41.
3. B. Boltwood, "The origin of radium" *Phil. Mag.*, 9 (Apr. 1905), 599–613.
4. *Ibid.*, p. 611.
5. R. J. Strutt, "On the radio-active minerals," *Proc. Roy. Soc. London*, A76 (24 May 1905), 88–101.
6. E. Rutherford letter to B. Boltwood, 9 Apr. 1905, BCY; printed in *Rutherford and Boltwood*, pp. 56–57.
7. B. Boltwood letter to E. Rutherford, 11 Apr. 1905, RCC; printed in *Rutherford and Boltwood*, pp. 57–61.
8. C. Baskerville, "Thorium; carolinium, berzelium," *J. Am. Chem. Soc.*, 26 (Aug. 1904), 923.
9. B. Boltwood, "On the ultimate disintegration products of the radio-active elements," *Am. J. Sci.*, 20 (Oct. 1905), 253–67.
10. O. Hahn, "A new radio-active element, which evolves thorium emanation," *Nature*, 71 (13 Apr. 1905), 574.
11. See note 7.
12. B. Boltwood letter to E. Rutherford, 4 May 1905, RCC; printed in *Rutherford and Boltwood*, pp. 69–72.
13. B. Boltwood letter to E. Rutherford, 22 Sept. 1905, RCC; printed in *Rutherford and Boltwood*, pp. 78–82.
14. B. Boltwood letter to E. Rutherford, 4 Oct. 1905, RCC; printed in *Rutherford and Boltwood*, pp. 87–89.
15. E. Rutherford letter to B. Boltwood, 10 Oct. 1905, BCY; printed in *Rutherford and Boltwood*, pp. 90–91.
16. E. Rutherford letter to B. Boltwood, 16 Apr. 1906, BCY; printed in *Rutherford and Boltwood*, pp. 131–32.
17. B. Boltwood letter to E. Rutherford, 21 Apr. 1906, RCC; printed in *Rutherford and Boltwood*, pp. 133–34.
18. E. Rutherford letter to B. Boltwood, 28 Apr. 1906, BCY; printed in *Rutherford and Boltwood*, pp. 134–35.
19. H. Schlundt and R. Moore, "The chemical separation of the radio-active types of matter in thorium compounds," *J. Phys. Chem.*, 9 (Nov. 1905), 682–706.
20. B. Boltwood letter to E. Rutherford, 18 Nov. 1905, RCC; printed in *Rutherford and Boltwood*, pp. 100–5.
21. E. Rutherford letter to B. Boltwood, 5 Dec. 1905, BCY; printed in *Rutherford and Boltwood*, pp. 106–8.
22. B. Boltwood letter to E. Rutherford, 10 Dec. 1905, RCC; printed in *Rutherford and Boltwood*, pp. 109–13.
23. *Ibid.*
24. B. Boltwood, "The radio-activity of thorium minerals and salts," *Am. J. Sci.*, 21 (June 1906), 415–26.
25. O. Hahn, "Ein neues Zwischenprodukt im Thorium," *Ber. Deut. Chem. Ges.*, 40 (1907), 1462–69.
26. O. Hahn letter to B. Boltwood, 6 May 1906, BCY.
27. B. Boltwood, "On the radio-activity of thorium salts," *Am. J. Sci.*, 24 (Aug. 1907), 93–100, quotation on p. 95. B. Boltwood letter to O. Hahn, 9 May 1906, BCY.
28. *Ibid.*, *Am. J. Sci.*, p. 95. O. Hahn, "Reminiscences," in *CPR*, vol. 1, p. 165.
29. E. Rutherford letter to B. Boltwood, 20 Oct. 1907, BCY; printed in *Rutherford and Boltwood*, pp. 170–71.

30. O. Hahn letter to B. Boltwood, 14 May 1907, BCY.
31. B. Boltwood letter to E. Rutherford, 15 May 1907, RCC; printed in *Rutherford and Boltwood*, pp. 156-57.
32. B. Boltwood (note 27), *Am. J. Sci.*
33. *Ibid.*, p. 99.
34. B. Boltwood letter to E. Rutherford, 23 Sept. 1907, RCC; printed in *Rutherford and Boltwood*, pp. 163-67. It may be noted that

radium	= Ra-226	thorium	= Th-232
thorium X	= Ra-224	radiothorium	= Th-228
mesothorium I	= Ra-228		

35. From a biographical sketch in the Center for the History of Physics, American Institute of Physics, N.Y.C. In addition to the data in this and the standard biographical reference works, I am deeply indebted to Professor Dadourian for much valuable information obtained during an interview on 13 Nov. 1963.
36. H. Dadourian, "The radio-activity of thorium," *Am. J. Sci.*, 21 (June 1906), 427-32.
37. H. McCoy and W. Ross, "The relation between the radio-activity and the composition of thorium compounds," *Am. J. Sci.*, 21 (June 1906), 433-43, quotation on p. 433.
38. *Ibid.*, p. 443.
39. H. McCoy and W. Ross, "The specific radioactivity of thorium and the variation of the activity with chemical treatment and with time," *J. Am. Chem. Soc.*, 29 (Dec. 1907), 1709-18.
40. This occurs after the maximum reached in a month, which is due to the growth of ThX from radiothorium. The radiothorium, which remained with the thorium, is now disintegrating, and since the mesothorium was removed, the former is not being replenished. The observed minimum is a function of the periods of both parent and daughter.
41. H. McCoy and W. Ross (note 39), p. 1718.
42. *Ind. Eng. Chem. News Ed.*, 15 (10 Feb. 1937), 47.

Chapter 9

1. F. O. Walkhoff, "Unsichtbare, photographisch wirksame Strahlen," *Photographische Rundschau* (Oct. 1900), 189-91; cited by H. Becquerel, "Recherches sur une propriété nouvelle de la matière. Activité radiante spontanée ou radioactivité de la matière," *Mémoires de l'Académie des Sciences, Paris*, 46 (1903), 263.
2. F. Giesel, "Ueber radioactive Stoffe," *Ber. Deut. Chem. Ges.*, 33 (1900), 3569-71.
3. H. Becquerel (note 1), pp. 263-66.
4. H. Becquerel and P. Curie, "Action physiologique des rayons du radium," *Compt. Rend.*, 132 (3 June 1901), 1289-91.
5. For a brief survey of this literature, see Harry H. Bowing and Robert E. Fricke, "Curie therapy," in Otto Glasser (ed.), *The Science of Radiology* (Springfield, Illinois: Thomas, 1933), p. 277.
6. *Ibid.* Also see George M. MacKee, "Cutaneous Roentgen and Curie therapy," in O. Glasser, *ibid.*, p. 293.
7. G. MacKee, *ibid.*, p. 291.
8. Carl von Noorden, quoted in *An Outline of Radium and its Emanations. A Complete Handbook for the Medical Profession* (New York: National Radium Products Co., 1924), p. 17.
9. Carroll Chase, "American literature on radium and radium therapy prior to 1906," *American Journal of Roentgenology*, 8 (1921), 766-78.
10. *Ibid.*, pp. 767-68.
11. *Ibid.*, p. 768.

12. W. Rollins, "Notes on x-light: Radio-active substances in therapeutics," *Boston Medical and Surgical Journal*, 146 (1902), 85.

13. C. Chase (note 9), p. 770.

14. R. Abbe, "The subtle power of radium," *Transactions of the American Surgical Association*, 22 (1904), 253-62.

15. A. Béclère, "The use of radium in medicine," *Smithsonian Institution Annual Report, 1924*, p. 207.

16. S. G. Tracy, "Thorium: A radioactive substance with therapeutical possibilities," *Medical Record*, (23 Jan. 1904), 126-28.

17. S. G. Tracy, "Radium: Induced radioactivity and its therapeutic possibilities," *New York Medical Journal* (9 Jan. 1904), 49-52.

18. Edith H. Quimby, "The background of radium therapy in the United States, 1906-1956," *American Journal of Roentgenology*, 75 (Mar. 1956), 443-50, quotation on p. 446.

19. A. G. Bell letter to Z. T. Sowers, printed in *American Medicine* (15 Aug. 1903), 261.

20. R. Abbe, "Radium and radio-activity," *Yale Medical Journal*, 10 (1904), 433-47.

21. H. Bowing and R. Fricke (note 5), p. 278.

22. A. Béclère (note 15), p. 209.

23. W. J. Morton, "Treatment of cancer by the x-ray, with remarks on the use of radium," *International Journal of Surgery*, 16 (1903), 289-94.

24. M. Einhorn, "Radium receptacles for the stomach, oesophagus, and rectum," *Medical Record*, 65 (5 Mar. 1904), 399-400. J. A. Storck, "Some facts concerning radium and the use of the intragastric radiode," *American Medicine*, 7 (21 May 1904), 820-22.

25. James T. Case, "The early history of radium therapy and the American Radium Society," *American Journal of Roentgenology*, 82 (1959), 579.

26. M. Einhorn, "Observations on radium," *Medical Record*, 66 (30 July 1904), 164-68.

27. N. S. Finzi, *Radium Therapeutics* (London: Henry Frowde, 1913), pp. 13-16.

28. Robley D. Evans, "Radium poisoning, a review of present knowledge," *American Journal of Public Health*, 23 (1933), 1017-23.

29. For this story see Daniel Lang, "A most valuable accident," *The New Yorker*, (2 May 1959), 49ff.

30. F. Soddy, "A method of applying the rays from radium and thorium to the treatment of consumption," *British Medical Journal* (25 July 1903), 197-99.

31. J. Storck (note 24).

32. N. Finzi (note 27), pp. 22-23. E. S. London, "Radium in biology and medicine; its effect on the normal and the diseased organism," *Sci. American Suppl.*, 72 (1911), 111-12.

33. N. Finzi, *ibid.*, pp. 16-26. E. London, *ibid.*

34. *Radiation Exposure of Uranium Miners*. Hearings Before the Subcommittee on Research, Development, and Radiation of the Joint Committee on Atomic Energy, Congress of the United States. Ninetieth Congress, First Session. Part 2 (9 May-10 Aug. 1967), pp. 960-61.

35. E. Quimby (note 18), pp. 446-47; "Fifty years of radium," *American Journal of Roentgenology*, 60 (Dec. 1948), 726. H. Bowing and R. Fricke (note 5), pp. 278-79. J. Case (note 25), p. 574. W. Duane, "On the extraction and purification of radium emanation," *Phys. Rev.*, 5 (Apr. 1915), 311-14.

36. F. van Beuren and H. Zinsser, "Some experiments with radium on bacteria," *American Medicine*, 6 (1903), 1021-22.

37. W. W. Keen commenting on paper by R. Abbe (note 14), pp. 261-62.

38. W. A. Pusey, "Radium and its therapeutic possibilities," *Journal of the American Medical Association*, (16 July 1904), 173-80.

39. Charles H. Viol, "History and development of radium-therapy," *Journal of Radiology*, 2 (1921), 29-34, esp. p. 32.

40. Miscellaneous papers in the Pegram collection, Physics Department, Columbia University.

41. James T. Case, "History of radiation therapy," in Franz Buschke (ed.), *Progress in Radiation Therapy* (New York: Grune and Stratton, 1958), p. 23.

42. C. Viol (note 39), pp. 32-33.
43. *Ibid.*, p. 34.
44. For more details of the medical rise of radioactive materials, in a book whose primary purpose is the description of x-ray therapy up to the present time, see Ruth and Edward Brecher, *The Rays. A History of Radiology in the United States and Canada* (Baltimore: Williams and Wilkins, 1969), esp. pp. 150-60, 271-89.

Chapter 10

1. For more details of uranium's early history, see L. Badash, "Chance favors the prepared mind: Henri Becquerel and the discovery of radioactivity," *Arch. Intern. Hist. Sci.,* 18 (Jan.-June 1965), 55-66, esp. pp. 61-63. T. W. Chalmers, *A Short History of Radio-Activity* (London: The Engineer, 1951), pp. 9-12.
2. Marie Curie, *Pierre Curie* (New York: Macmillan, 1923), pp. 98-102, 185-88.
3. Eve Curie, *Madame Curie* (Garden City, N.Y.: Doubleday, Doran, 1938), pp. 157-77.
4. M. Curie (note 2), pp. 111-13.
5. O. Hahn, "Friedrich Giesel," *Physik. Z.,* 29 (15 June 1928), 353-57.
6. T. Chalmers (note 1), pp. 13-14. Anon., "Carl Auer von Welsbach, the inventor of the incandescent light mantle," *Sci. American,* 106 (13 Apr. 1912), 327.
7. Edmund White, "Lecture on thorium and its compounds," printed as a booklet by the Institute of Chemistry of Great Britain and Ireland, 1912, p. 6.
8. Robert M. Keeney, "Radium," *Mineral Industry,* 25 (1916), 642.
9. *New York Tribune,* Illustrated Supplement, 26 Dec. 1897. Also see L. Badash, "Carnotite: What's in a name?," *Chem. in Britain,* 2 (June 1966), 240-41.
10. "Radium edition," *The Western Miner and Financier,* 10 (17 Mar. 1904), 6-7. Charles L. Parsons, "Our radium resources," *J. Ind. Eng. Chem.,* 5 (Nov. 1913), 944.
11. C. Friedel and E. Cumenge, "Sur un nouveau minerai d'urane, la carnotite," *Compt. Rend.,* 128 (27 Feb. 1899), 532-34.
12. C. Parsons (note 10), p. 944. Stephen T. Lockwood, "Radium research in America," a collection of Congressional hearings statements, letters, clippings, book extracts, and photographs (especially of himself), bound in silk and limp leather by Lockwood and deposited in the Princeton University Library. S. Lockwood letter to the author, 11 July 1963. S. Lockwood, "Radio-active elements," *Engineering and Mining Journal,* 74 (27 Sept. 1902), 417.
13. E. Curie (note 3), pp. 203-5.
14. A portion of this correspondence is preserved in the Buffalo Museum of Natural History, Buffalo, N.Y.: P. Curie to S. Lockwood, 22 Feb. 1903 and 6 Oct. 1903.
15. S. Lockwood, "Radium research" (note 12), p. 5. C. Parsons (note 10), p. 944.
16. S. Lockwood, "Radium research" (note 12), pp. 5-6.
17. F. Soddy letter to E. Rutherford, 4 Dec. 1903, RCC.
18. R. Abbe, "Radium and radio-activity," *Yale Medical Journal,* 10 (June 1904), 433-47.
19. B. Boltwood, "Radioactivity—Lecture IV, Radioactive change," undated (but ca. 1910) notes in BCY.
20. [A. S. Russell] letter to the editor of the *Manchester Guardian,* 6 July 1914.
21. *Ibid.* Also, unpublished information from Edward Brecher, co-author of *The Rays. A History of Radiology in the United States and Canada* (Baltimore: Williams and Wilkins, 1969). C. Parsons (note 10), p. 944.
22. S. Lockwood, "Radium research" (note 12), pp. 9-10.
23. Otto Brill, "Uranium in Colorado," *Radium,* 1 (1913), 11.
24. George E. Lees, of the Standard Chemical Co., letters to B. Boltwood, 6 May, 27 May, 14 June, 20 June, 24 June 1911, and Boltwood's replies of 1 June, 19 June, 23 June, 29 June 1911, BCY.
25. R. Bosworth letter to B. Boltwood, 31 Jan. 1912, BCY.
26. G. Lees letter to B. Boltwood, 11 Sept. 1911, BCY.

27. Ads in *Radium*, first published in April 1913 by the Radium Chemical Co., a subsidiary of the Standard Chemical Co. Report No. 214, from the Committee on Mines and Mining, to the House of Representatives, printed in *Chem. News*, 110 (27 Nov. 1914), 265.

28. C. Viol, "Production of radium in America," *Mining and Scientific Press*, 109 (19 Sept. 1914), 443–44. Frank L. Hess, "Radium, uranium, and vanadium," *Mineral Resources of the U.S., 1913*, part 1, pp. 363–64. Robert M. Keeney, "Radium," *Mineral Industry*, 27 (1918), 639–43.

29. C. Parsons (note 10), p. 944.

30. Several letters between H. A. Kelly and B. Boltwood, Feb. to June 1911; Julian R. Blackman letter to B. Boltwood, 13 Feb. 1911, and Boltwood's reply of 16 Feb. 1911; Thomas F. V. Curran letter to B. Boltwood, 17 May 1912, BCY.

31. Bob Considine, *That Many May Live* (New York: Memorial Center for Cancer and Allied Diseases, 1959), pp. 37–41.

32. R. Bosworth letter to B. Boltwood, 13 Nov. 1913; B. Boltwood letter to G. Pegram, 11 Dec. 1913, BCY.

33. G. D. van Arsdale letter to B. Boltwood, 12 Dec. 1911, BCY.

34. H. Kelly letter to B. Boltwood, 11 Oct. 1913, BCY.

35. R. Moore and K. Kithil, "A preliminary report on uranium, radium, and vanadium," *U.S. Bureau of Mines Bulletin*, No. 70 (1913). "America ignores her radium mines," *New York Times*, 5 May 1913. C. Parsons, "The uranium and radium situation," *J. Ind. Eng. Chem.*, 5 (May 1913), 356–57.

36. "Misquoted, says Dr. Moore. But repeats radium prices exceed four times cost of production," *New York Times*, 11 Jan. 1914.

37. "Urges that radium be nationalized," *New York Times*, 30 Nov. 1913.

38. "Plan to corner world's radium," *New York Times*, 6 Oct. 1913. "Radium plant here a trust prospect," *New York Times*, 7 Oct. 1913.

39. "Lane wants radium deposits protected," *New York Times*, 30 Dec. 1913. "U.S. to conserve radium deposits," *New York Times*, 15 Jan. 1914.

40. "Approves radium plan," *New York Times*, 3 Jan. 1914.

41. "U.S. to conserve radium deposits," *New York Times*, 15 Jan. 1914.

42. "Radium fulfilling Bible prophecies," *New York Times*, 9 Jan. 1914.

43. "Bremner's progress gratifying," *New York Times*, 3 Jan. 1914. "Bremner sends message to public," *New York Times*, 12 Jan. 1914. "Plea for radium cure," *New York Times*, 7 Feb. 1914.

44. See note 42.

45. "Wants U.S. to take his radium secret," *New York Times*, 14 Feb. 1914.

46. "Vexed with Dr. H. A. Kelly," *New York Times*, 11 Jan. 1914.

47. "Wants radium land free," *New York Times*, 4 Jan. 1914.

48. "Coloradoans to protest," *New York Times*, 10 Jan. 1914.

49. "Big radium owner threatens a strike," *New York Times*, 16 Jan. 1914.

50. "Against radium monopoly," *New York Times*, 12 Jan. 1914.

51. "Phipps millions for radium cure," *New York Times*, 22 Jan. 1914.

52. Editorial, *New York Times*, 25 Jan. 1914.

53. "Assails Dr. Howard A. Kelly," *New York Times*, 24 Jan. 1914.

54. "Rush to grab radium area," *New York Times*, 26 Jan. 1914. "Rush to radium fields," *New York Times*, 27 Jan. 1914.

55. "Federal control of radium doomed," *New York Times*, 23 Mar. 1914.

56. "Radium cure free to all," *New York Times*, 24 Oct. 1913. Archibald Douglas, "The National Radium Institute," *Mining and Scientific Press* (5 Jan. 1914), as reprinted in *Chem. News*, 110 (24 Dec. 1914), 311–12. C. Parsons, "Our radium resources," *J. Ind. Eng. Chem.*, 5 (Nov. 1913), 946; "Radium from carnotite," *Metallurgical and Chemical Engineering*, 14 (1 Jan. 1916), 52.

57. A. Douglas, *ibid.*, p. 312.

58. "Will give radium deposits to nation," *New York Times*, 19 Dec. 1913. "DuPont radium free," *New York Times*, 20 Feb. 1914.

59. "American radium shown," *New York Times*, 28 Jan. 1915. "Cost of radium re-

duced," *New York Times,* 29 July 1915. "Cut radium's cost fully one-third," *New York Times,* 22 Nov. 1915. "Radium production from Colorado carnotite ores by U.S. Bureau of Mines," *Chem. News,* 112 (24 Dec. 1915), 315-16.

60. "Experts disagree on cost of radium," *New York Times,* 23 Nov. 1915.

61. Dept. of Interior news release, 27 July 1915.

62. *Ibid.* "Cost of radium reduced," *New York Times,* 29 July 1915.

63. C. Viol, "Radium and radio-activity," *Chem. & Met. Eng.,* 19 (1 Dec. 1918), 752. R. Moore, "Radium," *Bull. Am. Inst. Mining Engrs.,* No. 140 (Aug. 1918), 1180.

64. R. Moore, *ibid.,* p. 1174. C. Viol (note 28), p. 444. C. Parsons (note 56), *Metallurgical and Chemical Engineering,* pp. 51-53. S. C. Lind letter to the author, 8 Feb. 1963. S. Lind, "Practical methods for the determination of radium. II—The emanation method," *J. Ind. Eng. Chem.,* 7 (Dec. 1915), 1024-29.

65. Robert M. Keeney, "Radium," *Mineral Industry,* 26 (1917), 616-18. James Paone, "Radium," *Mineral Facts and Problems, 1960 Edition,* p. 1.

66. C. Viol (note 28). Frank L. Hess, "Radium, uranium, and vanadium," *Mineral Resources of the U.S., 1914,* part 1, pp. 943-44.

67. T. Thorne Baker, "The industrial uses of radium" *Journal of the Royal Society of Arts,* 68 (16 Apr. 1915), 490-98.

68. From an advertisement for *An Outline of Radium and its Emanations: A Complete Handbook for the Medical Profession* (New York: National Radium Products Co., 1924).

69. T. Baker (note 67), pp. 490-93. "Radium edition" (note 10), p. 5. "Radium as a fertilizer," *J. Ind. Eng. Chem.,* 7 (Dec. 1915), 1081. R. Keeney (note 28), pp. 642-43.

70. "Radium cure for insanity," *New York Times,* 9 Jan. 1914. "Government warns against radioactive claims," *J. Ind. Eng. Chem.,* 18 (Aug. 1926), 807.

71. C. Viol and Glenn D. Kammer, "The application of radium in warfare," *Trans. Am. Electrochem. Soc.,* 32 (1917), 381-90.

72. R. Keeney (note 8), p. 643.

73. H. A. Mount, "The story of radium," *Sci. American,* 122 (24 Apr. 1920), 454.

74. T. Baker (note 67), p. 495. R. Keeney (note 8), p. 641; (note 65), p. 616; (note 28), p. 639. H. Mount, *ibid.,* pp. 454, 468. Wallace Savage, "Radio-active luminous materials," *Chem. & Met. Eng.,* 19 (28 Sept. 1918), 517.

75. R. Keeney (note 65), p. 616; (note 28), pp. 639-43.

76. H. Mount (note 73), p. 454.

77. H. E. Bishop, "The present situation in the radium industry," *Science,* 57 (23 Mar. 1923), 341.

78. Camille Matignon, "The manufacture of radium," *Smithsonian Institution Annual Report, 1925,* p. 223.

79. H. Bishop (note 77), pp. 341-42. U.S. Geological Survey Press Notice No. 13967, dated 17 Nov. 1922. "Radium at low prices promised physicians," *New York Times,* 14 Nov. 1922. "Radium $70,000 a gram," *New York Times,* 29 Nov. 1922. William H. Byler, Vice President of United States Radium Corporation, letter to author, 27 Mar. 1963.

80. C. Matignon (note 78), p. 232.

81. Paul M. Tyler, "Minor metals," a chapter from *Minerals Yearbook 1938* (U.S. Dept. Interior, Bureau of Mines, 1938), pp. 8-11. W. R. Jones, "Why radium is cheaper," *The Listener,* 20 (Dec. 1938), 1229.

82. Robert M. Keeney, "Radium," *Mineral Industry,* 24 (1915), 621; (note 28), p. 641.

Chapter 11

1. M. Curie and A. Debierne, "Sur le radium métallique," *Compt. Rend.,* 151 (5 Sept. 1910), 523-25.

2. E. Rutherford, *Radio-Activity* (Cambridge: University Press, 1904), pp. 154-58.

3. E. Rutherford, *ibid.,* pp. 332-33. Note that Rutherford made an arithmetical error here and reported the half-life as 1500 years.

4. E. Rutherford, *Radioactive Transformations* (New Haven: Yale University Press, and London: Oxford University Press, 1911), p. 149. Also see F. Soddy, *Radio-Activity* (London: The Electrician, 1904), p. 168, for a slightly different calculation, yielding 1700 years.

5. E. Rutherford, *ibid.*, pp. 149-50.

6. F. Soddy (note 4), pp. 168-69. Also see E. Rutherford (note 4), p. 150, for a variation of this method which yielded 1300 years. A. T. Cameron and W. Ramsay, "Some properties of radium emanation," *J. Chem. Soc.*, 91 (1907), 1266-82, reported a half-life of only 163 years, which was greeted with the skepticism other workers in radioactivity customarily gave to Ramsay's conclusions.

7. E. Rutherford, "Charge carried by the α and β rays of radium," *Phil. Mag.*, 10 (Aug. 1905), 193-208; reprinted in *CPR*, vol. 1, pp. 816-29.

8. E. Rutherford, "The mass and velocity of the α particles expelled from radium and actinium," *Phil. Mag.*, 12 (Oct. 1906), 348-71; reprinted in *CPR*, vol. 1, pp. 880-900.

9. E. Rutherford and H. Geiger, "An electrical method of counting the number of α-particles from radio-active substances," *Proc. Roy. Soc. London*, A81 (27 Aug. 1908), 141-61; reprinted in *CPR*, vol. 2, pp. 89-108.

10. E. Rutherford and H. Geiger, "The charge and nature of the α-particle," *Proc. Roy. Soc. London*, A81 (27 Aug. 1908), 162-73; reprinted in *CPR*, vol. 2, pp. 109-20.

11. E. Rutherford, J. Chadwick, and C. D. Ellis, *Radiations from Radioactive Substances* (Cambridge: University Press, 1930, reissued 1951), p. 18.

12. B. Boltwood, "On the life of radium," *Am. J. Sci.*, 25 (June 1908), 493-506; "The life of radium," *Science*, 42 (17 Dec. 1915), 851-59.

13. B. Keetman, "Über Ionium," *Jahrb. Radioakt. u. Elektronik*, 6 (1909), 265-74.

14. S. Meyer, "Über die Lebensdauer von Uran und Radium," *Sitzber. Akad. Wiss. Wien, Math. Naturw. Kl., Abt. IIa*, 122 (1913), 1085-94.

15. E. Rutherford, "The velocity and energy of the α particles from radioactive substances," *Phil. Mag.*, 13 (Jan. 1907), 110-17; reprinted in *CPR*, vol. 1, pp. 910-16.

16. H. Geiger and J. M. Nuttall, "The ranges of the α particles from various radioactive substances and a relation between range and period of transformation," *Phil. Mag.*, 22 (Oct. 1911), 613-21.

17. B. Boltwood letter to E. Rutherford, 12 Sept. [1913], RCC; printed in *Rutherford and Boltwood*, pp. 285-86.

18. E. Gleditsch letter to the author, 2 Oct. 1962.

19. E. Gleditsch, "The life of radium," *Am. J. Sci.*, 41 (Jan. 1916), 112-24.

20. B. Boltwood letter to E. Rutherford, 18 Jan. 1916, RCC; printed in *Rutherford and Boltwood*, pp. 315-18.

21. For a discussion of later developments, see T. P. Kohman, D. P. Ames, and J. Sedlet, "The specific activity of radium," in *The Transuranium Elements*, National Nuclear Energy Series (New York: McGraw-Hill, 1949), vol. 4, 14B, pp. 1675-99, as cited in H. W. Kirby, "Nuclear properties and genetic relationships of the naturally occurring radioactive series," U. S. Atomic Energy Commission Report MLM-2036 (15 Aug. 1973), p. 24.

22. B. Boltwood (note 12), *Science*. E. Rutherford, "Radium constants on the International Standard," *Phil. Mag.*, 28 (Sept. 1914), 320-27; reprinted in *CPR*, vol. 2, pp. 486-92.

23. E. Rutherford letter to B. Boltwood, 7 Mar. 1909, BCY; printed in *Rutherford and Boltwood*, pp. 210-13.

24. E. Rutherford, "The amount of emanation and helium from radium," *Nature*, 68 (20 Aug. 1903), 366-67; reprinted in *CPR*, vol. 1, pp. 609-10.

25. E. Rutherford (note 8). E. Rutherford and H. Geiger (note 9).

26. E. Rutherford and T. Royds, "The nature of the α particle," *Mem. Proc. Manchester Lit. & Phil. Soc.*, 53:1 (1908), 1-3; "The nature of the α paticle from radioactive substances," *Phil. Mag.*, 17 (Feb. 1909), 281-86; reprinted in *CPR*, vol. 2, pp. 134-35 and 163-67.

27. J. Dewar, "The rate of production of helium from radium," *Proc. Roy. Soc. London*, A81 (11 Sept. 1908), 280-86; "Long-period determination of the rate of production of helium from radium," *Proc. Roy. Soc. London*, A83 (22 Mar. 1910), 404-8.

28. E. Rutherford and H. Geiger (note 10).

29. E. Rutherford and B. Boltwood, "Production of helium by radium," *Mem. Proc. Manchester Lit. & Phil. Soc.*, 54:6 (1909), 1-2; reprinted in *CPR*, vol. 2, p. 177.

30. B. Boltwood letter to E. Rutherford, 18 Nov. 1905, RCC, and E. Rutherford letter to B. Boltwood, 5 Dec. 1905, BCY; printed in *Rutherford and Boltwood*, pp. 100–5, 106–8.
31. Isaac Newton, *Principia*, Cajori edition (Berkeley: University of California Press, 1960), p. 398.
32. B. Boltwood and E. Rutherford, "Die Erzeugung von Helium durch Radium," *Sitzber. Akad. Wiss. Wien, Math. Naturw. Kl., Abt. IIa*, 120 (1911), 313–36; "Production of helium by radium," *Phil. Mag.*, 22 (Oct. 1911), 586–604; the first paper is reprinted in *CPR*, vol. 2, pp. 221–37. With Dewar's radium recalibrated against Rutherford's standard, and then adjusted in terms of the International Standard prepared by Marie Curie in 1912, his value for helium was 172 cubic millimeters per year. Boltwood and Rutherford's adjusted value was 163 cubic millimeters per year. E. Rutherford, J. Chadwick, and C. D. Ellis (note 11), p. 164.
33. E. Rutherford and B. Boltwood, "The relative proportion of radium and uranium in radio-active minerals," *Am. J. Sci.*, 22 (July 1906), 1–3; reprinted in *CPR*, vol. 1, pp. 856–58.
34. Joseph J. Katz and Eugene Rabinowitch, *The Chemistry of Uranium* (New York: Dover, 1961), p. 244.
35. B. Boltwood, "On the radio-activity of uranium minerals," *Am. J. Sci.*, 25 (Apr. 1908), 269–98.
36. E. Rutherford (note 22).
37. H. McCoy and W. Ross, "The specific radioactivity of uranium," *J. Am. Chem. Soc.*, 29 (Dec. 1907), 1698–1709.
38. J. Danne, "Sur un nouveau minéral radifère," *Compt. Rend.*, 140 (23 Jan. 1905), 241–43.
39. H. McCoy and W. Ross (note 37), pp. 1702–3.
40. E. Gleditsch, "Sur le radium et l'uranium contenus dans les minéraux radioactifs," *Compt. Rend.*, 148 (1 June 1909), 1451–53; "Sur le rapport entre l'uranium et le radium dans les minéraux radioactifs," *Compt. Rend.*, 149 (26 July 1909), 267–68.
41. F. Soddy, "Radioactivity," *Ann. Repts. on Progr. Chem., Chem. Soc. London*, 8 (1911), 294–95. Also see E. Rutherford, *Radioactive Substances and their Radiations* (Cambridge: University Press, 1913), pp. 460–65.
42. Review of John Joly, *The Birth-Time of the World, and Other Scientific Essays* (London: Unwin, 1915), by Arthur Holmes, *Geol. Mag.*, 3 (1916), 176–78.
43. B. Heimann and W. Marckwald, "Über den Radiumgehalt von Pechblenden," *Jahrb. Radioakt. u. Elektronik*, 10 (1913), 299–323. A. Becker and P. Jannasch, "Radioaktive und chemische Analyse der Uranpechblende von Joachimsthal," *Jahrb. Radioakt. u. Elektronik*, 12 (1915), 1–34.
44. S. Lind and L. Roberts, "New determination of the absolute value of the radium:uranium ratio," *J. Am. Chem. Soc.*, 42 (June 1920), 1170–77. Also see S. Lind and C. Whittemore, "The radium:uranium ratio in carnotites," *J. Am. Chem. Soc.*, 36 (Oct. 1914), 2066–82. H. Schlundt, "Some experiments on the quantitative determination of radium," *Trans. Am. Electrochem. Soc.*, 26 (1914), 163–79.

Chapter 12

1. B. Boltwood, "On ionium, a new radio-active element," *Am. J. Sci.*, 25 (May 1908), 365–81.
2. B. Boltwood, *ibid.*, p. 379.
3. F. Soddy, "The relation between uranium and radium. Part IX. The period of ionium," *Phil. Mag.*, 12 (Nov. 1931), 939–45. Only this last article is here cited, as the rest of this series is listed therein.
4. J. Chadwick letter to F. Soddy, 21 Oct. 1942, file 140, Soddy Collection, Bodleian Library, Oxford.
5. E. Rutherford, *Radio-Activity* (Cambridge: At the University Press, 1904), p. 308.

6. H. McCoy and W. Ross, "The specific radioactivity of uranium," *J. Am. Chem. Soc.*, 29 (Dec. 1907), 1698-1709. B. Boltwood, "Radium and its disintegration products," *Nature*, 75 (3 Jan. 1907), 223.

7. B. Boltwood, "On the radio-activity of uranium minerals," *Am. J. Sci.*, 25 (Apr. 1908), 269-98; "The origin and life of radium," *Phys. Rev.*, 26 (May 1908), 413-14.

8. J. H. L. Johnstone and B. Boltwood, "The relative activity of radium and the uranium with which it is in radioactive equilibrium," *Am. J. Sci.*, 50 (July 1920), 1-19, esp. p. 9.

9. H. McCoy and E. Leman, "The relation between the alpha-ray activities and ranges of radium and its short-lived products," *Phys. Rev.*, 6 (Sept. 1915), 184-91. B. Boltwood, "The radio-activity of the salts of radium," *Am. J. Sci.*, 21 (June 1906), 409-14.

10. B. Boltwood (note 7), *Am. J. Sci.*, pp. 282-83.

11. B. Boltwood (note 6).

12. H. McCoy and W. Ross (note 6), p. 1707.

13. B. Boltwood (note 7), *Am. J. Sci.*, p. 298.

14. B. Boltwood (note 6). R. Moore and H. Schlundt, "Some new methods of separating uranium X from uranium," *Phil. Mag.*, 12 (Oct. 1906), 393-96.

15. H. McCoy and W. Ross, "The relation between uranium and radium," *Phys. Rev.*, 24 (Jan. 1907), 124-25; "The variation of the alpha-ray ionization of radioactive solids with the thickness of the layer," *Phys. Rev.*, 1 (May 1913), 395.

16. B. Boltwood letters to E. Rutherford, 3 May 1908 and 11 Oct. 1908, RCC; printed in *Rutherford and Boltwood*, pp. 181-83, 189-97. The quotation is in the second letter.

17. E. Rutherford letter to B. Boltwood, 7 Mar. 1909, BCY; printed in *Rutherford and Boltwood*, pp. 210-13.

18. H. Geiger and E. Rutherford, "The number of α particles emitted by uranium and thorium and by uranium minerals," *Phil. Mag.*, 20 (Oct. 1910), 691-98; reprinted in *CPR*, vol. 2, pp. 196-202.

19. H. Geiger and J. M. Nuttall, "The ranges of the α particles from uranium," *Phil. Mag.*, 23 (Mar. 1912), 439-45.

20. E. Marsden and T. Barratt, "The probability distribution of the time intervals of α particles with application to the number of α particles emitted by uranium," *Proc. Phys. Soc. London*, 23 (1911), 367-73.

21. E. Rutherford, *Radioactive Substances and their Radiations* (Cambridge: University Press, 1913), pp. 447-50.

22. H. McCoy and W. Ross (note 6). Also see chapter 6, and H. McCoy and G. C. Ashman, "The preparation of urano-uranic oxide, U_3O_8, and a standard of radioactivity," *Am. J. Sci.*, 26 (Dec. 1908), 521-30.

23. S. Meyer and F. Paneth, "Über die Intensität der α-Strahlung von Uran," *Sitzber. Akad. Wiss. Wien, Math. Naturw. Kl., Abt. IIa*, 121 (1912), 1403-12.

24. J. H. L. Johnstone and B. Boltwood (note 8), p. 13.

25. F. Soddy and J. Cranston, "The parent of actinium," *Proc. Roy. Soc. London*, A94 (1 June 1918), 384-405.

26. E. Rutherford, "The succession of changes in radioactive bodies," *Phil. Trans. Roy. Soc. London*, A204 (1904), 169-219; reprinted in *CPR*, vol. 1, pp. 671-722, esp. p. 721.

27. B. Boltwood, "On the ultimate disintegration products of the radioactive elements," *Am. J. Sci.*, 20 (Oct. 1905), 258; (note 7), *Am. J. Sci.*, p. 270. About a decade later Karl Fussler, at the University of Pennsylvania, confirmed Boltwood's work using a different mineral. K. Fussler, "The actinium-uranium ratio in Colorado carnotite," *Phys. Rev.*, 9 (Feb. 1917), 142-47.

28. E. Rutherford and B. Boltwood, "The relative proportion of radium and uranium in radio-active minerals," *Am. J. Sci.*, 20 (July 1905), 55-56, quotation on p. 56; reprinted in *CPR*, vol. 1, pp. 801-2.

29. See chapters 6 and 7.

30. H. McCoy and W. Ross (note 6), pp. 1708-9.

31. M. Levin, "Über einige radioaktive Eigenschaften des Uraniums," *Physik. Z.*, 7 (1906), 692-96.

32. B. Boltwood letter to E. Rutherford, 11 Oct. 1908, RCC; printed in *Rutherford and Boltwood*, pp. 189-97.

33. B. Boltwood, "Report on the separation of ionium and actinium from certain residues and on the production of helium by ionium," *Proc. Roy. Soc. London,* A85 (14 Mar. 1911), 77-81.

34. F. Soddy, "Multiple atomic disintegration, a suggestion in radioactive theory," *Phil. Mag.,* 18 (Nov. 1909), 739-44.

35. F. Soddy, "Attempts to detect the production of helium from the primary radio-elements," *Phil. Mag.,* 16 (Oct. 1908), 513-30, esp. pp. 516-17; (note 34), p. 741; "Radioactivity," *Ann. Repts. on Progr. Chem., Chem. Soc. London,* 6 (1909), 259.

36. K. Fajans, "Über die komplexe Natur von Radium C," *Physik. Z.,* 12 (1911), 369-78. F. Soddy, "Radioactivity," *Ann. Repts. on Progr. Chem., Chem. Soc. London,* 8 (1911), 287.

37. E. Marsden and T. Barratt, "The α particles emitted by the active deposits of thorium and actinium," *Proc. Phys. Soc. London,* 24 (1911), 50-61.

38. G. N. Antonoff, "The disintegration products of uranium," *Phil. Mag.,* 22 (Sept. 1911), 419-32.

39. E. Marsden and P. Perkins, "The transformations in the active deposit of actinium," *Phil. Mag.,* 27 (Apr. 1914), 690-703.

40. H. McCoy and E. Leman, "The relation between alpha-ray activities and ranges in the actinium series, with notes on the period and range of radioactinium," *Phys. Rev.,* 4 (Nov. 1914), 409-19.

41. H. McCoy and E. Leman, *ibid.*; "Über die Zerfalls-konstante von Aktinium X," *Physik. Z.,* 14 (1913), 1280-82. A. Kovarik, "Bemerkung über den Wert der Zerfallsperiode des *Act C*," *Physik. Z.,* 12 (1911), 83. P. Perkins, "A determination of the periods of transformation of thorium and actinium emanation," *Phil. Mag.,* 27 (Apr. 1914), 720-31.

42. M. Curie, "Sur la variation avec le temps de l'activité de quelques substances radioactives," *Le Radium,* 8 (Oct. 1911), 353-54.

43. F. Soddy (note 34), p. 742. Essentially the reverse possibility, that thorium was the parent of uranium, had been proposed by Strutt in 1905. See chapter 8.

44. See chapter 8.

45. G. Ashman, "The specific radio-activity of thorium and its products," *Am. J. Sci.,* 27 (Jan. 1909), 65-72.

46. H. McCoy and W. Ross, "The specific radioactivity of thorium and the variation of the activity with chemical treatment and with time," *J. Am. Chem. Soc.,* 29 (Dec. 1907), 1709-18. See chapter 8.

47. O. Hahn, "Ein neues Zwischenprodukt im Thorium," *Ber. Deut. Chem. Ges.,* 40 (1907), 1462-69.

48. H. McCoy and C. Viol, "The chemical properties and relative activities of the radio-products of thorium," *Phil. Mag.,* 25 (Mar. 1913), 333-59.

49. F. Soddy, *The Chemistry of the Radio-Elements* (London and New York: Longmans, Green, 1911).

50. H. Geiger, "The transformation of the actinium emanation," *Phil. Mag.,* 22 (July 1911), 201-4. E. Rutherford and H. Geiger, "Transformation and nomenclature of the radioactive emanations," *Phil. Mag.,* 22 (Oct. 1911), 621-29; reprinted in *CPR,* vol. 2, pp. 255-61.

51. O. Hahn and L. Meitner, "Eine neue Methode zur Herstellung radioaktiver Zerfallsprodukte; Thorium D, ein kurzlebiges Produkt des Thoriums," *Verhandl. Deut. Physik. Ges.,* 11 (15 Feb. 1909), 55-62.

52. E. Rutherford, *Radio-Activity* (2nd edit., Cambridge: University Press, 1905), pp. 360-61.

53. M. Leslie, "Le thorium et ses produits de desagrégation," *Le Radium,* 8 (Oct. 1911), 356-63.

54. G. Blanc, "Die Zerfallskonstante des Radiothoriums," *Physik. Z.,* 8 (1907), 321-24.

55. F. von Lerch, "Über das Th X und die induzierte Thoraktivität," *Sitzber. Akad. Wiss. Wien, Math. Naturw. Kl., Abt. IIa,* 114 (1905), 553-83.

56. O. Hahn, "Ein kurzlebiges Zwischenprodukt zwischen Mesothor und Radiothor," *Physik. Z.,* 9 (1908), 246-48.

57. F. von Lerch, "Beitrag zur Kenntnis der Thoriumzerfallsprodukte," *Sitzber. Akad. Wiss. Wien, Math. Naturw. Kl., Abt. IIa,* 116 (1907), 1443-50.

58. F. von Lerch, *ibid.*

59. H. McCoy, "The periods of transformation of uranium and thorium," *Phys. Rev.*, 1 (May 1913), 401-4.

60. H. McCoy and C. Viol (note 48), pp. 355-58.

61. E. Marsden and T. Barratt (note 37). T. Barratt, "Sur les nombres et les parcours des particules α émises par l'émanation et le dépôt actif du thorium," *Le Radium*, 9 (Mar. 1912), 81-84.

62. O. Hahn, "On some properties of the α rays of radiothorium. (II.)," *Phil. Mag.*, 12 (July 1906), 82-93.

63. H. McCoy and C. Viol (note 48), pp. 353-55, 358.

64. H. McCoy and Lawrence M. Henderson, "The ratio of mesothorium to thorium," *J. Am. Chem. Soc.*, 40 (Sept. 1918), 1316-26. G. H. Cartledge, "A study of the emanation method of determining thorium," *J. Am. Chem. Soc.*, 41 (Jan. 1919), 42-50. H. McCoy and G. H. Cartledge, "The gamma ray activity of thorium D," *J. Am. Chem. Soc.*, 41 (Jan. 1919), 50-53. P. Perkins (note 41).

65. E. Rutherford (note 21), pp. 596-99.

66. A. Smithells, "Presidential address to Section B," *Nature*, 76 (8 Aug. 1907), 352-57.

Chapter 13

1. This story is told in A. Romer (ed.), *Radiochemistry and the Discovery of Isotopes* (New York: Dover, 1970), pp. 76ff.

2. O. Hahn, "A new radio-active element which evolves thorium emanation. Preliminary communication," *Proc. Roy. Soc. London*, A76 (24 May 1905), 115-17.

3. B. Boltwood, "The radio-activity of thorium minerals and salts," *Am. J. Sci.*, 21 (June 1906), 415-26. H. M. Dadourian, "The radio-activity of thorium," *Am. J. Sci.*, 21 (June 1906), 427-32. H. McCoy and W. Ross, "The relation between the radio-activity and the composition of thorium compounds," *Am. J. Sci.*, 21 (June 1906), 433-43.

4. O. Hahn, "Ein neues Zwischenprodukt im Thorium," *Ber. Deut. Chem. Ges.*, 40 (1907), 1462-69.

5. B. Boltwood, "On the radio-activity of thorium salts," *Am. J. Sci.*, 24 (Aug. 1907), 99. B. Boltwood letter to E. Rutherford, 23 Sept. 1907, RCC; printed in *Rutherford and Boltwood*, pp. 163-67. B. Boltwood letters to H. S. Miner, 19 June 1907 and 8 July 1907, BCY.

6. H. McCoy and W. Ross, "The specific radioactivity of thorium and the variation of the activity with chemical treatment and with time," *J. Am. Chem. Soc.*, 29 (Dec. 1907), 1709-18.

7. F. Soddy, "Radioactivity," *Ann. Repts. on Progr. Chem.*, *Chem. Soc. London*, 4 (1907), 325.

8. F. Soddy, "The origins of the conception of isotopes," *Les Prix Nobel en 1921-1922* (Stockholm: Imprimerie Royale, P. A. Norstedt, 1923), p. 6.

9. B. Boltwood, "On the life of radium," *Am. J. Sci.*, 25 (June 1908), 497.

10. B. Boltwood letter to E. Rutherford, 21 Feb. 1909, RCC; printed in *Rutherford and Boltwood*, pp. 208-10.

11. B. Boltwood letter to E. Rutherford, 30 May 1909, RCC; printed in *Rutherford and Boltwood*, pp. 216-18.

12. A. S. Russell, "Lord Rutherford: Manchester, 1907-19: a partial portrait," *Proc. Phys. Soc. London*, 64 (1 Mar. 1951), 224.

13. Rutherford had lately calculated the alpha particle's atomic weight as 3.9. Thus, U (238.5) $\xrightarrow{\alpha}$ UX(230.7) $\xrightarrow{\beta}$ Io (230.7).

14. B. Keetman, "Uber Ionium," *Jahrb. Radioakt. u. Elektronik*, 6 (1909), 265-74.

15. B. Szilard, "Étude sur le radioplomb," *Le Radium*, 5 (Jan. 1908), 1-5. H. Herchfinkel, "Sur le radioplomb," *Le Radium*, 7 (July 1910), 198-200.

16. See chapter 12.

17. O. Hahn, "Ein kurzlebiges Zwischenprodukt zwischen Mesothor und Radiothor," *Physik. Z.*, 9 (1908), 246-48.

18. W. Marckwald, "Zur Kenntnis des Mesothoriums," *Ber. Deut. Chem. Ges.*, 43 (1910), 3420-22.

19. F. Soddy, "The chemistry of mesothorium," *J. Chem. Soc.*, 99 (1911), 72-83; "Radioactivity," *Ann. Repts. on Progr. Chem., Chem. Soc. London*, 7 (1910), 285-86.

20. F. Soddy, contribution to "Discussion on isotopes," *Proc. Roy. Soc. London*, A99 (2 May 1921), 97.

21. E. Rutherford and H. Brooks, "The new gas from radium," *Trans. Roy. Soc. Canada*, 7 (1901), 21-25; reprinted in *CPR*, vol. 1, pp. 301-5.

22. H. A. Bumstead and L. P. Wheeler, "On the properties of a radio-active gas found in the soil and water near New Haven," *Am. J. Sci.*, 17 (Feb. 1904), 97-111.

23. P. B. Perkins, "A determination of the molecular weight of radium emanation by the comparison of its rate of diffusion with that of mercury vapor," *Am. J. Sci.*, 25 (June 1908), 461-73. B. Boltwood letter to E. Rutherford, 3 May 1908, RCC; printed in *Rutherford and Boltwood*, pp. 181-83.

24. D. Strömholm and T. Svedberg, "Untersuchungen über die Chemie der radioaktiven Grundstoffe," *Z. Anorg. Chem.*, 61 (1909), 338-46, and 63 (1909), 197-206.

25. F. von Lerch, "Über die induzierte Thoraktivität," *Ann. Physik*, 12 (27 Oct. 1903), 745-66; "Trennungen des Radiums C vom Radium B," *Ann. Physik*, 20 (1 June 1906), 345-54.

26. F. von Lerch, "Die physikalischen und chemischen Eigenschaften der Umwandlungsprodukte des Thoriums," *Jahrb. Radioakt. u. Elektronik*, 2 (1905), 461-75. R. Lucas, "Über das elektrochemische Verhalten der radioaktiven Elemente," *Physik. Z.*, 7 (1906), 340-42.

27. F. von Lerch and E. von Wartburg, "Über das Thorium D," *Sitzber. Akad. Wiss. Wien, Math. Naturw. Kl., Abt. IIa*, 118 (1909), 1575-90.

28. G. von Hevesy, "The electrochemistry of radioactive bodies," *Phil. Mag.*, 23 (Apr. 1912), 628-46.

29. G. von Hevesy, "Über den Zusammenhang zwischen den chemischen Eigenschaften der Radioelemente und der Reihenfolge radioaktiver Umwandlungen," *Physik. Z.*, 13 (1912), 672-73.

30. G. von Hevesy, "The valency of the radioelements," *Phil. Mag.*, 25 (Mar. 1913), 390-414; "Die Valenz der Radioelemente," *Physik. Z.*, 14 (1913), 49-62. G. von Hevesy letters to E. Rutherford, 7 Dec. 1912 and 3 Jan. 1913, RCC.

31. L. B. Loeb letter to the author, 30 Nov. 1962.

32. B. Boltwood, "Report on the separation of ionium and actinium from certain residues and on the production of helium by ionium," *Proc. Roy. Soc. London*, A85 (14 Mar. 1911), 77-81. F. Soddy, "Radioactivity," *Ann. Repts. on Progr. Chem., Chem. Soc. London*, 9 (1912), 320-21.

33. A. S. Russell and R. Rossi, "An investigation of the spectrum of ionium," *Proc. Roy. Soc. London*, A87 (13 Dec. 1912), 478-84. Independently and slightly earlier, F. Exner and E. Haschek, "Spektroskopische Untersuchung des Joniums," *Sitzber. Akad. Wiss. Wien, Math. Naturw. Kl., Abt. IIa*, 121 (1912), 1075-77, also developed negative evidence of ionium's spectrum, using a sample prepared by Auer von Welsbach for the same type of separations from thorium as attempted by Keetman.

34. F. Soddy, *The Chemistry of the Radio-Elements* (London: Longmans, Green, 1911).

35. F. Soddy, *ibid.*, pp. 26-29.

36. F. Soddy, *ibid.*, p. 29.

37. F. Soddy, *ibid.*, pp. 29-30.

38. Correspondence between B. Boltwood and A. Fleck, 18 May 1914 and 9 June 1914, BCY.

39. A. Fleck, "The chemical nature of uranium X, radio-actinium, and thorium B," *Chem. News*, 106 (13 Sept. 1912), 128; "The chemical nature of some radioactive disintegration products," *J. Chem. Soc.*, 103 (1913), 381-99.

40. E. Rutherford and F. Soddy, "Radioactive change," *Phil. Mag.*, 5 (May 1903), 576-91; reprinted in *CPR*, vol. 1, pp. 596-608.

41. W. Marckwald (note 18).
42. C. Baskerville, "Some recent transmutations," *Popular Sci. Monthly,* 72 (Jan. 1908), 46–51, quote on p. 49.
43. A. S. Russell (note 12), pp. 224–25.

Chapter 14

1. A. T. Cameron, "The position of the radio-active elements in the periodic table," *Nature,* 82 (18 Nov. 1909), 67–68.
2. A. S. Russell letter to E. Rutherford, 14 Sept. 1912, RCC. A. S. Russell, "The periodic system and the radio-elements," *Chem. News,* 107 (31 Jan. 1913), 49–52.
3. A. S. Russell letter to E. Rutherford, 14 Sept. 1912, RCC. A. S. Russell conversation with the author, 20 Apr. 1970. James Chadwick conversation with the author, 19 Feb. 1970.
4. G. von Hevesy letter to E. Rutherford, 7 Dec. 1912, RCC.
5. A. S. Russell conversation with the author, 20 Apr. 1970.
6. A. S. Russell (note 2), *Chem. News.*
7. K. Fajans, "Über eine Beziehung zwischen der Art einer radioaktiven Umwandlung und dem elektrochemischen Verhalten der betreffenden Radioelemente," *Physik. Z.,* 14 (1913), 131–36; "Die Stellung der Radioelemente im periodischen System," *ibid.,* pp. 136–42.
8. K. Fajans letter to the author, 27 July 1974.
9. K. Fajans, "Die radioaktiven Umwandlungen und die Valenzfrage vom Standpunkte der Struktur der Atome," *Verhandl. Deut. Physik. Ges.,* 15 (1913), 240–59. A. Fleck, "Frederick Soddy," *Biog. Mem. Fellows Roy. Soc. London,* 3 (Nov. 1957), 208.
10. F. Soddy, "The radio-elements and the periodic law," *Chem. News,* 107 (28 Feb. 1913), 97–99.
11. A. Fleck, "The periodic system and the radio-elements," *Chem. News,* 107 (21 Feb. 1913), 95.
12. F. Soddy, "The origins of the conception of isotopes," *Lex Pris Nobel en 1921–1922* (Stockholm: Imprimerie Royale, 1923), p. 17.
13. In a letter to Fajans on 2 Apr. 1913 (Fajans' personal collection, now in the University of Michigan library), Rutherford seems to imply this to be the case. The charge is made explicit by I. M. Frank, "The sixtieth anniversary of the discovery of the law of displacement: on the lecture of Professor K. Fajans" (in Russian), *Priroda,* no. 10 (1973), 70–73.
14. References to the original literature are omitted in this section as the material is well known and is here presented only as background to the main story in radioactivity. For an overview, see Samuel Glasstone, *Sourcebook on Atomic Energy* (3rd ed., Princeton: Van Nostrand, 1967).
15. Ruth Moore, *Niels Bohr* (New York: Knopf, 1966).
16. J. L. Heilbron, *H. G. J. Moseley. The Life and Letters of an English Physicist, 1887–1915* (Berkeley: University of California Press, 1974).
17. F. Soddy letter to W. A. Noyes, 22 Feb. 1936, Soddy collection, Bodleian Library, Oxford.
18. A. Fleck, "The chemical nature of some radioactive disintegration products. Part II," *Chem. News,* 107 (6 June 1913), 273; same title, *J. Chem. Soc.,* 103 (1913), 1052–61.
19. K. Fajans and O. Göhring, "Über die komplexe Natur des Ur X," *Naturwissenschaften,* 14 (4 Apr. 1913), 339; "Über das Uran X₂—das neue Element der Uranreihe," *Physik. Z.,* 14 (1913), 877–84. K. Fajans, "Discovery and naming of the isotopes of the element 91," remarks prepared for the Third International Protactinium Conference, 15 Apr. 1969.
20. K. Fajans conversation with the author, 16–17 July 1966. K. Fajans letter to James B. Conant, 30 Oct. 1969. T. W. Richards letter to B. Boltwood, 10 Jan. 1917, BCY.

21. Obituary notice of T. W. Richards, *Proc. Roy. Soc. London*, A121 (1928), xxix-xxxiv. Gregory P. Baxter, "Theodore William Richards," *Dictionary of American Biography*. James B. Conant, "Theodore William Richards and the periodic table," *Science*, 168 (24 Apr. 1970), 425-28. Harold Hartley, "Theodore William Richards Memorial Lecture," *Chem. Soc. London, Memorial Lectures*, 3 (1929), 131-63. E. Rutherford letter to B. Boltwood, 20 Nov. 1911, BCY; printed in *Rutherford and Boltwood*, pp. 257-59.

22. T. W. Richards and M. Lembert, "The atomic weight of lead of radioactive origin," *Science*, 39 (5 June 1914), 831-32; same title, *J. Am. Chem. Soc.*, 36 (July 1914), 1329-44.

23. T. W. Richards and M. Lembert, *ibid.* Richards letter to B. Boltwood, 30 Apr. 1914, and Boltwood letter to Richards, 1 May 1914, BCY.

24. F. Soddy and Henry Hyman, "The atomic weight of lead from Ceylon thorite," *J. Chem. Soc.*, 105 (1914), 1402-8.

25. *Le Matin*, 9 June 1914. Maurice Curie, "Sur les écarts de poids atomiques obtenus avec le plomb provenant de divers minéraux," *Compt. Rend.*, 158 (8 June 1914), 1676-79.

26. T. W. Richards letters to O. Hönigschmid, 17 Dec. 1912 and 24 Feb. 1913, Harvard University Archives.

27. O. Hönigschmid and S. Horovitz, "Sur le poids atomique du plomb de la pechblende," *Compt. Rend.*, 158 (15 June 1914), 1796-98. F. Soddy, "Radioactivity," *Ann. Repts. on Progr. Chem., Chem. Soc. London*, 11 (1914), 266-69.

28. Samuel Glasstone letter to the author, 5 Sept. 1966.

29. J. B. Conant (note 21), pp. 425-26.

30. O. Hönigschmid and S. Horovitz, "Über das Atomgewicht des 'Uranbleis'," *Monatsh. Chem.*, 36 (1915), 335-80.

31. T. W. Richards and C. Wadsworth, "Further study of the atomic weight of lead of radioactive origin," *J. Am. Chem. Soc.*, 38 (Dec. 1916), 2613-22.

32. K. Fajans, *Radioactivity* (London: Methuen, 1923), p. 49.

33. O. Hönigschmid, "Neuere Atomgewichtsbestimmungen," *Z. Elektrochem.*, 25 (1 Apr. 1919), 91-96. For a brief survey of this subject see A. F. Kovarik and L. W. McKeehan, "Radioactivity," *Bulletin of the National Research Council (U.S.)*, no. 51 (Mar. 1929), 25-28. Another useful source is Arthur Holmes, *The Age of the Earth* (London and New York: Harper, 1927), p. 62.

34. J. H. L. Johnstone and B. Boltwood, "The relative activity of radium and the uranium with which it is in radioactive equilibrium," *Am. J. Sci.*, 50 (July 1920), 1.

35. K. Fajans (note 32), p. 48.

36. O. Göhring, "Einige Versuche zur Frage des Aktiniumursprunges," *Physik. Z.*, 15 (1914), 642-45.

37. F. Soddy, "The origin of actinium," *Nature*, 91 (21 Aug. 1913), 634-35.

38. O. Hahn and L. Meitner, "Die Muttersubstanz des Actiniums, ein neues radioaktives Element von langer Lebensdauer," *Physik. Z.*, 19 (1918), 208-18.

39. F. Soddy and J. Cranston, "The parent of actinium," *Proc. Roy. Soc. London*, A94 (1 June 1918), 384-405.

40. A. Piccard, "L'Hypothèse de l'existence d'un troisième corps simple radioactif dans la pléiade uranium," *Archives des Sciences Physiques et Naturelles* (Geneva), 44 (Sept. 1917), 161-64.

41. F. W. Aston, "The mass-spectrum of uranium lead and the atomic weight of protactinium," *Nature*, 123 (2 Mar. 1929), 313. E. Rutherford, "Origin of actinium and age of the earth," *ibid.*, pp. 313-14; reprinted in *CPR*, vol. 3, pp. 216-17.

42. O. Hönigschmid and S. Horovitz, "Revision des Atomgewichtes des Thoriums. Analyse des Thoriumbromids," *Monatsh. Chem.*, 37 (1916), 305-34; "Zur Kenntnis des Atomgewichtes des Ioniums," *ibid.*, pp. 335-45.

43. Samuel Lind, "The atomic weight of radium emanation," *Science*, 43 (31 Mar. 1916), 464-65.

44. F. Soddy (note 12), p. 2.

45. T. W. Richards, "The problem of radioactive lead," *Science*, 49 (3 Jan. 1919), 1-11. F. W. Aston, *Mass Spectra and Isotopes* (2nd ed., London: Arnold, 1942), chapter 14-15. S. Glasstone (note 14), chapter 6. P. W. Bridgman, "A comparison of certain electrical prop-

erties of ordinary and uranium lead," *Proc. Nat. Acad. Sci. U.S.*, 5 (1919), 351-53. W. Duane and T. Shimizu, "On the x-ray absorption wave lengths of lead isotopes," *Proc. Nat. Acad. Sci. U.S.*, 5 (1919), 198-200. Rather than arbitrarily cite a few of Harkins' many papers as examples, his own survey of his career is preferable. This appears as his Willard Gibbs Medal Address, "Surface structure and atom building," *Science*, 70 (8 and 15 Nov. 1929), 433-42 and 463-70. Note that his significant research contributions continued after this date, such as early experimental work on the neutron.

46. A. S. Russell (note 5).

47. K. Fajans (note 32), p. xii.

48. K. Fajans (note 9).

49. F. Soddy, "Intra-atomic charge," *Nature*, 92 (4 Dec. 1913), 399-400.

50. H. Creighton, M. Jermain, and A. S. McKenzie, "The influence of radium on the decomposition of hydriodic acid," *Am. Chem. J.*, 39 (Apr. 1908), 474-93.

51. H. N. McCoy and Herbert H. Bunzel, "The speed of oxidation, by air, of uranous solutions, with a note on the volumetric determination of uranium," *J. Am. Chem. Soc.*, 31 (Mar. 1909), 367-73.

52. W. Duane and G. L. Wendt, "A chemically active modification of hydrogen produced by alpha rays," *Phys. Rev.*, 7 (June 1916), 689-91; "A reactive modification of hydrogen produced by alpha-radiation," *Phys. Rev.*, 10 (Aug. 1917), 116-28.

53. H. S. Taylor, "The interaction of hydrogen and chlorine under the influence of alpha particles," *J. Am. Chem. Soc.*, 37 (Jan. 1915), 24-38, and 38 (Feb. 1916), 280-85.

54. The best source of information about S. C. Lind's work is his own survey of the field, *The Chemical Effects of Alpha Particles and Electrons* (New York: Chemical Catalog Co., 1921; 2nd ed., 1928).

55. This definition restricts the subject to chemical studies of the radioelements. S. C. Lind, *ibid.*, 2nd ed., p. 19, offers an expanded definition which encompasses all reactions involving radiant energy.

Chapter 15

1. R. A. Millikan, review in *Science*, 38 (4 July 1913), 29-30.

2. F. K. Richtmyer, "The romance of the next decimal place," *Science*, 75 (1 Jan. 1932), 1-5.

3. There is a modest literature concerning the nature of American science, though the question has generally been posed as basic vs applied research. Recent work has tended to place in proper perspective the earlier overstatements to the effect that it was almost entirely applied. My distinction of experimental versus theoretical is by no means congruent with the above categories, but does nevertheless benefit from the discussion of this topic. Note that an experimental-observational emphasis was probably true of most countries at most times, the few exceptions that come to mind being ancient Greece, Czarist Russia, and Germany between the world wars. Even in the latter two countries the impression may be due more to known accomplishments than to overall activity. See Richard H. Shryock, "American indifference to basic science during the nineteenth century," *Arch. Intern. Hist. Sci.*, 2 (1948-1949), 50-65. George H. Daniels, *American Science in the Age of Jackson* (New York: Columbia University Press, 1968), pp. 19, 237-38. Nathan Reingold, "American indifference to basic research: A reappraisal," in George H. Daniels (ed.), *Nineteenth-Century American Science* (Evanston, Ill.: Northwestern University Press, 1972), pp. 38-62. Daniel Kevles, "On the flaws of American physics: A social and institutional analysis," also in Daniels, *op. cit.*, pp. 133-51.

4. E. L. Nichols, "Henry Augustus Rowland, physicist," *Phys. Rev.*, 13 (July 1901), 62.

5. J. McKeen Cattell, "The distribution of American men of science in 1927," in *American Men of Science* (4th ed., New York: Science Press, 1927), pp. 1118-29. For an interesting analysis of the institutions that trained the men who trained the next generation of physicists—a worthwhile measure of quality—see L. S. Reich, "The structure of American physics, 1876-1916," an unpublished Johns Hopkins paper, dated March 1974.

6. P. Forman, J. Heilbron, and S. Weart, "Physics *circa* 1900," *Historical Studies in the Physical Sciences,* 5 (1975), 30.

7. J. McK. Cattell (note 5), p. 1126.

8. Theodore Lyman, "An explanation of the false spectra from diffraction gratings," *Phys. Rev.,* 16 (May 1903), 257-66. W. S. Franklin, "Derivation of equation of decaying sound in a room and definition of open window equivalent of absorbing power," *Phys. Rev.,* 16 (June 1903), 372-74. Eugene C. Woodruff, "A study of the effects of temperature upon a tuning fork," *Phys. Rev.,* 16 (June 1903), 325-55.

9. S. J. Allen, "On the range and total ionization of the α particle," *Phys. Rev.,* 27 (Oct. 1908), 294-321.

10. H. Bumstead letter to E. Rutherford, 23 June 1908, RCC.

11. B. Kučera and B. Mašek, "Über die Strahlung des Radiotellurs. II," *Physik. Z.,* 7 (15 Sept. 1906), 630-40. W. H. Bragg, "The influence of the velocity of the α particle upon the stopping power of the substance through which it passes," *Phil. Mag.,* 13 (Apr. 1907), 507-16.

12. E. Rutherford, "Retardation of the α particle from radium in passing through matter," *Phil. Mag.,* 12 (Aug. 1906), 134-46; reprinted in *CPR,* vol. 1, pp. 859-69. R. K. McClung, "The absorption of α rays," *Phil. Mag.,* 11 (Jan. 1906), 131-42. M. Levin, "Über die Absorption der α-Strahlen des Poloniums," *Physik. Z.,* 7 (1 Aug. 1906), 519-21.

13. T. S. Taylor, "On the retardation of 'alpha rays' by metal foils, and its variation with the speed of the alpha particles," *Am. J. Sci.,* 26 (Sept. 1908), 169-79.

14. T. S. Taylor, "The retardation of alpha rays by metals and gases," *Phys. Rev.,* 28 (June 1909), 465-66; "On the retardation of alpha rays by metals and gases," *Am. J. Sci.,* 28 (Oct. 1909), 357-72; this latter paper appeared also in the *Phil. Mag.,* 18 (Oct. 1909), 604-19.

15. H. Geiger, "The ionisation produced by an α-particle. Part I," *Proc. Roy. Soc. London,* A82 (31 July 1909), 486-95; "The ionisation produced by an α-particle. Part II. Connection between ionisation and absorption," *Proc. Roy. Soc. London,* A83 (14 Apr. 1910), 505-15.

16. T. S. Taylor, "On the ionization of different gases by the alpha particles from polonium and the relative amounts of energy required to produce an ion," *Am. J. Sci.,* 31 (Apr. 1911), 249-56; same title, *Phil. Mag.,* 21 (Apr. 1911), 571-79; "On the ionization of gases by the alpha particles from polonium," *Phys. Rev.,* 32 (Feb. 1911), 236.

17. T. S. Taylor, "A determination of the number of ions produced by an alpha particle from polonium," *Phil. Mag.,* 23 (Apr. 1912), 670-76.

18. T. S. Taylor, "A determination of the ionization curve for the alpha rays from polonium in mercury vapour," *Phil. Mag.,* 24 (Aug. 1912), 296-301.

19. E. Rutherford letter to B. Boltwood, 15 Aug. 1912, BCY; printed in *Rutherford and Boltwood,* pp. 275-77.

20. E. Rutherford, "The scattering of α and β particles by matter and the structure of the atom," *Phil. Mag.,* 21 (May 1911), 669-88; reprinted in *CPR,* vol. 2, pp. 238-54. C. G. Darwin, "A theory of the absorption and scattering of the α rays," *Phil. Mag.,* 23 (June 1912), 901-20. N. Bohr, "On the theory of the decrease of velocity of moving electrified particles on passing through matter," *Phil. Mag.,* 25 (Jan. 1913), 10-31.

21. E. Marsden and T. S. Taylor, "The decrease in velocity of α-particles in passing through matter," *Proc. Roy. Soc. London,* A88 (1 July 1913), 443-54. E. Rutherford, J. Chadwick, and C. D. Ellis, *Radiations from Radioactive Substances* (Cambridge: University Press, 1930, reissued 1951), p. 102.

22. T. S. Taylor, "The range and ionization of the alpha particle in simple gases," *Phil. Mag.,* 26 (Sept. 1913), 402-10.

23. H. N. McCoy and C. H. Viol, "The chemical properties and relative activities of the radio-products of thorium," *Phil. Mag.,* 25 (Mar. 1913), 333-59, esp. pp. 355-58. H. N. McCoy, "The variation of the alpha-ray ionization of radioactive solids with the thickness of the layer," *Phys. Rev.,* 1 (May 1913), 393-400, esp. p. 393.

24. H. N. McCoy and E. D. Leman, "The relation between alpha-ray activities and ranges in the actinium series, with notes on the period and range of radioactinium," *Phys. Rev.,* 4 (Nov. 1914), 409-19; "The relation between the alpha-ray activities and ranges of radium and its short-lived products," *Phys. Rev.,* 6 (Sept. 1915), 184-91.

25. A. F. Kovarik, "Range of α-particles in air at different temperatures," *Phys. Rev.*, 3 (Feb. 1914), 148–49.

26. H. A. Bumstead, "On the emission of electrons by metals under the influence of alpha rays," *Am. J. Sci.*, 32 (Dec. 1911), 403–17.

27. H. A. Bumstead and A. G. McGougan, "On the emission of electrons by metals under the influence of alpha rays," *Am. J. Sci.*, 34 (Oct. 1912), 309–28; same title, *Phil. Mag.*, 24 (Oct. 1912), 462–83.

28. L. Wertenstein, "Sur l'ionisation par projections radioactives," *Le Radium*, 9 (Jan. 1912), 6–19.

29. H. A. Bumstead and A. G. McGougan (note 27).

30. H. A. Bumstead, "On the velocities of delta rays," *Am. J. Sci.*, 36 (Aug. 1913), 91–108; same title, *Phil. Mag.*, 26 (Aug. 1913), 233–51.

31. A. G. McGougan, "Some properties of metals under the influence of alpha rays," *Phys. Rev.*, 12 (Aug. 1918), 122–29.

32. W. H. Bragg and R. D. Kleeman, "On the recombination of ions in air and other gases," *Phil. Mag.*, 11 (Apr. 1906), 466–84.

33. R. D. Kleeman, "On the recombination of ions, made by α, β, γ, and x rays," *Phil. Mag.*, 12 (Oct. 1906), 273–97.

34. M. Moulin, "L'ionisation des gaz par les rayons α et l'hypothèse de la recombinaison initiale," *Le Radium*, 5 (May 1908), 136–41.

35. F. E. Wheelock, "On the nature of the ionization produced by α rays," *Am. J. Sci.*, 30 (Oct. 1910), 233–55.

36. W. R. Barss, "Note on measurements of radio-activity by means of alpha rays," *Am. J. Sci.*, 33 (June 1912), 546–50. The question of the accuracy of ionization measurements was also examined by E. Regener, "Über den Einfluss der Kondensatorform auf den Verlauf der α-Strahlen-Sättigungsstromkurven," *Verhandl. Deut. Physik. Ges.*, 13 (1911), 1065–73.

37. E. M. Wellisch and J. W. Woodrow, "Experiments on columnar ionization," *Phil. Mag.*, 26 (Sept. 1913), 511–28. E. M. Wellisch and H. L. Bronson, "The distribution of the active deposit of radium in an electric field," *Phil. Mag.*, 23 (May 1912), 714–29.

38. For a review, see E. Rutherford, J. Chadwick, and C. D. Ellis (note 21), pp. 147–52.

39. H. A. Bumstead, "On the ionization of gases by alpha rays," *Phys. Rev.*, 8 (Dec. 1916), 715–20.

Chapter 16

1. S. J. Allen letter to E. Rutherford, 28 Feb. 1904, RCC.

2. S. J. Allen, "The velocity and ratio e/m for the primary and secondary rays of radium," *Phys. Rev.*, 23 (Aug. 1906), 65–94.

3. A. S. Eve, "On the secondary radiation caused by the β and γ rays of radium," *Phil. Mag.*, 8 (Dec. 1904), 669–85.

4. H. F. Dawes, "On the secondary radiation excited in different metals by the γ rays from radium," *Phys. Rev.*, 20 (Mar. 1905), 182–85.

5. J. A. McClelland, "On secondary radiation," *Phil. Mag.*, 9 (Feb. 1905), 230–43; "Secondary β-rays," *Proc. Roy. Soc. London*, A80 (10 June 1908), 501–15. The second paper lists several others by McClelland. See also H. W. Schmidt, "Über Reflexion und Absorption von β-Strahlen," *Ann. Physik*, 23 (1907), 671–97.

6. E. Rutherford, *Radioactive Transformations* (New Haven: Yale University Press, and London: Oxford University Press, 1911—although noted in the book itself as published in 1906).

7. C. Davisson, "Note on radiation due to impact of β-particles upon solid matter," *Phys. Rev.*, 28 (June 1909), 469–70.

8. W. B. Huff, "Typical cases of secondary radiation excited by uranium-X," *Phys. Rev.*, 30 (Apr. 1910), 482–91.

9. C. Harrison Dwight, *The First Seventy-Five Years of the Physics Department (1883–1958)* (Privately printed by the University of Cincinnati, 1969).

10. S. J. Allen, "On the secondary radiation produced from solids, solutions, and pure liquids, by the β rays of radium," *Phys. Rev.*, 29 (Sept. 1909), 177-211, quotes on pp. 210-11.

11. S. J. Allen, *ibid.*, p. 211.

12. W. H. Bragg, "The secondary radiation produced by the beta rays of radium," *Phys. Rev.*, 30 (May 1910), 638-40.

13. S. J. Allen (note 10).

14. S. J. Allen, "On the secondary β radiation from solids, solutions and liquids," *Phys. Rev.*, 30 (Feb. 1910), 276-79.

15. S. J. Allen, "On the secondary β radiation from solids and liquids," *Phys. Rev.*, 32 (Feb. 1911), 201-10. E. Rutherford, *Radioactive Substances and Their Radiations* (Cambridge: University Press, 1913), pp. 212-18.

16. E. Rutherford, *ibid.*, pp. 211, 610-11. E. Rutherford, J. Chadwick, and C. D. Ellis, *Radiations From Radioactive Substances* (Cambridge: University Press, 1930), pp. 215-16. For beta scattering in relation to the construction of atomic models, see John Heilbron, "The scattering of α and β particles and Rutherford's atom," *Arch. Hist. Exact Sci.*, 4 (1968), 247-307, esp. pp. 265-80.

17. A. F. Kovarik, "The effect of changes in the pressure and temperature of gases upon the mobility of the negative ions produced by ultra-violet light," *Phys. Rev.*, 30 (Apr. 1910), 415-45.

18. O. Hahn and L. Meitner, "Über die Absorption der β-Strahlen einiger Radioelemente," *Physik. Z.*, 9 (15 May 1908), 321-33.

19. W. Wilson, "On the absorption of homogeneous β-rays by matter, and on the variation of the absorption of the rays with velocity," *Proc. Roy. Soc. London*, A82 (18 Sept. 1909), 612-28.

20. H. W. Schmidt, "Bericht über den Durchgang der β-Strahlen durch feste Materie," *Jahrb. Radioakt. u. Elektronik*, 5 (1908), 451-92. J. A. Crowther, "On the transmission of β-rays," *Proc. Cambridge Phil. Soc.*, 15 (1910), 442-58.

21. A. F. Kovarik, "Absorption and reflexion of the β-particles by matter," *Phil. Mag.*, 20 (Nov. 1910), 849-66.

22. A. F. Kovarik and W. Wilson, "On the reflexion of homogeneous β-particles of different velocities," *Phil. Mag.*, 20 (Nov. 1910), 866-70.

23. H. Geiger and A. F. Kovarik, "On the relative number of ions produced by the β-particles from the various radioactive substances," *Phil. Mag.*, 22 (Oct. 1911), 604-13.

24. A. F. Kovarik, "Absorption of the β-particles by gases," *Phys. Rev.*, 34 (Feb. 1912), 142-43; "Absorption of β-particles by gases," *Phys. Rev.*, 3 (Feb. 1914), 150-51; "Absorption of the β-particles from some of the radioactive substances by air and carbon dioxide," *Phys. Rev.*, 6 (Dec. 1915), 419-25. A. F. Kovarik letters to E. Rutherford, 17 Aug. 1912 and 8 June 1914, RCC.

25. W. Wilson, "The decrease of velocity of the β-particles on passing through matter," *Proc. Roy. Soc. London*, A84 (28 July 1910), 141-50.

26. W. B. Huff, "Reflection of β rays by thin metal plates," *Phys. Rev.*, 35 (Sept. 1912), 194-202.

27. A. F. Kovarik and L. W. McKeehan, "Counting the transmitted and reflected β-particles," *Phys. Rev.*, 3 (Feb. 1914), 149-50; "Messung der Absorption und Reflexion von β-Teilchen durch direkte Zählung," *Physik. Z.*, 15 (1914), 434-40.

28. C. T. R. Wilson, "On an expansion apparatus for making visible the tracks of ionising particles in gases and some results obtained by its use," *Proc. Roy. Soc. London*, A87 (19 Sept. 1912), 277-92. L. W. McKeehan, "On the passage of cathode particles through gases at low pressure," *Phys. Rev.*, 4 (Aug. 1914), 140-44.

29. A. F. Kovarik and L. W. McKeehan, "Distribution of transmitted and reflected β-particles determined by the statistical method," *Phys. Rev.*, 6 (Dec. 1915), 426-36.

30. E. Rutherford, J. Chadwick, and C. D. Ellis (note 16), pp. 341-52.

31. A. F. Kovarik and L. W. McKeehan, "The magnetic spectra of the β-rays of radium DE and of radium and its products, determined by the statistical method," *Phys. Rev.*, 8 (Nov. 1916), 574-78.

32. F. Sanford, "Electrical density and absorption of beta rays," *Science*, 42 (23 July 1915), 130-31.

33. F. Sanford, "Is the Einstein radiation factor *h* a constant?," *Phys. Rev.*, 15 (Jan. 1920), 67-72. E. Rutherford, "The connexion between the β and γ ray spectra," *Phil. Mag.*, 28 (Sept. 1914), 305-19; reprinted in *CPR*, vol. 2, pp. 473-85. E. Rutherford, "Penetrating power of the x radiation from a Coolidge tube," *Phil. Mag.*, 34 (Sept. 1917), 153-62; reprinted in *CPR*, vol. 2, pp. 538-46.

34. H. A. Wilson, "On the scattering of β-rays," *Proc. Roy. Soc. London*, A102 (2 Oct. 1922), 9-20.

35. H. A. Erikson, "The ionization of gases at high pressures," *Phys. Rev.*, 27 (Dec. 1908), 473-91.

36. H. A. Erikson, "The absorption of gamma rays of radium by air at different pressures," *Phys. Rev.*, 34 (Mar. 1912), 231.

37. A. S. Eve (note 3).

38. For this subject and an exhaustive study of related developments, see Roger H. Stuewer, *The Compton Effect* (New York: Science History Publications, 1975).

39. D. C. H. Florance, "Primary and secondary γ rays," *Phil. Mag.*, 20 (Dec. 1910), 921-38.

40. F. Soddy and A. S. Russell, "The γ-rays of uranium and radium," *Phil. Mag.*, 18 (Oct. 1909), 620-49. F. Soddy, W. M. Soddy, and A. S. Russell, "The question of the homogeneity of γ-rays," *Phil. Mag.*, 19 (May 1910), 725-57.

41. S. J. Allen, "On the passage of γ rays of radium through matter," *Phys. Rev.*, 34 (Apr. 1912), 296-310. Preliminary announcements are "On the absorption of the γ rays of radium by solids," *Phys. Rev.*, 32 (Feb. 1911), 222-24; "On the absorption of γ rays of radium by liquids," *Phys. Rev.*, 32 (Feb. 1911), 225-26. For an example of his later work, see S. J. Allen and L. M. Alexander, "The effects of previous filtering upon the absorption coefficients of high frequency x-rays," *Phys. Rev.*, 9 (Mar. 1917), 198-204. S. J. Allen letter to E. Rutherford, 9 May 1912, RCC. A. Kovarik letter to E. Rutherford, 20 Jan. 1916, RCC.

42. J. A. Gray, "The scattering and absorption of the γ rays of radium," *Phil. Mag.*, 26 (Oct. 1913), 611-23.

43. E. Rutherford and E. N. da C. Andrade, "The wavelength of the soft γ rays from radium B," *Phil. Mag.*, 27 (May 1914), 854-68; reprinted in *CPR*, vol. 2, pp. 432-44. Also, "The spectrum of the penetrating γ rays from radium B and radium C," *Phil. Mag.*, 28 (Aug. 1914), 263-73; reprinted in *CPR*, vol. 2, pp. 456-65.

44. A. H. Compton, "The corpuscular properties of light," *Naturwissenschaften*, 17 (28 June 1929), 507-15. Also see R. H. Stuewer (note 38).

45. R. A. Millikan and Harvey Fletcher, "On the question of valency in gaseous ionization," *Phil. Mag.*, 21 (June 1911), 753-70. R. A. Millikan, V. H. Gottschalk, and M. J. Kelly, "The nature of the process of ionization of gases by alpha rays," *Phys. Rev.*, 15 (Mar. 1920), 157-77.

Chapter 17

1. O. Hahn and L. Meitner, "Eine neue Methode zur Herstellung radioaktiver Zerfallsprodukte; Thorium D, ein kurzlebiges Produkt des Thoriums," *Verhandl. Deut. Physik. Ges.*, 11 (Feb. 1909), 55-62; "Nachweis der komplexen Natur von Radium C," *Physik. Z.*, 10 (15 Oct. 1909), 697-703. S. Russ and W. Makower, "The expulsion of radio-active matter in the radium transformations," *Proc. Roy. Soc. London*, A82 (6 May 1909), 205-24. E. Rutherford, J. Chadwick, and C. D. Ellis, *Radiations From Radioactive Substances* (Cambridge: University Press, 1930), pp. 153-58, 557-58.

2. E. M. Wellisch, "Über die Vorgänge beim Transport des aktiven Niederschlages," *Verhandl. Deut. Physik. Ges.*, 13 (1911), 159-71.

3. A. F. Kovarik letter to E. Rutherford, 17 Aug. 1912, RCC.

4. A. F. Kovarik, "Recoil atoms in ionized air," *Phil. Mag.*, 24 (Nov. 1912), 722–27.

5. L. W. McKeehan, "The diffusion of actinium emanation and the range of recoil from it," *Phys. Rev.*, 10 (Nov. 1917), 473–82.

6. E. M. Wellisch and H. L. Bronson, "The distribution of the active deposit of radium in an electric field," *Phys. Rev.*, 34 (Feb. 1912), 151–52; same title, *Phil. Mag.*, 23 (May 1912), 714–29; same title, *Am. J. Sci.*, 33 (May 1912), 483–98.

7. E. M. Wellisch, "The distribution of the active deposit of radium in an electric field. II," *Phil. Mag.*, 26 (Oct. 1913), 623–35; same title, *Am. J. Sci.*, 36 (Oct. 1913), 315–27; "Experiments on the active deposit of radium," *Phil. Mag.*, 28 (Oct. 1914), 417–39; same title, *Am. J. Sci.*, 38 (Oct. 1914), 283–304. A. N. Lucian, "The distribution of the active deposit of actinium in an electric field," *Am. J. Sci.*, 38 (Dec. 1914), 539–55.

8. A. F. Kovarik, "The effect of changes in the pressure and temperature of gases upon the mobility of the negative ions produced by ultra-violet light," *Phys. Rev.*, 30 (Apr. 1910), 415–45.

9. A. F. Kovarik, "Mobility of the positive and negative ions in gases at high pressures," *Proc. Roy. Soc. London*, A86 (31 Jan. 1912), 154–62.

10. H. A. Erikson, "The mobility of ions at different temperatures and constant gas density," *Phys. Rev.*, 3 (Feb. 1914), 151–52.

11. H. A. Erikson, "Size and aging of ions produced in air," *Phys. Rev.*, 17 (Mar. 1921), 400; "Factors affecting the nature of ions in air," *Phys. Rev.*, 34 (15 Aug. 1929), 635–43.

12. James A. Crowther, *Ions, Electrons, and Ionizing Radiations* (8th ed., New York: Longmans, Green, and London: Edward Arnold, 1949), pp. 20–34.

13. M. Curie, "Rayons émis par les composés de l'uranium et du thorium," *Compt. Rend.*, 126 (12 Apr. 1898), 1101–3.

14. W. W. Strong, "On the possible radioactivity of erbium, potassium and rubidium," *Phys. Rev.*, 19 (Aug. 1909), 170–73.

15. E. Rutherford, J. Chadwick, and C. D. Ellis (note 1), pp. 537–44. National Bureau of Standards, *Nuclear Data* (NBS Circular 499, Washington, D. C.: Government Printing Office, 1950).

16. F. C. Brown, "Evidence that sodium belongs to a radioactive series of elements," *Science*, 37 (10 Jan. 1913), 72–76; "Sur une preuve que le sodium appartient à une série radioactive d'éléments," *Le Radium*, 9 (Oct. 1912), 352–55.

17. T. T. Quirke and L. Finkelstein, "Measurements of the radioactivity of meteorites," *Am. J. Sci.*, 44 (Sept. 1917), 237–42.

18. R. J. Strutt, "On the distribution of radium in the earth's crust, and on the earth's internal heat," *Proc. Roy. Soc. London*, A77 (14 May 1906), 472–85.

19. M. B. Snyder, "Radium in spiral nebulae and in star clusters," *Science*, 29 (28 May 1909), 865–69. E. Rutherford, *Radioactive Substances and Their Radiations* (Cambridge: University Press, 1913), p. 656. A brief review of "Cosmical radioactivity" is given on pp. 653–56 of this book.

20. B. Boltwood letter to David Todd, 20 Apr. 1920, BCY.

21. The following is a sample of such papers: S. J. Allen, "Radioactivity of a smoke-laden atmosphere," *Phys. Rev.*, 26 (Feb. 1908), 206–8; same title, *Phys. Rev.*, 26 (June 1908), 483–96; "The radioactive deposit from the atmosphere on an uncharged wire," *Phys. Rev.*, 7 (Jan. 1916), 133–38. H. M. Dadourian, "On the constituents of atmospheric radioactivity," *Am. J. Sci.*, 25 (Apr. 1908), 335–42. W. W. Strong, "Variation of the penetrating radiation," *Phys. Rev.*, 26 (June 1908), 518–19. Frederic A. Harvey (a graduate student under E. P. Lewis at Berkeley at the time of the first paper; on the Syracuse University faculty for the second), "Atmospheric radioactivity in California and Colorado and the range of the α-particles from radium B," *Phys. Rev.*, 28 (Mar. 1909), 188–216; "The half-value of the radioactive deposit collected in the open air," *Phys. Rev.*, 35 (Aug. 1912), 120–27. Oliver H. Gish (of Westinghouse Research Laboratory), "Absorption coefficient of the penetrating radiation," *Phys. Rev.*, 13 (Feb. 1919), 155–56.

22. The following is a sample of such papers: J. C. Sanderson (a graduate student at Yale; then on the Minnesota faculty), "The probable influence of the soil on local atmospheric radioactivity," *Am. J. Sci.*, 32 (Sept. 1911), 169–84; "Radio-active content of certain Minnesota soils," *Am. J. Sci.*, 39 (Apr. 1915), 391–97. Stewart J. Lloyd and John Cunning-

ham (of the University of Alabama), "The radium content of some Alabama coals," *Am. Chem. J.*, 50 (1913), 47–51. R. B. Moore, "The radioactivity of some type soils of the United States," *J. Ind. Eng. Chem.*, 6 (May 1914), 370–74. H. Schlundt and R. B. Moore, "Radioactivity of the thermal waters of the Yellowstone National Park," *Bulletin of the U. S. Geological Survey*, No. 395 (1909). John C. Hemmeter and Ernest Zueblin (of the University of Maryland), "The radioactivity of the mineral water of hot springs, warm springs and healing springs in Hot Springs, Va.," *Archives of Internal Medicine*, 15 (1915), 188–203. R. B. Moore and C. F. Whittemore, "The radioactivity of the waters of Saratoga Springs, New York," *J. Ind. Eng. Chem.*, 6 (July 1914), 552–53. P. B. Perkins, "Radioactivity of underground waters in Providence and the vicinity," *Science*, 42 (3 Dec. 1915), 806–8.

23. J. Satterly, "On the amount of radium emanation in the lower regions of the atmosphere and its variation with the weather," *Phil. Mag.*, 20 (July 1910), 1–36. A. S. Eve, "On the ionization of the atmosphere due to radioactive matter," *Phil. Mag.*, 21 (Jan. 1911), 26–40.

24. L. J. Lassalle, "The diurnal variation of the earth's penetrating radiation at Manila, Philippine Islands," *Phys. Rev.*, 5 (Feb. 1915), 135–48. J. R. Wright and O. F. Smith, "The variation with meteorological conditions of the amount of radium emanation in the atmosphere, in the soil gas, and in the air exhaled from the surface of the ground, at Manila," *Phys. Rev.*, 5 (June 1915), 459–82.

25. H. L. Bronson, "Radio-active measurements by a constant deflection method," *Am. J. Sci.*, 19 (Feb. 1905), 185–87. S. J. Allen, "A null instrument for measuring ionization," *Phil. Mag.*, 14 (Dec. 1907), 712–23. Also see E. N. da C. Andrade, *Rutherford and the Nature of the Atom* (Garden City, New York: Doubleday, 1964), p. 36, and E. Rutherford (note 19), pp. 97–100.

26. The relative sensitivity of electroscope compared to electrometer actually involved a degree of experimental skill. In the hands of a Boltwood, the electroscope was a superior device. A recent author claims, however, that except for the torsion type of electroscope, the ordinary electrometer is more sensitive. He adds that it is inconvenient to use a null method of reading with the electroscope, and difficult to vary its sensitivity. But electroscopes enjoy popularity because of their freedom from external changes of temperature and humidity, freedom from changes in battery potential and resistance, freedom from drift, need for only one potential, ease of setting up and operating, portability, and low cost. John Strong, *Procedures in Experimental Physics* (New York: Prentice-Hall, 1945), p. 254.

27. K. Fajans conversation with the author, 16–17 July 1966.

28. C. T. R. Wilson, "On a sensitive gold-leaf electrometer," *Proc. Cambridge Phil. Soc.*, 12 (1903), 135–39. For an earlier improvement by Wilson, see "On the ionisation of atmospheric air," *Proc. Roy. Soc. London*, 68 (1901), 151–61. Also see E. Rutherford (note 19), pp. 88–96.

29. J. Zeleny, "A lecture electroscope for radioactivity," *Phys. Rev.*, 32 (Feb. 1911), 255–56; "A lecture electroscope for radioactivity and other ionization experiments," *Phys. Rev.*, 32 (June 1911), 581–84.

30. H. A. Bumstead, "On the emission of electrons by metals under the influence of alpha rays," *Phil. Mag.*, 22 (Dec. 1911), 907–22; same title, *Am. J. Sci.*, 32 (Dec. 1911), 403–17.

31. S. C. Lind, "Practical methods for the determination of radium. I. Interchangeable electroscope and its use," *J. Ind. Eng. Chem.*, 7 (May 1915), 406–10.

32. T. H. Leaming, H. Schlundt, and J. Underwood, "Comparison of the ionization currents due to equal quantities of radium emanation in different types of electroscopes," *Trans. Am. Electrochem. Soc.*, 30 (1916), 365–78. O. C. Lester, "On the calibration and the constants of emanation electroscopes," *Am. J. Sci.*, 44 (Sept. 1917), 225–36.

33. W. Duane, "Sur une méthode photographique d'enregistrement des particules α," *Compt. Rend.*, 151 (18 July 1910), 228–30.

34. H. A. Bumstead letter to E. Rutherford, 23 June 1908, RCC.

35. R. C. Hills, "A Denver-made spinthariscope," *Proceedings of the Colorado Scientific Society*, 11 (1917), 209–14, as noted in *Chem. Abst.*, 11 (1917), 2303.

36. E. Rutherford and H. Geiger, "An electrical method of counting the number of α-particles from radio-active substances," *Proc. Roy. Soc. London*, A81 (27 Aug. 1908), 141–61; reprinted in *CPR*, vol. 2, pp. 89–108. H. Geiger and E. Rutherford, "Photo-

graphic registration of α particles," *Phil. Mag.*, 24 (Oct. 1912), 618–23; reprinted in *CPR*, vol. 2, pp. 288–91.

37. H. J. Vennes, "The retardation of alpha particles by metals," *Am. J. Sci.*, 44 (July 1917), 69–72.

38. H. Geiger, "Über eine einfache Methode zur Zählung von α- und β-Strahlen," *Ber. Deut. Physik. Ges.*, 15 (1913), 534–39.

39. J. E. Shrader, "The Geiger apparatus for the photographic registration of alpha and beta particles," *Phys. Rev.*, 6 (Oct. 1915), 292–303.

40. A. F. Kovarik and L. W. McKeehan, "Distribution of transmitted and reflected β-particles determined by the statistical method," *Phys. Rev.*, 6 (Dec. 1915), 426–36.

41. A. Kovarik, "New methods for counting the alpha and the beta particles," *Phys. Rev.*, 9 (June 1917), 567–68.

42. A. F. Kovarik, "A device for the automatic registration of the α- and β-particles and γ-ray pulses," *Phys. Rev.*, 13 (Feb. 1919), 153–54; "On the automatic registration of α-particles, β-particles and γ-ray and x-ray pulses," *Phys. Rev.*, 13 (Apr. 1919), 272–80.

43. A. F. Kovarik, "Some experiments bearing on the nature of γ-rays," *Phys. Rev.*, 14 (Aug. 1919), 179–80; "A statistical method for studying the radiations from radioactive substances and the x-rays and its application to some γ-ray problems," *Proc. Nat. Acad. Sci. U. S.*, 6 (Mar. 1920), 105–7. A. F. Kovarik letter to E. Rutherford, 20 July 1919, RCC.

44. A. Zeleny, "The dependence of progress in science on the development of instruments," *Science*, 43 (11 Feb. 1916), 185–93, quotation on p. 186.

45. Merle Randall, "The Boltwood standard of radioactivity," *Trans. Am. Electrochem. Soc.*, 21 (1912), 463–97.

46. M. Randall, *ibid.*, p. 463. For the McCoy number, see chapter 12 above, and the *International Critical Tables* (Washington, D. C.: National Research Council, 1926), vol. 1, p. 368.

47. F. Soddy, "Radioactivity," *Ann. Repts. on Progr. Chem.*, *Chem. Soc. London*, 6 (1909), 261.

48. M. Randall (note 45), pp. 463–64, 496.

49. J. Moran, "A comparison of radium standard solutions," *Phil. Mag.*, 30 (Oct. 1915), 660–64; "A comparison of radium standard solutions. Continued," *Transactions of the Royal Society of Canada*, 10 (Dec. 1916), 77–84.

50. B. Boltwood letter to O. Hahn, 1 Nov. 1910, BCY.

51. B. Boltwood letter to N. Ernest Dorsey, 20 Jan. 1921, BCY. M. Curie unsigned letter to E. Rutherford [Sept. 1910], RCC. E. Rutherford, "Radium standards and nomenclature," *Nature*, 84 (6 Oct. 1910), 430–31; reprinted in *CPR*, vol. 2, pp. 193–95. Samuel Glasstone, *Sourcebook on Atomic Energy* (3rd ed., Princeton: Van Nostrand, 1967), p. 670, notes that in 1930 the definition of the curie was extended to embrace the quantity of any product in equilibrium with one gram of radium. Since one gram of radium undergoes 3.7×10^{10} disintegrations each second, by 1948 it was seen to be more convenient to label this rate of activity of any substance the curie. At the same time it was proposed to call 10^6 disintegrations per second a "rutherford," but this unit has not been widely adopted. G. B. Cook and J. F. Duncan, *Modern Radiochemical Practice* (Oxford: Clarendon Press, 1952), p. 47.

52. E. Rutherford, J. Chadwick, and C. D. Ellis (note 1), p. 568. E. Rutherford letter to B. Boltwood, 22 Apr. 1912, BCY; printed in *Rutherford and Boltwood*, pp. 269–72. F. A. Paneth, "Radioactive standards and units," *Nature*, 116 (2 Dec. 1950), 931–33. Charles H. Viol, "History and development of radium-therapy," *Journal of Radiology*, 2 (1921), 30–31.

53. C. Chamié, "Sur le nouvel étalon international de radium," *J. Phys. Radium*, 1 (1940), 319–21.

Chapter 18

1. H. W. Webb, autobiographical notes, Center for History of Physics, American Institute of Physics, New York City.

2. Leonard Loeb letter to the author, 30 Nov. 1962.

3. Otto Hahn, *A Scientific Autobiography* (New York: Scribner's, 1966), p. 50.

4. L. Badash, "How the 'newer alchemy' was received," *Sci. American*, 215 (Aug. 1966), 88–95.

5. O. Hahn (note 3), pp. 50–51.

6. A. Smithells, "Presidential address to Section B," *Nature*, 76 (8 Aug. 1907), 352–57, offers a good survey of chemists' attitudes at this time.

7. F. W. Aston, "The atoms of matter; their size, number, and construction," *Nature*, 110 (25 Nov. 1922), 702–5.

8. T. S. Kuhn, *The Structure of Scientific Revolutions* (Chicago: University of Chicago Press, 1962, revised 1970).

9. L. Badash (note 4).

10. D. J. de S. Price, *Science Since Babylon* (New Haven, Connecticut: Yale University Press, 1961, revised 1975), chapter 8. Diana Crane, *Invisible Colleges. Diffusion of Knowledge in Scientific Communities* (Chicago: University of Chicago Press, 1972), p. 172. The number of papers on radioactivity each year was obtained from the author's rather complete file.

11. A. L. Kroeber, *Style and Civilization* (Ithaca, New York: Cornell University Press, 1957), as cited by D. Crane, *ibid.*, pp. 27–28.

12. Nathan Reingold, "American indifference to basic research: a reappraisal," in George H. Daniels (ed.), *Nineteenth-Century American Science, a Reappraisal* (Evanston, Illinois: Northwestern University Press, 1972), pp. 38–62, esp. pp. 54–55, argues that America was a small but mature scientific country by about 1875, and a peer of European nations by the turn of the century. P. Forman, J. Heilbron, and S. Weart, "Physics *circa* 1900," *Historical Studies in the Physical Sciences*, 5 (1975), show that American science was in several respects quantitatively on a par with European science around 1900. Edward H. Beardsley, *The Rise of the American Chemical Profession, 1850–1900* (Gainesville: University of Florida Press, 1964), suggests that the American chemical profession stood on its own feet by about 1900. But I would maintain that, despite quantitative vigor, Americans still felt that this country had a long way to go to equal the accomplishments of Germany and England. That Americans continued to publish in European journals hints at where the major action was. That I. I. Rabi found his Hamburg hosts in 1927 held the *Physical Review* in such low esteem that they purchased it in the yearly volume rather than by monthly issue, to economize on postage, hints at how long the major action remained there. The Rabi information is from a biographical sketch by Jeremy Bernstein, "Profiles. Physicist-I," *The New Yorker* (13 Oct. 1975), 86.

13. A survey of the social development of science in the late nineteenth and early twentieth centuries is provided in Daniel J. Kevles, "The study of physics in America, 1865–1916," unpublished Ph.D. dissertation, Princeton University, 1964. The late nineteenth century is discussed in chapters 3–4. See the revision and extention of this dissertation: *The Physicists* (New York: Knopf, 1978). Also see George H. Daniels, *Science in American Society* (New York: Knopf, 1971), chapters 12–13.

14. H. S. Carhart, "The humanistic element in science," *Science*, 4 (31 July 1896), 124–30, quotation on p. 125.

15. D. Kevles (note 13), chapters 5–7.

16. D. Kevles (note 13), chapters 5–6.

17. D. Kevles (note 13), chapter 5 and appendices 5–7.

18. D. Kevles (note 13), chapter 5.

19. W. Huggins, "Address delivered by the President," *Proc. Roy. Soc. London*, A76 (22 Apr. 1905), 1–29, quotation on p. 21.

20. D. Kevles (note 13), chapters 5–6. AAAS and APS data are from Kevles, appendices 3–4. The number of original APS members (69) is taken from A. G. Webster, "Some practical aspects of the relations between physics and mathematics," *Phys. Rev.*, 18 (Apr. 1904), 297–318, esp. p. 297; Kevles gives 59. ACS figures were supplied by the ACS. For a survey of twentieth century developments, see G. H. Daniels (note 13), chapter 14.

21. For relative numbers of scientists, see Robert W. Lawson, "Part played by different countries in the development of the science of radioactivity," *Scientia*, 30 (1921), 257–70. For each country he gives the number of authors who contributed four or more original papers, and the total number of authors who contributed any noteworthy papers

on radioactivity. His results are: British Empire 45/171; Germany 28/210; France 18/70; Austria 10/76; America 9/89; Poland 4/14; Switzerland 3/19; Sweden 3/9; Italy 2/21; Norway 2/20; Holland 2/12; Hungary 2/7; Russia 1/13; Japan 1/12; Denmark 1/4; Rumania 0/4; Spain 0/1.

22. For Soddy, see E. O. Lovett letter to J. J. Thomson, 28 Feb. 1912, Thomson collection, Cambridge University Library.

23. A. G. Webster (note 20). David Starr Jordan, "The making of a Darwin," *Science,* 32 (30 Dec. 1910), 929-42. F. C. Brown, "The predicament of scholarship in America and one solution," *Science,* 39 (24 Apr. 1914), 587-95.

24. S. J. Allen, "On the secondary radiation produced from solids, solutions, and pure liquids by the β rays of radium," *Phys. Rev.,* 29 (Sept. 1909), 177-211. T. W. Richards, "The problem of radioactive lead," *Science,* 49 (3 Jan. 1919), 1-11.

25. B. Boltwood letter to E. Rutherford, 11 Oct. 1908, RCC; printed in *Rutherford and Boltwood,* p. 193.

26. J. Zeleny letter to E. Rutherford, 8 Apr. 1902, RCC.

27. Eve Curie, *Madame Curie* (Garden City, New York: Doubleday, Doran, 1938), p. 275. Robert M. Lester, *Forty Years of Carnegie Giving* (New York: Scribner's, 1941), p. 7.

28. H. Schlundt letter to B. Boltwood, 4 Nov. 1907, BCY.

29. H. A. Bumstead letter to E. Rutherford, 30 Sept. 1905, RCC.

30. For a discussion of such conditions in a different science, see Nicholas Mullins, "The development of a scientific specialty: The phage group and the origins of molecular biology," *Minerva,* 10 (Jan. 1972), 51-82.

31. G. Holton, "Striking gold in science: Fermi's group and the recapture of Italy's place in physics," *Minerva,* 12 (Apr. 1974), 159-98, esp. p. 186.

32. E. Gleditsch letter to the author, 2 Oct. 1962.

33. L. Badash, "Radioactivity before the Curies," *Am. J. Phys.,* 33 (Feb. 1965), 128-35.

Index

Library of Congress Cataloging in Publication Data

Badash, Lawrence.
    Radioactivity in America.

    1. Radioactivity—United States—History.   I. Title.
QC794.98.B33   539.7'5'0973   78–20525
ISBN 0–8018–2187–8